ENGINEERING TOOLS, TECHNIQUES AND TABLES

STRUCTURAL STEEL AND CASTINGS: SHAPES AND STANDARDS, PROPERTIES AND APPLICATIONS

ENGINEERING TOOLS, TECHNIQUES AND TABLES

Additional books in this series can be found on Nova's website under the Series tab.

Additional E-books in this series can be found on Nova's website under the E-book tab.

ENGINEERING TOOLS, TECHNIQUES AND TABLES

STRUCTURAL STEEL AND CASTINGS: SHAPES AND STANDARDS, PROPERTIES AND APPLICATIONS

LENA M. BECKER
EDITOR

Nova Science Publishers, Inc.
New York

LIBRARY OF CONGRESS CATALOGING-IN-PUBLICATION DATA

Structural steel and castings : shapes and standards, properties, and
applications / editor, Lena M. Becker.
p. cm.
Includes bibliographical references and index.
ISBN 978-1-61728-104-4 (hardcover)
1. Steel, Structural. 2. Steel castings. I. Becker, Lena M.
TS350.S77 2010
624.1'821--dc22

2010017743

Published by Nova Science Publishers, Inc. † New York

CONTENTS

PREFACE

Structural steel is steel construction material, a profile, formed with a specific shape or cross section and certain standards of chemical composition and mechanical properties. Casting is a manufacturing process by which a liquid material is usually poured into a mold, which contains a hollow cavity of the desired shape, and then allowed to solidify. This book reviews research in the fields of structural steel and castings including recent advances on the design of shear connections, research on multiaxial fatigue in structural steel connections, thermosetting polymer nanocomposites castings based on polyurethane and epoxy polymers and gas emission from molds with synthetic additives during metal castings.

From 1988 to 1994, the Uniform Building Code (UBC) prescribed a beam-to-column connection to be used in steel special moment frames. This connection type is known as Pre-Northridge connection. Numerous tests have been done until 1994 to investigate the ductility level of the Pre-Northridge connections. Most of these tests were successful, with a number of specimens sustaining total rotation larger than 0.04 rad. Despite of the design expectations and the successful tests, the 1994 Northridge earthquake showed that the level of ductility of pre-Northridge connections is lower than the required levels.

Research results showed that higher ductility can be achieved by using smooth weld access holes, high fracture toughness weld metal E70-TGK2 and removing the bottom beam flange backing bar. This type of modified pre-Northridge connections are known as post-Northridge connections. However, these modifications may not lead to achieving minimum required ductility levels. To improve the ductility of post-Northridge connections, researchers started to change the connection configuration by keeping the previously mentioned modifications. One of the such modified connections is reduced beam web (RBW) connection. In RBW connection the beam web is penetrated by a number of circular voids. The size and spacing between the voids are designed as such to cause the shear yielding of the beam web along the beam span and to keep the connection in elastic region. Experimental tests of RBW connections indicated that these connections could achieve adequate strength and ductility. Chapter 1 aimed to use the above mentioned design criterion to investigate whether connections of beams deeper than those previously tested could also achieve adequate strength and ductility. In addition, the effect of moment yielding and shear-moment yielding criterions on the behaviour of RBW connection were also investigated. Hence, using finite element method, three pre-tested post-Northridge connections with deep beams of different depths were modeled. Results showed that web shear yielding criterion can lead to higher connection ductility. Furthermore, multiple voids of same size at the beam web area could only increase the total rotation of the shallowest beam specimen (600 mm beam depth) to the minimum required ductility. Therefore, a void arrangement of different void sizes was proposed for deep beams in terms of the beam depth.

Due to its robust post-buckling strength, substantial ductility, stable hysteretic characteristics and high initial stiffness, the steel plate shear wall (SPSW) system is now considered to be an appealing alternative to conventional lateral load resisting systems used for earthquake resistant design of structures. SPSW systems, when compared to a traditional reinforced concrete shear wall, offer lighter structures, increased floor area, faster speed of construction, considerable economy and better quality control. Although research work on SPSW has been going on since the early 1980s, the satisfactory structural performance of SPSW systems in the Northridge, USA (1994) and Kobe, Japan (1995) earthquakes led researchers and practicing engineers to study and implement the SPSW system to a greater extent. Experimental and analytical research works, conducted primarily in Canadian, US and UK universities on SPSW considered various aspects of the seismic behavior of SPSW and developed the fundamental guidelines for its implementation as an effective lateral load resisting system. Chapter 2 presents a comprehensive review of the stages of the development in SPSW research and implementation over the past three decades. The primary focus of this review is on the recent experimental and analytical research works on unstiffened thin steel plate shear walls carried out all over the world. The existing design guidelines for the SPSW system in current US and Canadian codes are also discussed. An overview of the ongoing research activities across the globe is presented with an emphasis to the development of design methodologies for SPSW systems conforming to the performance-based seismic design (PBSD) philosophy. The final sections of this chapter deal with the research areas needing development in the immediate future, roadblocks in the process of implementation, standardization of design guidelines and a roadmap for the future aiming at a wide acceptance of this system in the field of earthquake engineering.

Chapter 3 reports some recent studies on the analysis and design of shear connections. These new developments include: 1) the formulae for computing the moment of various shear connections (such as double-angle, single-angle, end-plate, and clipped end-plate shear connections) corresponding to the plastic limit state of its supported beam; 2) an experimental study on the double-angle shear connections with shortened outstanding-legs; 3) design method for bolted single-angle connections; and 4) resilience analysis of shear connections to resist progressive collapse under a double-span condition.

The effect of stress concentration in connections between primary members of structures was found to be one of the main reasons for fatigue damage. Most of these connections are subjected to multiaxial fatigue in both high cycle and low cycle ranges. In case of high cycle fatigue, the equivalent effective stress with Miner's rule is the most popular multiaxial fatigue criterion in service life estimation of structures. However, under many variable amplitude-loading conditions, Miner's predictions have been found to be unreliable since it does not properly take into account the loading sequence effect. Recently, a number of damage models have been proposed to capture the loading sequence effect of variable amplitude loads more precisely.

The scale of metal grains of a metallic aggregate is generally defined to be within the mesoscopic scale (grain scale). Many of the more precise high cycle fatigue theories involve grain scale than macroscopic scale. In the high cycle fatigue region, some grains undergo local plasticity while the rest of the matrix behaves elastically. Recently, one mesoscopic scale fatigue model was developed to obtain more precise estimation to multiaxial fatigue and it has been found that this model is applicable to common structures since the associated material parameter determination procedures are dependent on commonly available fatigue test data. However, experimental comparisons of this theory (model) have exhibited a certain amount of deviation for multiaxial fatigue life due to variable amplitude loading conditions since the considered damage law is Miner's rule.

Therefore the main objective of this chapter is to propose a new mesoscopic scale (grain scale) fatigue model to obtain a more precise estimation to multiaxial high cycle fatigue life for variable amplitude proportional loading. The accumulated inelastic meso-strain is considered as the damage variable in this model, and a new damage indicator is defined as an alternative to the damage factor of the previous model. Initially, Chapter 4 describes the existing multiaxial high cycle fatigue models and their shortcomings. Then, the chapter describes the new fatigue model. The verification of the proposed fatigue model is performed by comparing the experimental fatigue lives with theoretical predictions. The proposed model is applied to estimate the fatigue life of a riveted bridge connection and obtained results are compared with previous estimations based on other known models. Finally, it is shown that the proposed multiaxial fatigue model gives a much more realistic fatigue life to variable amplitude proportional loading situation where detailed stress histories are known.

In Chapter 5, thermosetting polymer nanocomposites castings based on polyurethane and epoxy polymers have been described. The goal was to develop these materials with high gas barrier properties for achieving their use in the packaging applications. The synthesis methodology for the generation of these castings and the analysis of subsequent parameters like thickness of the coatings, density of the composite etc. have been explained. The optimization of the curing parameters like curing time, temperature, crosslinker amount etc. was obtained by using differential scanning calorimetry. The filler used in the study have been modified with different surface modifications in order to quantify the effect of the interaction of the filler surface modification with the polymer matrices on the composite microstructure and properties. It was concluded that the exfoliated platelets contribute maximum to the barrier properties of the castings. Also, it was the interactions of the filler modifications with the polymer, rather than initial high basal plane spacing of the filler by exchanging long alkyl chains, which leads to exfoliation of the filler platelets and hence enhancement of the barrier properties.

Chapter 6 analyses the main features of gas emission from molds during metal casting. A set of experiments on determination of the parameters of the gas emission from molding sands was carried out. The thickness of the casting, the thickness of the mold wall, the process time, and the initial temperature of the molten metal were varied. The investigated molding sands were prepared from quartz sand and synthetic binders: M19-62 (urea resin), FF-1C (furan resin), and PK-104 (phenol-formaldehyde resin). It was shown that the highest rate of thermal decomposition has urea resin. A mathematical model of gas emission from molds in the stage of solidification and cooling of metal was constructed. The model includes heat and mass transfer in the mold's capillary-porous medium containing various synthetic resins as a binder, the kinetics of the resin's thermal decomposition, and secondary reactions between products formed. It was shown that the model gives a fairly good description of the process. The kinetic parameters of the thermal decomposition of considered resins were determined on the base of fitting of calculated results to experimental data.

The continuous casting technology appeared almost more than 40 years. The most attractive advantage of continuous casting machines is the shortest process from the melt liquid metal to the near—end—products. Further the energy consuming can be remarkably reduced. With the development of socio—economic, the qualitative requirements of metal products are getting stricter. Thanks to the deep scientific researches of magnetohydrodynamics (MHD), the applications of MHD are enjoying more and more extensions. For example, the electromagnetic continuous casting (EMCC) is the most spotlighted issue during the past double decades.

Magnetic fields can provide energy into media without any contact, and the energy supplement can be controlled locally and directionally. According to the types of magnetic fields produced and applied in the continuous casting process, the applications of electromagnetic fields in EMCC can be classified into the following categories (a) Electromagnetic braking (EMBr), (b) Electromagnetic soft contact (EMSC), (c) Electromagnetic stirring (EMS).

Based on the latest results of EMCC, a comprehensive review is given in Chapter 7. The experiments, numerical simulations, together with some theoretical analysis and industrial applications are presented and discussed in detail.

Chapter 8 advances the state of the art in two main ways. First, the authors propose a methodology to adapt machine learning classifiers to the prediction of microshrinkages or ultimate tensile strength and they describe the method for training them. Second, the authors evaluate the classifiers with a historical dataset from a real foundry process in order to compare the accuracy and suitability of each method.

The remainder of this chapter is organised as follows. Section 2 discusses related work. Section 3. details the foundry process of iron castings. Section 5 introduces and describes the *machine-learning algorithms* the authors tailor to iron foundries. Section 6 describes the experiments performed and examines the obtained results and explains feasible enhancements. Finally, section 7 concludes and outlines the avenues of future work.

Nanocasting refers to the special fabrication technology that can be used to produce nanoscale features. This technology has the capability of generating well-ordered, very fine entities such as nanofibers, coaxial nanotubes, and porous structures. In Chapter 9, an overview of the process of nanocasting technology will be provided followed by a description of the mechanism of generating controlled nanostructures. Different types of molds used for nanocasting will also be introduced. Some details on how to use nanocasting to make nanoscale hollow shells, nanofibers with core structures and nanoscale networks will be presented. Finally, the applications of nanocasting for preparing high surface area, electrochemically and photoelectrochemically active nanomaterials and structures will be discussed.

In: Structural Steel and Castings
Editor: Lena M. Becker, pp. 1-56
ISBN: 978-1-61728-104-4

Chapter 1

REDUCED BEAM WEB (RBW) CONNECTIONS WITH CIRCULAR OPENINGS

Amir A. Hedayat and Murude Celikag*
Civil Engineering Department, Eastern Mediterranean University,
Gazimagusa, North Cyprus, Mersin 10 Turkey

Abstract

From 1988 to 1994, the Uniform Building Code (UBC) prescribed a beam-to-column connection to be used in steel special moment frames. This connection type is known as Pre-Northridge connection. Numerous tests have been done until 1994 to investigate the ductility level of the Pre-Northridge connections. Most of these tests were successful, with a number of specimens sustaining total rotation larger than 0.04 rad. Despite of the design expectations and the successful tests, the 1994 Northridge earthquake showed that the level of ductility of pre-Northridge connections is lower than the required levels.

Research results showed that higher ductility can be achieved by using smooth weld access holes, high fracture toughness weld metal E70-TGK2 and removing the bottom beam flange backing bar. This type of modified pre-Northridge connections are known as post-Northridge connections. However, these modifications may not lead to achieving minimum required ductility levels. To improve the ductility of post-Northridge connections, researchers started to change the connection configuration by keeping the previously mentioned modifications. One of the such modified connections is reduced beam web (RBW) connection. In RBW connection the beam web is penetrated by a number of circular voids. The size and spacing between the voids are designed as such to cause the shear yielding of the beam web along the beam span and to keep the connection in elastic region. Experimental tests of RBW connections indicated that these connections could achieve adequate strength and ductility. This chapter aimed to use the above mentioned design criterion to investigate whether connections of beams deeper than those previously tested could also achieve adequate strength and ductility. In addition, the effect of moment yielding and shear-moment yielding criterions on the behaviour of RBW connection were also investigated. Hence, using finite element method, three pre-tested post-Northridge connections with deep beams of different depths were modeled. Results showed that web shear yielding criterion can lead to higher connection ductility. Furthermore, multiple voids of same size at the beam web area could only increase the total rotation of the shallowest beam specimen (600 mm beam depth) to the

* E-mail address: murude.celikag@emu.edu.tr, Tel: +903926301159, Fax: +903926302869. (Corresponding author)

minimum required ductility. Therefore, a void arrangement of different void sizes was proposed for deep beams in terms of the beam depth.

1. Introduction

Seismic design of welded steel moment frame (WSMF) construction is based on the assumption that, in a severe earthquake, frame members will be stressed beyond the elastic limit. Inelastic action is permitted in frame members (normally beams or girders) because it is presumed that they will behave in a ductile manner thereby dissipating energy. It is assumed that welds and bolts, being considerably less ductile, will not fracture. Thus, the design philosophy requires that sufficient strength is provided in the connection to allow inelastic yielding and deformation of the beam and/or column panel zones [SEAOC 1990].

The Uniform Building Code (UBC) [ICBO 1994] is adopted as the standard for seismic design by nearly all jurisdictions in California. From 1988 to 1994 the UBC prescribed a beam-to-column connection that was deemed to satisfy the above mentioned strength requirements. This connection type is known as Pre-Northridge connection. Typical pre-Northridge welded moment connections were often constructed by using large rolled steel sections of grade A36 for beam and grade A572 for column and were designed to transfer the flexural moment and shear force of beams to columns. Figure 1 shows the configuration of a typical moment connection used in moment-resisting frames. A bolted shear tab is commonly used to transfer the shear force, and a complete joint penetration (CJP) groove weld is employed in the field to join the beam flange to the column flange. The weld metal was chosen to overmatch the base metal. For example, 483 MPa (70 ksi) weld metal was commonly specified to overmatch the nominal A36 steel beam. There were no requirements on the weld metal toughness and no restrictions on which welding process and practice to choose. In this connection the E70T-4 weld metal with standard access hole was used and backup bars were left in place (Figure 2).

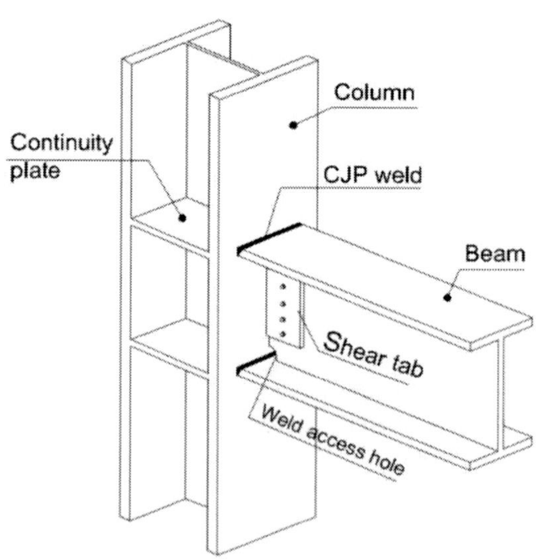

Figure 1. Typical pre-Northridge beam-to-column moment connection.

Although this connection detail was first prescribed by the UBC in 1988, it has been widely used since the early 1970's. A lot of tests have been done until 1994 to investigate the ductility level of the Pre-Northridge connections [Beedle et al. 1973, Parfitt et al. 1976, Chen el al. 1981, Carpenter et al. 1973, Popov et al. 1969, Popov et al. 1970, Popov et al. 1972, Popov et al. 1973]. Most of these tests were successful, with a number of specimens sustaining total rotation larger than 0.04 rad. It must be noted that, in all cases, the beam size was relatively small (typically W18×50 and the largest was W24×76).

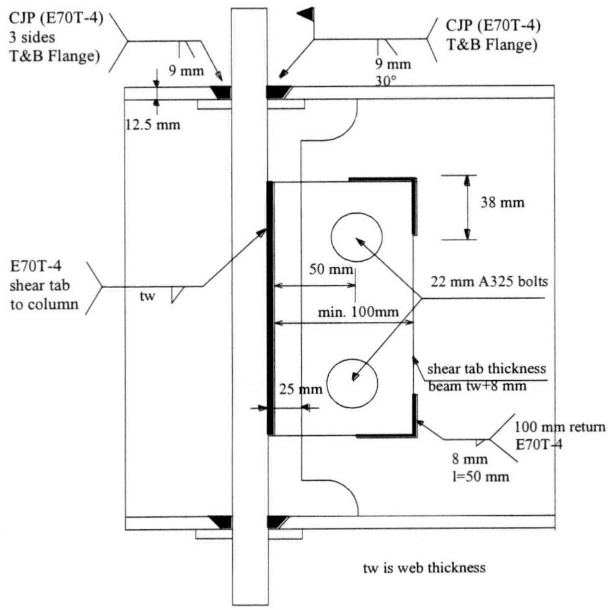

Figure 2. SAC pre-Northridge connection. [Lee et al., 2000]

Despite of the design expectations and successful tests, the 1994 Northridge earthquake caused widespread damage to moment-resisting frames [Gates et al., 1994, Naeim et al., 1994, Green et al., 1994 and Hajjar et al., 1994]. Various brittle fractures were found in beam-to-column welded moment connections. This prompted the initiation of new research programs to investigate the causes of these fractures and to propose the necessary changes in design procedures. In these connections, the beam-to-column interface is the critical section that has the maximum flexural moment when moment frames are subjected to earthquake-induced forces.

The fractures of welded moment connections in the Northridge Earthquake exhibited a variety of origins and paths. In general, fracture was initiated at the root of the beam flange CJP weld and propagated either through the beam flange, the column flange, or the weld itself. In some instances, fracture extended through the column flange and into the column web. A schematic view of the typical fracture paths is shown in Figure 3 [Popov et al., 1998]. The failure modes are catastrophic because they fracture at extremely high speeds without exhibiting prior ductile behaviour. Therefore, they prevented the welded moment connections from exhibiting the inelastic behaviour expected to resist earthquake loading. A welded connection is ductile when it can reach the plastic rotation capacity of 0.03 radians [FEMA 355D, 2000 and AISC-LRFD, 1999].

The poor performance of pre-Northridge moment connections was verified in laboratory testing conducted under SAC Program to Reduce Earthquake Hazards in Steel Moment-Resisting Frame Structures (Phase 1) [SAC, 1996]. Cyclic loading tests were conducted on 12 specimens constructed by using W30x99 and W36x150 beams. These specimens used the same connection details and welding practices that were used prior to the Northridge earthquake. Most of the 12 specimens failed in a brittle manner with little or no ductility. A number of specimens failed at zero plastic rotation, and at a moment well below the plastic moment capacity of the beam.

2. Cause of Failures

Many factors are believed to contribute to the occurrence of brittle fracture at the pre-Northridge connections. A study was done by Hedayat and Celikag [2009a] to summarize these factors and to determine the reliability of using this type of connection in seismic regions. These factors are excessive beam strength [Coons, 1999, Miller, 1998, Chi et al., 1997 and El-Tawil et al., 1998], weld defects and notch effects [Chi et al., 2000; Azuma et al., 2000; Righiniotis et al., 2002 and Kiral et al., 2005], fracture toughness of deposited weld metal [Lu et al., 2000; Kaufmann et al., 1995; Burdekin et al., 2000; Ricles et al., 2002; Chi, 2000b], low temperature [Popov et al., 1998], high stress triaxiality due to severe structural restraint conditions [Lee et al., 1999 and 2006], residual stresses [Zhang et al., 2000 & Matos et al., 2000], geometry of access holes [El-Tawil et al., 1998; Lu et al., 2000] and stress distribution on the flange of beam.

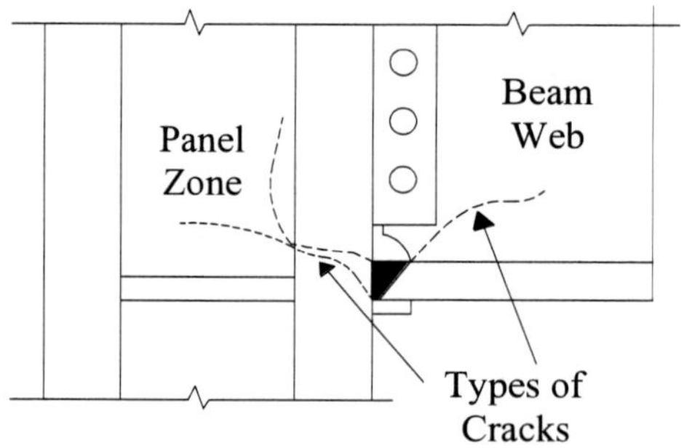

Figure 3. Typical failure modes of the welded beam-to-column connection. [Popov et al., 1998]

In order to improve the behaviour of pre-Northridge connections, various changes were applied to these connections to obtain the post-Northridge connections. The research results showed that higher ductility can be achieved by using smooth weld access holes, high fracture toughness weld metal E70-TGK2 and removing bottom beam flange backing bar. However, these modifications may not lead to achieving minimum required plastic rotation, 3 percent, as required by the seismic codes.

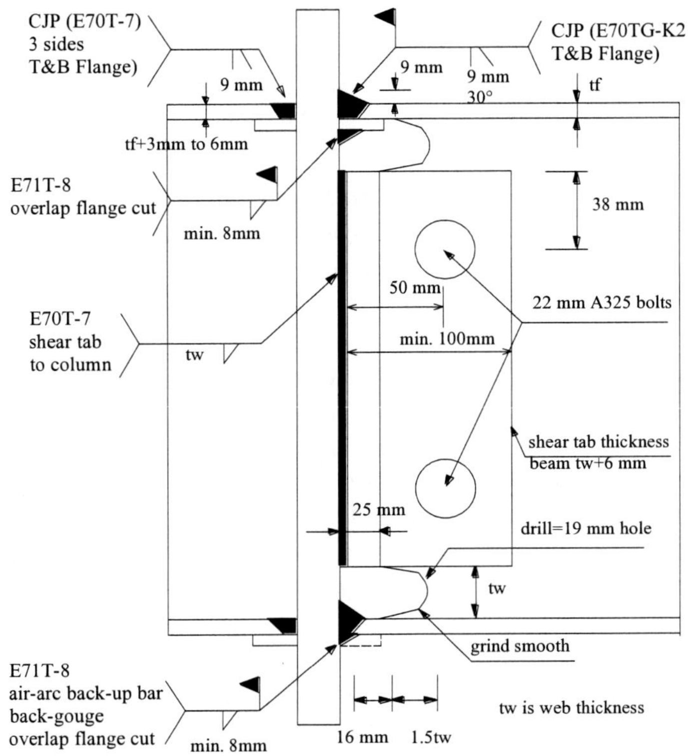

Figure 4. SAC post-Northridge connection. [Lee et al., 2000]

Hedayat and Celikag [2009a] studied the test results of some pre-Northridge and post-Northridge connections to find the cumulative distribution function (CDF) for the normalized fracture moment, M_f / M_p (M_f is moment at failure and M_p is beam plastic moment), and plastic rotation (θ_p) of these connections. This study showed that the mean values of the normalized fracture moment of pre and post-Northridge connections are 0.935 and 0.989 respectively. The mean values of plastic rotation of pre and post-Northridge connections were 0.833 percent and 1.539 percent radian respectively which are less than the minimum required ductility.

3. Modification Methods

As it was discussed in section 2, according to current design codes, the plastic rotation capacity developed by the modified pre-Northridge connections (post-Northridge connections) is not sufficient to qualify this type of connections for use in the special moment resisting frames. A variety of approaches may facilitate improved seismic performance of existing or new welded steel moment frames. These approaches include:

- Modifying the lateral force resisting system to reduce deformation demands at the connections and/or provide alternative load paths. For example, addition of a bracing system; such as reinforced concrete or steel plate shear walls, or new moment resisting bays.

- Modifying existing simple (pinned) beam-to-column connections to behave as partially-restrained connections. For example, the addition of a seat angle at the web cleat connection.
- Reducing the force and deformation demands at the post-Northridge connections through the use of base isolation or supplemental damping devices.
- Modifying the post-Northridge connections by changing the connection configuration.

Any one or a combination of the above approaches may be appropriate for a given project. To improve seismic performance, the choice of the modification strategy should carefully consider the seismic hazard at the building site, the performance goals of the modification and the cost of the modification. Economic considerations of modification include the cost of the structural works, the cost associated with the removal of architectural finishes and other non-structural elements to permit access to the structural frame and the subsequent restoration of these elements, and the costs associated with the disruption to the building function and occupants.

Out of the four approaches listed above for the modification of welded steel moment frames, this chapter deals only with the last one, i.e., methods to modify post-Northridge connections by changing the connection configuration.

4. Changing the Connection Configuration

So far numerous efforts have been made to find a better connection configuration to improve the behaviour of the post-Northridge moment connection. The improvements are mainly based on reinforcing the connection or weakening the beam section. In general terms, the purpose of connection reinforcement or beam section reduction is to achieve a connection that is stronger than the beam. These methods are intended to force the plastic hinge away from the face of the column, to avoid premature fractures that could occur at the connection.

4.1. Reinforcing the Connection

Reinforcement of the connection can be done by using one of the following methods: cover plates, triangular/straight haunches, upstanding ribs, lengthened ribs, bolted brackets and side plates.

4.1.1. Use of Cover Plate

Cover plates are considered as one of the most common alternatives since the Northridge earthquake (Figure 5). The bottom flange of the beam is reinforced by using a rectangular cover plate wider than the beam flange and the top flange is reinforced by using a tapered cover plate thinner than the beam flange. This enables down hand welding (welding from above) to be carried out, in both cases leading to a stronger weld. If the bottom plate is welded to the column before the beam is connected to the column then it can act as a construction seat speeding the erection process.

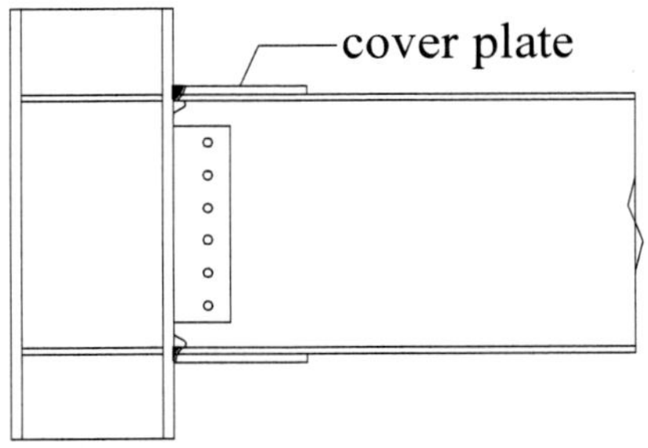

Figure 5. Strengthening of the connection using top and bottom cover plates.

Prior to the Northridge earthquake, a number of investigators conducted cyclic loading tests on steel moment connections reinforced with cover plates [Engelhardt. et al., 1989 & Tsai et al., 1993]. These early tests showed that very high levels of cyclic ductility could be achieved by reinforced connections. However, these earlier tests were very limited in number and were conducted on relatively shallow beams. A total of 12 large scale specimens were constructed with cover plated connections and subjected to cyclic loading tests by Engelhardt [1998]. Out of the 12 specimens tested, 10 showed excellent performance, achieving high levels of plastic rotation. However, two of them showed poor performance, experiencing brittle failures at low levels of plastic rotation.

4.1.2. Use of Haunches

Another method for strengthening the steel beams near the column face area is using triangular or straight haunches beneath the beam bottom flange (Figure 6). Reinforcing the beam with a welded haunch can be viewed as a means of increasing the section modulus of the beam at the face of the column. Therefore, the strengthened section of the beam would remain primarily elastic during plastic hinging of the beam, thereby limiting the stress in the welds.

Experimental investigations of beam-column subassemblies conducted by SAC [1996] showed that welding a tapered haunch to the beam's bottom flange is very effective for enhancing the cyclic performance of damaged moment connections or connections for new construction [Noel and Uang 1996; Uang, 1998]. The cyclic performance can be further improved when haunches are welded to both top and bottom flanges of the beam [SAC 1996], although such a scheme requires the removal of the concrete floor slab in existing buildings which is time consuming and costly.

The tapered haunch is usually cut from a structural tee or wide flange section although it could be fabricated from plate. The haunch flange is groove welded to the beam and column flanges. The haunch web is then fillet welded to the beam and column flanges (Figure 6.a). However, the labor cost for fitting a triangular haunch connection is expensive. In addition, the complete joint penetration groove weld at both ends of the haunch flange with an inclined angle requires a significant amount of overhead welding. To minimize the construction cost, the use of a straight haunch with one free end (Figure 6.b) has been proposed by the SAC

Joint Venture [SAC 1996]. However, the force transfer mechanism of the straight haunch differs from that of the tapered haunch because a direct strut action does not exist [Lee 2001].

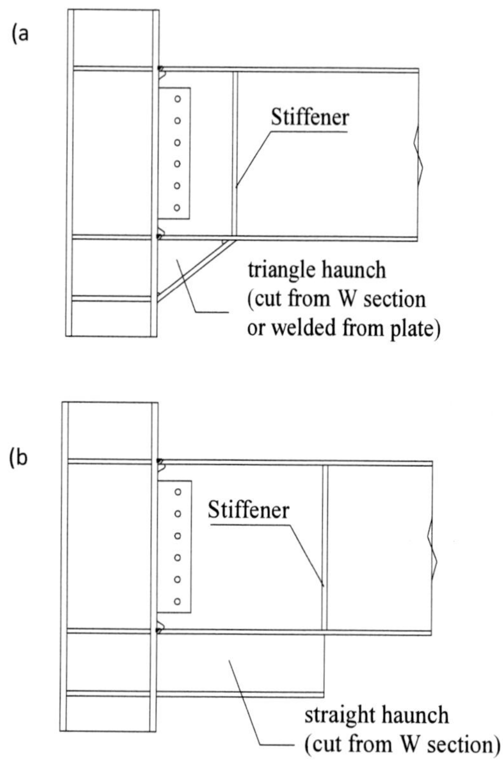

Figure 6. Welded haunches: (a) Triangular; (b) Straight

Experimental tests conducted by Lee [2003] showed that the straight haunch is still a viable solution if the stress concentration at the free end of the haunch, which tends to unzip the weld between the haunch web and beam flange, can be alleviated. A sloped edge combined with a drilled hole near the haunch tip, or a pair of stiffeners partially or fully extended from the beam web, successfully prevented crack initiation at the haunch tip (Figure 7).

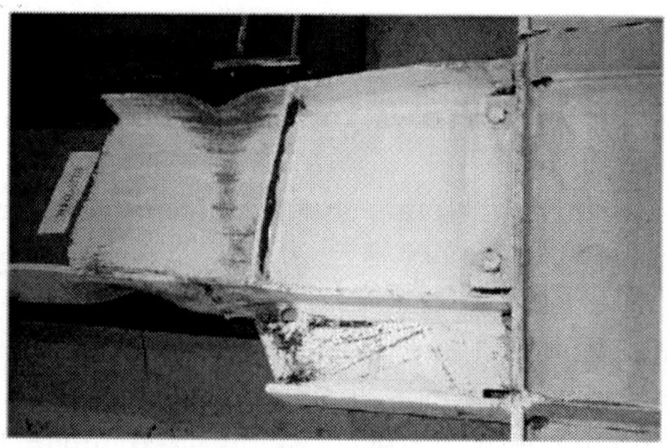

Figure 7. Plastic hinging of straight haunch specimen. [Lee 2003]

When triangular or straight haunches are used they can be added beneath the beam's bottom flange which does not require breaking of the concrete slab. However, this method of stiffening sometimes leads to the achievement of weak column-strong beam behaviour [Chia et al., 2006] and it becomes necessary to break the concrete slab so that stiffeners can be added on the beam top flange.

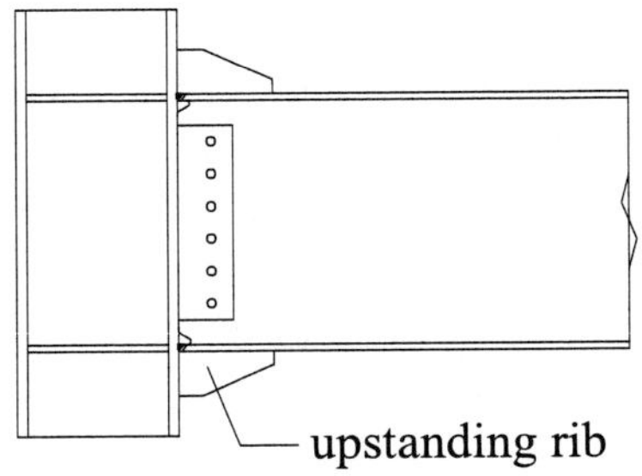

Figure 8. Strengthening of connection using upstanding ribs.

4.1.3. Use of Upstanding Ribs

Among the various strengthening schemes, it is common to use vertical rib plates to improve the seismic performance of steel moment connections. Such ribs generally have the form of a tapered triangular plate. The ribs welded to the top and bottom beam flanges close to the column face are used to reduce the stresses at the beam flange groove weld and to move the critical section away from the column face (Figure 8).

Experimental investigations of reinforced beam-column connections showed that welding two upstanding ribs to each beam flange is effective for enhancing the cyclic performance of moment connections [Engelhardt et al., 1995; Anderson and Duan, 1998 and Zekioglu et al., 1997]. However, there are some reports about the unsatisfactory performance of this type of connection strengthening where the early beam fracture at the tip of ribs prevented the connection to reach adequate plastic rotation [Popov and Tsai, 1989 & Chen et al., 2005].

To gain more insight into the behaviour of the connections strengthened by using upstanding ribs, an analytical study using finite element method was carried out by Chen [2005]. In this study two finite element models were constructed. The first model used two separate rib plates on the top and bottom beam flanges (DR specimen). The other finite element model used only a single rib welded to the centerline of the beam flange. The single rib was located in the same plane as the column web to enhance the transfer of the force from the beam to the column. This single-rib (SR) connection used a rib plate of the same size as those used for the double-rib (DR) connection. The results showed that the von Mises stress in the DR connection was maximum at the access hole region, while that in the SR connection was maximum at the beam flange at the rib tip. Also, finite element analysis proved that the rib-reinforced connection can reduce the stress concentration in the access hole region as well as the stress demand in the beam CJP groove weld. Furthermore, a single rib was more effective than the double spaced ribs for reducing the localized stress concentration near the weld access hole.

In order to eliminate the damage associated with the rib plate, such as the fracture of the beam flange at the rib tip and the fracture of the weld between the rib and the column flange, a lengthened rib was proposed by Chen et al. [2003a & 2003b]. This rib is welded to the centerline of the top and bottom beam flanges and comprises a main reinforced part, a curved part, and an extension (Figure 9). The main reinforced part of the rib was designed to reduce the demand on the full penetration groove weld joining the beam and column flanges. The rib extension was proposed to mitigate the potential stress concentration on the beam flange at the tip of the rib and consequent tearing of the beam flange.

Test results of connections reinforced with lengthened ribs and having I-shape [Chen et al., 2003a & 2003b] and welded box-shape columns [Chen et al., 2004] showed that this type of connection is able to sustain desired inelastic deformation (3% plastic rotation) without any sign of brittle fracture. Despite the good performance of this type of connection, strengthening of existing buildings using lengthened ribs becomes very costly since a significant portion of concrete slab at vicinity of column face should be removed to connect the lengthened rib to the top beam flange.

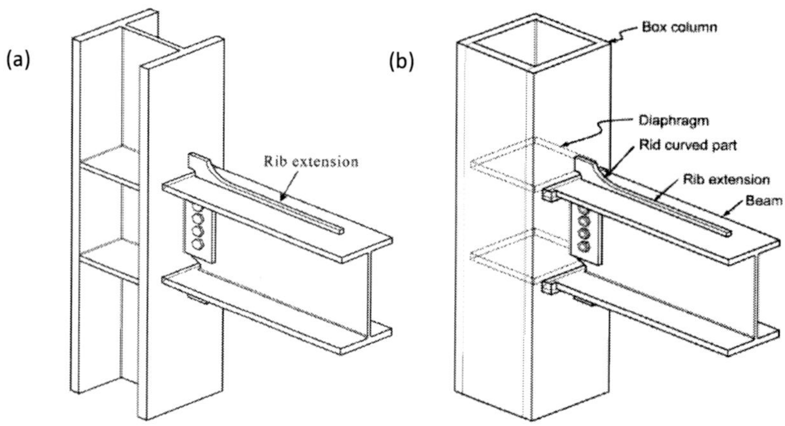

Figure 9. Lengthened flange rib strengthened connection: (a) with I-shape column [Chen, 2003a & 2003b]; (b) with welded box-shape column [Chen et al., 2004]

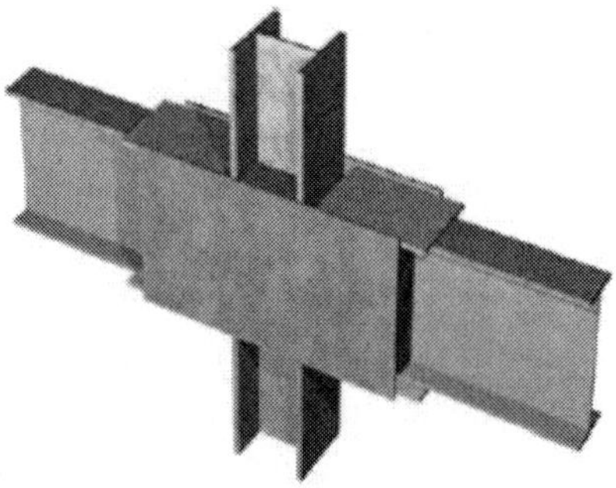

Figure 10. SidePlate Moment connection. [Uang 1995]

4.1.4. Use of Side Plates

The SidePlate connection system (Figure 10) is another post-Northridge connection design that distinguished itself from the other post-Northridge designs by using a totally different connection geometry. The technical prowess of SidePlate connection geometry centers on the intentional physical separation (commonly referred to as a 'gap') between the face of the column flange and the end of the girder by means of parallel full-depth side plates which inherently eliminate the highly-restrained condition and the high-order tri-axial strain concentrations that are intrinsic to the basic geometry of the other moment connection systems comprising the Pre-Northridge connection and its post-Northridge derivatives.

Unlike the 'traditional' connection geometries, all the moments and the loads transfer from the girder to the column reverts back to simple statics and is accomplished with plates and shop fillet welds using predictable equivalent force couples and basic engineering principles. The parallel full depth side plates are designed with adequate strength and stiffness to force all significant plastic behaviour of the connection system into the beam, in the form of ductile hinging at a predictable distance away from the end of the side plates.

Four full-scale cyclic testing of the SidePlate moment connection system were conducted by Uang [1995 & 1996] at the University of California, San Diego to determine the performance of plastic rotation and to verify the predicted strength of connection elements, including welds. The connections of all tested specimens demonstrated robustness and reliability in sustaining multiple inelastic cyclic rotations with no reduction in strength in any of the connection plates, weld elements, or column, while permitting the beam to develop its full strain-hardened plastic flexural strength.

The development of a practical design application for use in blast mitigation, including the prevention of progressive floor collapse after a terrorist attack, the torsional strengthening and hardening of column and girder ends and the resistance of the bomb blast pressures near the column-to-girder joint was made possible by the connection geometry of the SidePlate system [Houghton and Karns, 2001 & Crawford et al., 2002]. Mitigation of ensuing progressive floor collapse using SidePlate connection technology is achieved through positive means by way of catenary action (i.e., a cable-like action - Figure 11), inherently made possible by two parallel side plates which provide a structural ***beam-to-beam continuity link*** which, by definition, is ***free to act independent of the failed column,*** thus allowing the moment frame beams on either side of the failed column to continue to resist tributary gravity floor loads from above [Houghton and Karns, 2001 & Crawford et al., 2002].

The resulting catenary can span two bay lengths between the two columns that are located on either side of the failed column. SidePlate's two full-depth vertical side plates connected to the failed column act as ***effective continuity elements*** to transfer axial load in ***catenary action***, using horizontal fillet welds loaded in shear along their length to transfer the axial load to the adjacent beams of the moment frame. The resulting catenary system can sustain the formation of one or more plastic hinges along its length at each column connection location, while remaining structurally stable.

4.1.5. Use of Bolted Brackets

The bolted bracket is an alternative to the welded haunch and has the added advantage that no field welding is required. Instead, high strength bolts are used to attach the bracket to both the beam and column as shown in Figure 12. Installation of the bolted bracket eliminates the problems associated with welding, such as, venting of welding fumes, supply of fresh air and the need for fire protection.

As with the welded haunch, the bolted bracket forces inelastic action in the beam outside the reinforced region. Tests have shown this approach to be an effective repair and modification technique producing a rigid connection with stable hysteresis loops and high ductility [Kasai et al., 1997 & 1998].

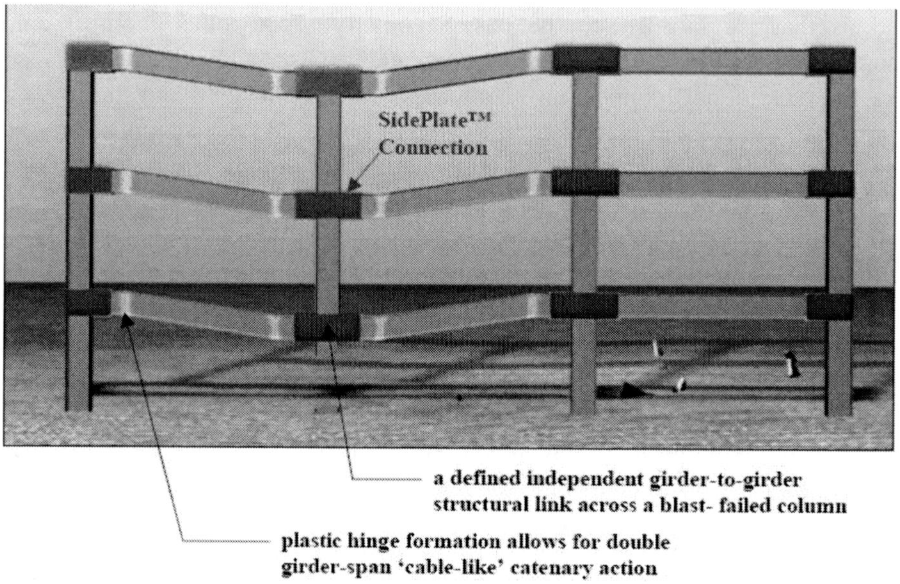

Figure 11. SidePlate mitigates progressive floor collapse through catenary action and independent beam-to-beam continuity links. [Houghton 2001 & Crawford 2002]

Various types of bolted brackets have been developed. The haunch bracket (Figure 12) consists of a shop-welded horizontal leg, vertical leg, and vertical stiffener. The two legs are bolted to the beam and column flanges. The pipe bracket (Figure 13) consists of pipes which are shop welded to a horizontal plate. The plate and pipes are bolted to the beam and column flanges, respectively. The angle bracket (Figure 14) uses an angle section cut from a relatively heavy wide flange section with the flange forming the vertical leg and the web forming the horizontal leg. For light beams, hot rolled angle sections may be sufficient.

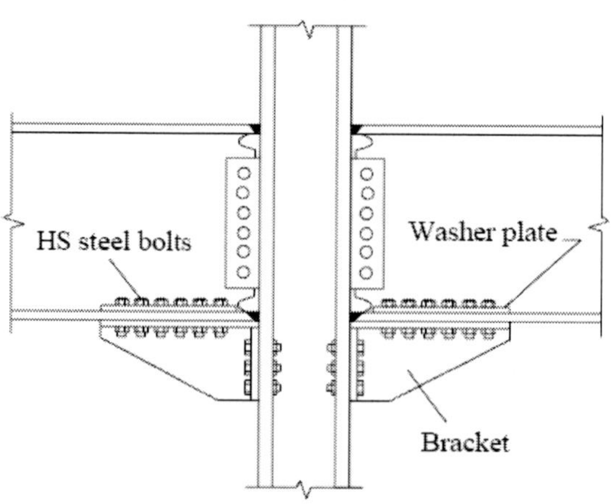

Figure 12. Bolted Bracket connection.

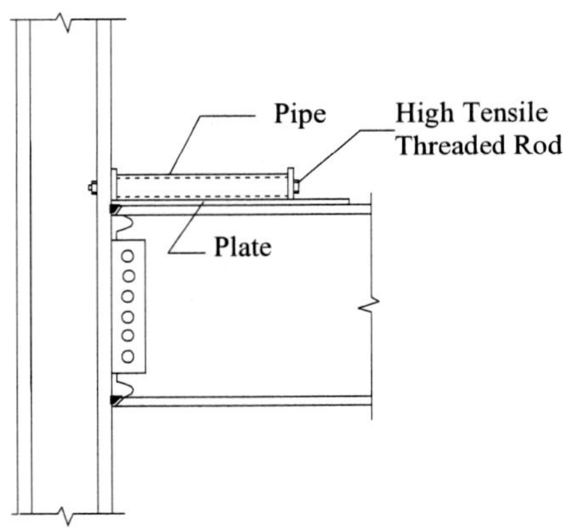

Figure 13. Pipe Bracket connection.

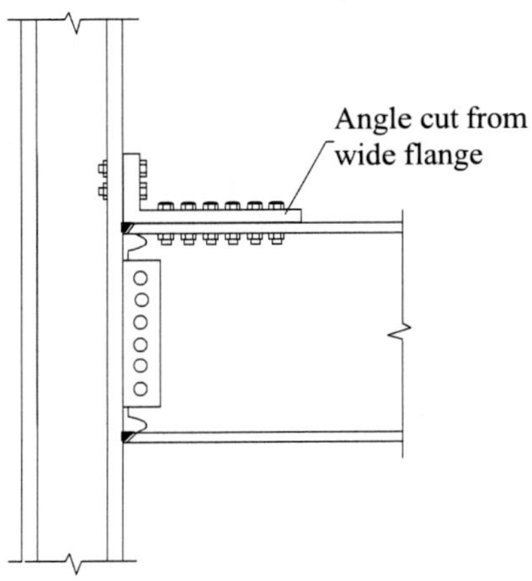

Figure 14. Angle Bracket connection.

Both pipe and angle brackets have the advantage of smaller dimension compared to the haunch bracket and can therefore be embedded in the concrete floor slab. However, for heavy beam sections, it may be necessary to place a pipe or angle bracket on both sides of the beam flange which may make fabrication and erection more costly than would be the case for the haunch bracket. When attaching the bracket to only one side of the beam flange, the use of a horizontal washer plate on the opposite side of the flange (Figure 12) has been shown to enhance connection ductility. It prevents propagation of flange buckling into the flange net area that otherwise may cause early fracture of the net area. Also, the use of a thin brass plate between the bracket and beam flange has been found to be effective in preventing both noise and galling associated with interface slip.

4.2. Weakening of the Beam Section

Weakening of the beam section can be done either by cutting a portion of the beam flange, reduced beam section connections or the beam web (wedge beam connections and reduced beam web connections).

4.2.1. Reduced Beam Section

The reduced beam section (or RBS) technique is illustrated in Figure 15. The beam flange is reduced in cross section thereby weakening the beam in flexure. Various profiles have been tried for the reduced beam section as illustrated in Figure 16. Other profiles are also possible. The intent is to force a plastic hinge to form in the reduced section. By introducing a structural "fuse" in the reduced section, the force demand that can be transmitted to the CJP flange welds is also reduced. The reduction in beam strength is, in most cases, acceptable since drift requirements frequently govern moment frame design and the members are larger than needed to satisfy strength requirements.

This technique has been shown to be quite promising in experimental tests in both with and without considering the effect of concrete slab [Chen 1996 and 2001, Engelhardt 1996 & Uang 2000]. The RBS plays a role quite similar to that of connection reinforcement schemes such as cover plates, ribs, and haunches. Both the RBS and connection reinforcement move the plastic hinge away from the face of the column and reduce stress levels in the vicinity of the CJP flange welds. The main problem with the stiffening of the joint is that stiffness attracts moment and increasing the moment at the column could induce failure in the column (especially in the case of a retrofit). Column failure is potentially more serious than the failure of the unimproved connection because it is more likely to lead to a collapse mechanism. In addition connection reinforcement often requires welds that are difficult and costly to make and inspect. These problems are lessened with the RBS, which is relatively easier to construct.

On the other hand, a greater degree of stress reduction can be achieved with connection reinforcement. For example, the size of haunches can be increased to achieve any desired level of stress reduction. With the RBS, on the other hand, there is a practical limit to the amount of flange material which can be removed. Consequently, there is a limit to the degree of stress reduction that can be achieved with the RBS. The reduced beam section therefore appears attractive for the modification of existing connections because of its relative simplicity and because it does not increase demands on the column and panel zone.

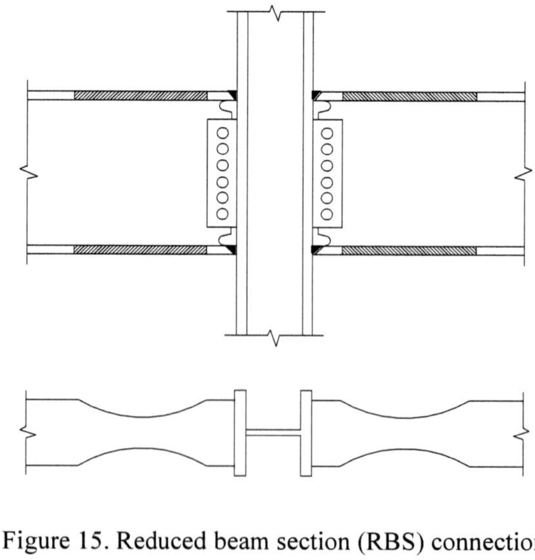

Figure 15. Reduced beam section (RBS) connection.

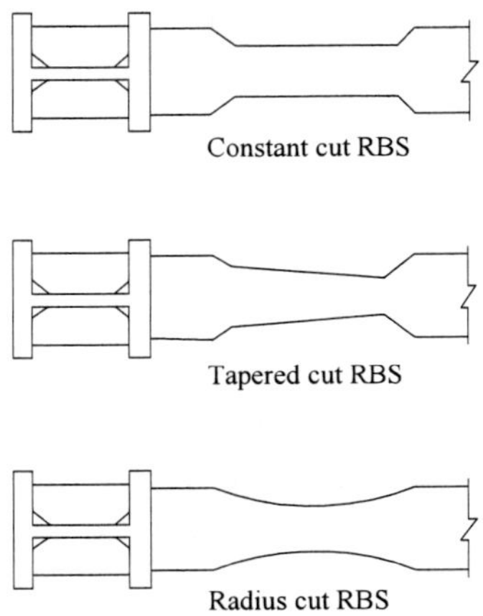

Figure 16. Various types of reduced beam section (RBS) connection.

For new construction, RBS cuts are typically provided in both the top and bottom beam flanges. However, this type of connection becomes relatively costly due to the cutting of flanges at four locations at each end, especially in the presence of floor slabs for rehabilitation purposes. Also in these connections cutting of flanges reduces the beam stability and increases the probability of beam lateral torsional buckling. This has been proven to be the case and "the reduced beam section usually experiences web local buckling first, followed by lateral torsional buckling and finally flange local buckling" [Naeim, 2001].

4.2.2. Slotted Web Connection

Figure 17 schematically shows the SlottedWeb connection proposed by Allen [1998]. It is similar to the popular field welded–field bolted beam-to-column moment frame connection, shown in the AISC LRFD and ASD steel design manuals, that has become known as the "pre-Northridge" connection.

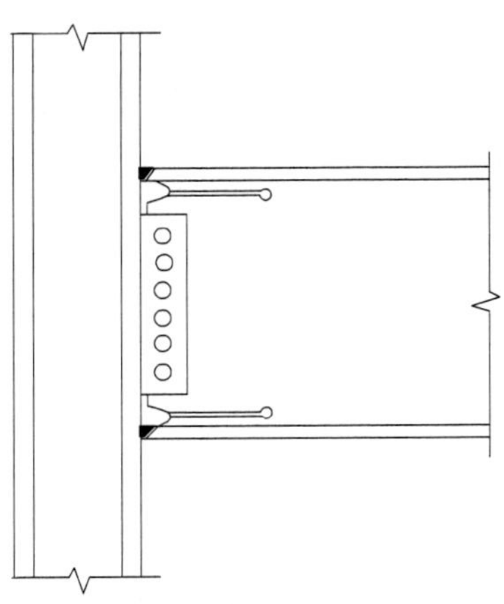

Figure 17. Proprietary SlottedWeb Connection. [Allen 1998]

The beam flanges are separated from the beam web in the region of the connection and the beam web is welded to the column flange, which changes the force, stress and strain distributions in this connection as explained below [Allen 1998]:

1. The beam web resists the entire beam vertical shear and its share of the beam moment whereas the flanges resist only moment.
2. The beam web separation from the beam flange reduces the large stress and strain gradients across and through the beam flanges and at the weld access hole by permitting the flanges to flex out of plane.
3. The separation of the beam web and the flange results in biaxial rather than triaxial stress and strain states in the region of the connection, which increases its fatigue life. Moreover, residual weldment stresses are significantly reduced because the slots also provide a long structural separation between the vertical web and horizontal flange weldments.

Ten SlottedWeb connections were tested by Allen [1998]. Experimental tests showed that all specimens are able to develop the full plastic moment capacity of the connected beam without any reduction in the elastic stiffness of the beam and also achieving the 3 percent plastic rotation.

4.2.3. Wedge Design Connection

Figure 18 shows the wedge design connection proposed by Wilkinsona [2006]. In this connection, the joint detail has the form of a beam with a wedge removed from the web and the flange is reattached to form a profile. The geometry of proposed wedge design connection by Wilkinsona [2006] was such that the plastic hinge was formed at a distance equal to the beam depth (D) away from the column face (Figure 18). The beam depth at hinge point can be calculated by equating the applied moment and the beam plastic moment capacity at the hinge point.

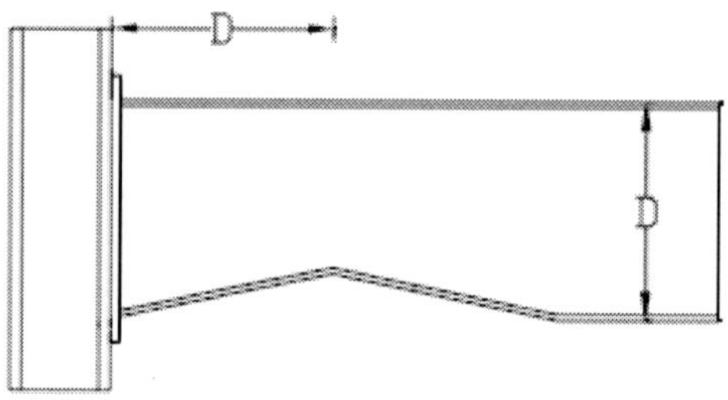

Figure 18. Geometry of the wedge detail. [Wilkinsona, 2006]

Two identical wedge design specimens were tested by Wilkinsona [2006] and connections easily reached to more than 3 percent plastic rotation. In these tests, the beam used was a shallow beam (UB254×102×22, beam depth: 254 mm) and the beam was connected to the column using an end-plate. A study was done by Hedayat and Celikag [2010] to investigate the ability of the wedge design concept to provide adequate joint strength and ductility in relation to welded flange-bolted web connections for deep beams.

This study showed inadequate connection ductility when deep beams were used. Therefore, they proposed to use double wedge with proper stiffener arrangements to achieve both strength and ductility.

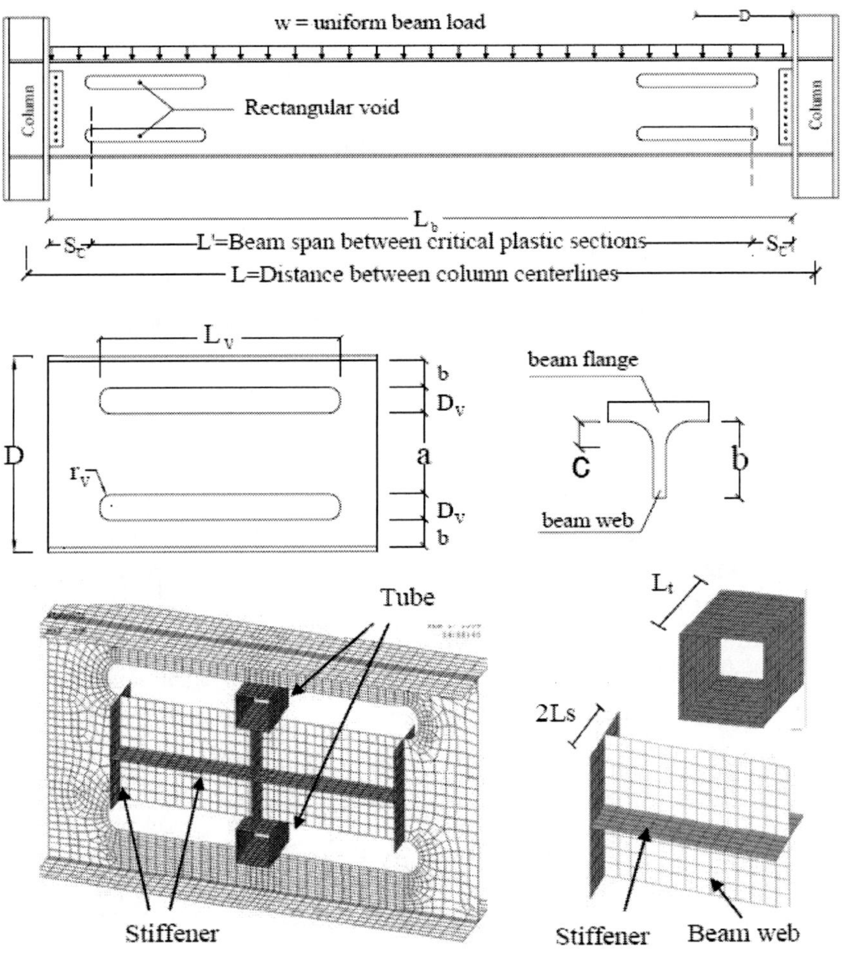

Figure 19. Details of proposed BEC by Hedayat and Celikag. [2009b]

4.2.4. Rbw Connections with Longitudinal Voids

In the study done by Hedayat and Celikag [2009b], in order to enhance the strength and ductility of post-Northridge connections with beam depths (varying from 450 mm to 912 mm), two parallel horizontal long voids were opened on their beams web (Figure 19). Results showed that the proposed beam end configuration (BEC) is effective in moving the plastic hinge away from the column face. This study also showed that adding web stiffeners and two tubes at the center of voids are effective in preventing excessive beam flange/web buckling (Figure 19). Based on the analytical results a step by step design procedure is proposed in this reference to determine the most suitable geometry for the BEC to achieve adequate connection strength and ductility.

4.2.5. Reduced Beam Web (Rbw) Connections with Circular Voids

Several design procedures have been developed for structural steel beams with web openings over the last three decades. One of them which provide a single technique that applies to all types of structural steel beams with web openings, that is, composite and noncomposite members, reinforced and unreinforced openings is presented in reference [ASCE, 1998]. This specification is intended to supplement the provisions for beams and other flexural members in the Load and Resistance Factor Design Specification for Structural Steel Buildings [AISC, 1993]. These design procedures may be classified into the following two types of construction:

1. Hot-rolled steel beams with single web openings: In these beams a rectangular or circular opening is used to pass water pipes. Welded steel plates might be provided above and below the web openings to reinforce the perforated sections.
2. Castellated beams: Castellated steel beams were fabricated from universal beams by cutting the universal beams and re-welding to create a new profile with higher section modulus of elasticity.

To improve the design guides presented by ASCE [1998] several valuable finite element and experimental studies were done by Liu and Chung [2003]. But it should be noted that in all mentioned documents, basic design procedures involved determination of maximum nominal flexural and shear capacity and interaction of flexure and shear. The main concern in these design procedures which often considered simply supported beams was to achieve adequate beam strength rather than enough connection rotational capacity to prevent early fracture at beam flanges at column face.

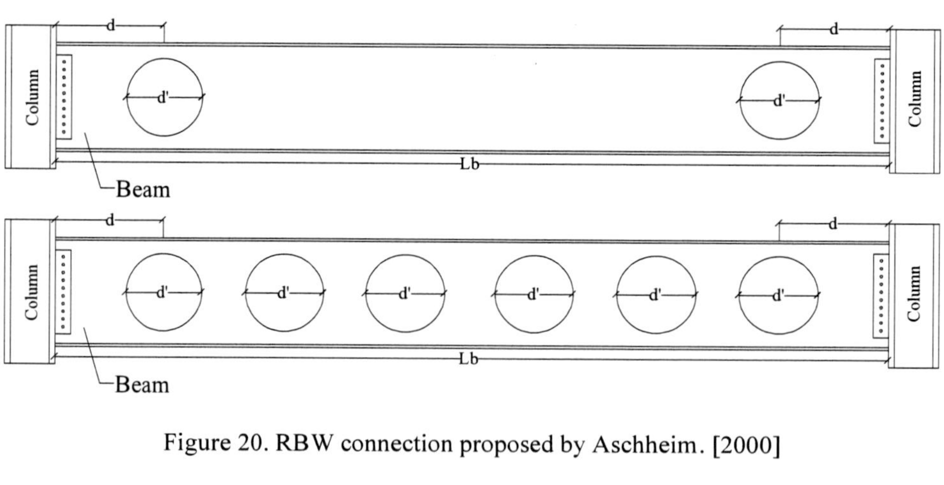

Figure 20. RBW connection proposed by Aschheim. [2000]

Among all design procedures with beam web openings, only the design procedure proposed by Aschheim [2000] is based on the achievement of adequate connection ductility for deep beams when used in seismic regions. Figure 20 shows the proposed RBW connections by Aschheim [2000]. In these connections the beam web is penetrated by a number of circular voids. The size and spacing between the voids are designed such as to cause the shear yielding of the beam web along the beam span and to keep the connection in elastic region. However, since the work presented by Aschheim [2000] is a US patent then there is limited information available. Five RBW connections of W21x68 Grade 50 beams were tested under cyclic loading by Aschheim [2000] and adequate inter-story drifts of 6% were obtained [FEMA350, 2000].

5. Strength and Ductility of Rbw Connections with Circular Voids

The study presented in this section is part of the PhD thesis on "Behaviour of reduced beam web connections" done by Hedayat [2009c]. This study aimed to investigate the ability of RBW connections with circular voids to provide adequate ductility in the case of using relatively deep beams. Also the design criterion used by Aschheim [2000] was compared with two other design criterions, which were named moment yielding and shear-moment yielding criterions. For these purposes, three post-Northridge connections previously tested by Lee [2000], W24x68, W30x99 and W36x150 beam (beam depths of 600, 750 and 910 mm respectively) were modeled by using finite element method. Then, the three design criterions were applied on these three connections and the strength, ductility and initial stiffness of these connections were investigated. In this process, a parametric study was done to find out the best arrangement of the voids on the beam web to achieve adequate strength, ductility and initial connection stiffness.

6. Finite Element Modelling

Proposed design procedures to modify pre and post Northridge connections are usually based on three beam sections: W36x150, W30x99 and W24x68. Therefore, three pre-tested post-Northridge connections (Figure 21), specimens SAC3 (beam: W24x68; column: W14x120), SAC5 (beam: W30x99; column: W14x176) and SAC7 (beam: W36x150; column: W14x257) used by Lee [2000], were modelled using the general purpose finite element program ANSYS [2007]. As stated by Lee and Stojadinovic [2000], specimens SAC3, SAC5 and SAC7 represent three conventional specimen sizes, small, medium and large respectively which were also tested in Phase 1 of SAC Steel Projects [1996]. The connection modification procedures presented by Engelhardt and Uang [2003] also were based on the experimental results of specimens with beams W36x150 and W30x99.

Figures 22 and 23 show the meshes of the finite element models for a typical non-modified SAC specimen and its modified configuration using single and multi circular voids respectively. There are no reports about the failure of pre-Northridge connections due to the fracture of bolts in all experimental tests and investigations done after Northridge earthquake. Therefore, in finite element modeling, the bolt shanks were not exactly modeled but shear tab, bolt holes and interaction between the shear tab and beam web were modeled to achieve a realistic model.

Both welds and base metals were modelled using shell elements (SHELL43, SHELL91) and associated material property was defined for each one (Figure 22). SHELL43 and SHELL91 are one-layer four-node and multi-layer eight-node shell elements respectively. Both of them have six (three translation and three rotation) degrees of freedom at each node and have plasticity, large deflection, and large strain capabilities. In the case of using SHELL91, each element was separated into five layers across the thickness. The number of layers was selected based on the finite element study carried out by Gilton and Uang [2002].

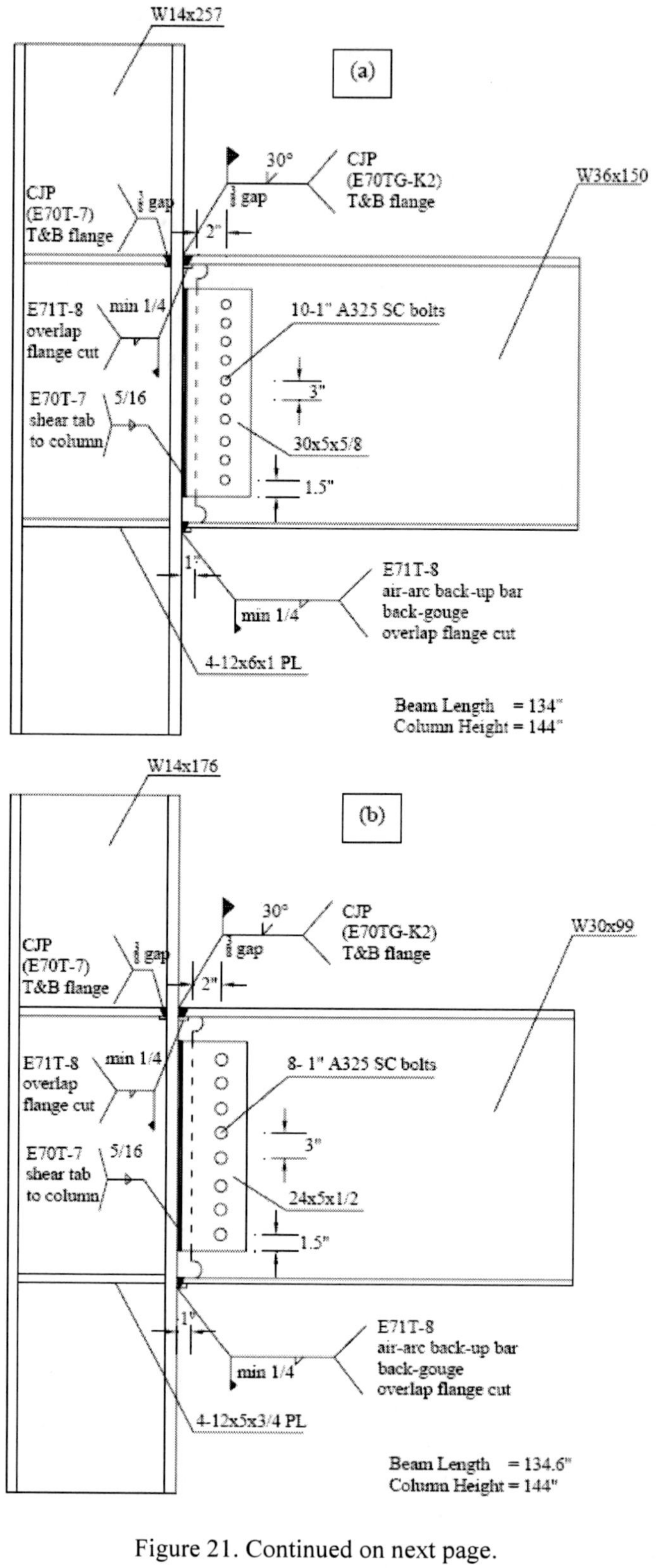

Figure 21. Continued on next page.

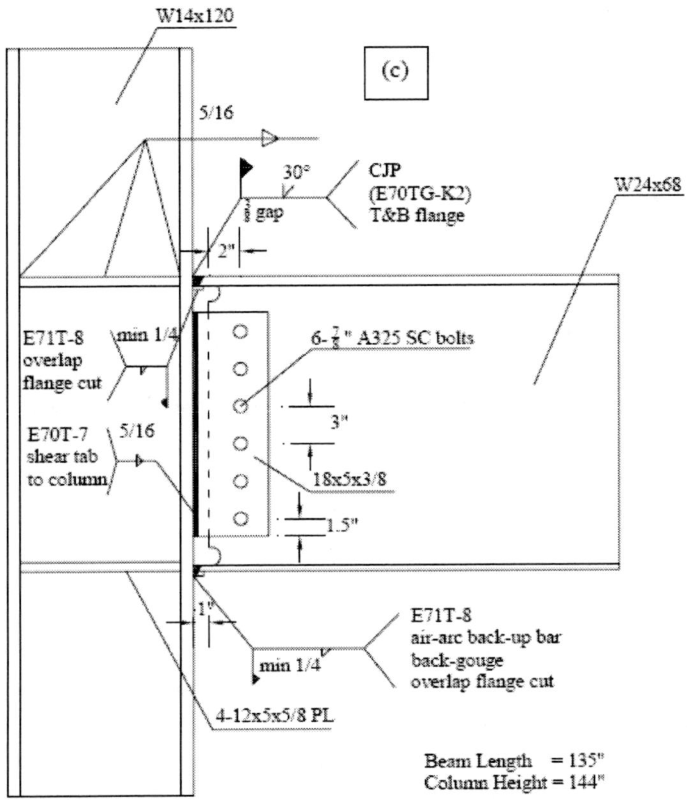

Figure 21. SAC specimens utilized by Lee [2000]: (a) SAC7; (b) SAC5; (c) SAC3.

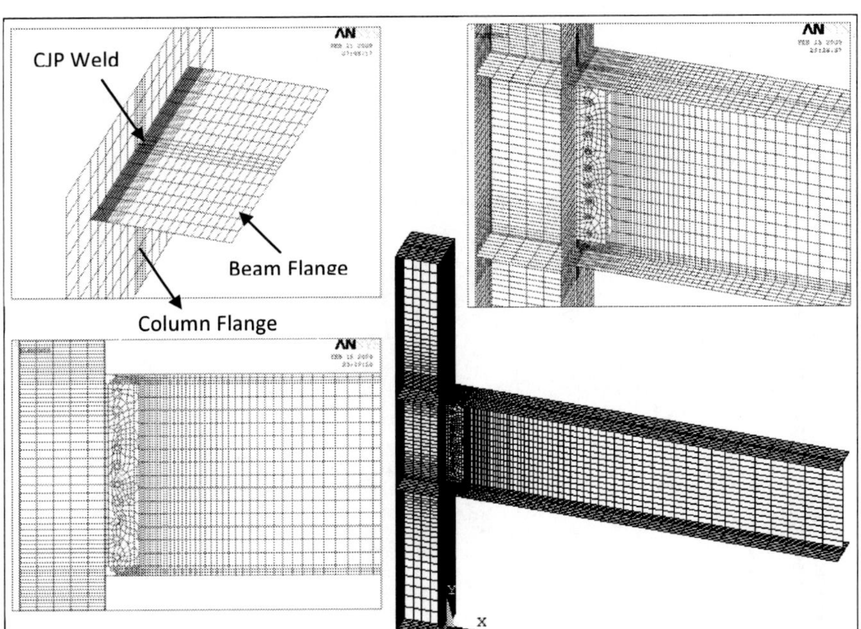

Figure 22. Finite element mesh of a typical SAC specimen.

In order to determine the appropriate mesh density, a mesh sensitivity study was done for both modified and non-modified specimens based on the recommendation given by ANSYS program and by comparing the analytical results with experimental results of other references [Lee et al., 2000; Wilkinsona et al., 2006). Based on this study, it was decided to carry out the model analyses for non-modified specimens (SAC3, SAC5, SAC7) with around 8500 elements, where 3000 elements were located at a region equal to the beam depth from the column face. For modified specimens, the total numbers of elements for the whole global model was around 13000 and at the void area between 3500 and 7000 for the largest and smallest void diameter respectively.

To perform the material nonlinearity analyses, plastic behaviour was based on the von-Mises yielding criteria and the associated flow rule. Isotropic hardening was assumed for the monotonic analysis, whereas kinematic hardening was assumed for the cyclic analysis as used by Mao et al. [2001] and Ricles et al. [2003]. A bilinear material response was used for base metals based on the material properties given by Lee et al., [2000], whilst for weld metals; a multi-linear material response based on material property given by Mao et al. [2001] and Ricles et al. [2003] was used.

After yielding, an accurate prediction of large deformations at void area was achieved via the consideration of the geometric nonlinearities through a small strain, large displacement formulation. The monotonic analyses were conducted by applying a monotonic vertical displacement load to the beam tip until achieving more than 4 percent total rotation at column web center, whereas the load history recommended by FEMA350 [2000] was utilized for cyclic analyses. When applied loads are in the vertical direction only, then the out of plane deformations (normal to the web) may not occur. Therefore, in order to ensure that buckling would occur when the model becomes unstable, the imperfect model was used to analyse under cyclic or monotonic loadings. In this study, in order to determine the imperfect model, firstly the buckling mode shapes were computed in a separate buckling analysis and then were implemented to perturb the original perfect geometry of the model as it was used by Moslehi Tabar [2005] and Zahrai [2006]. However, the imperfect model can also be obtained by applying small perturbing out of plane forces to the original perfect model at locations which are determined by either using experimental results or buckling analyses [Hennessey et al., 2004].

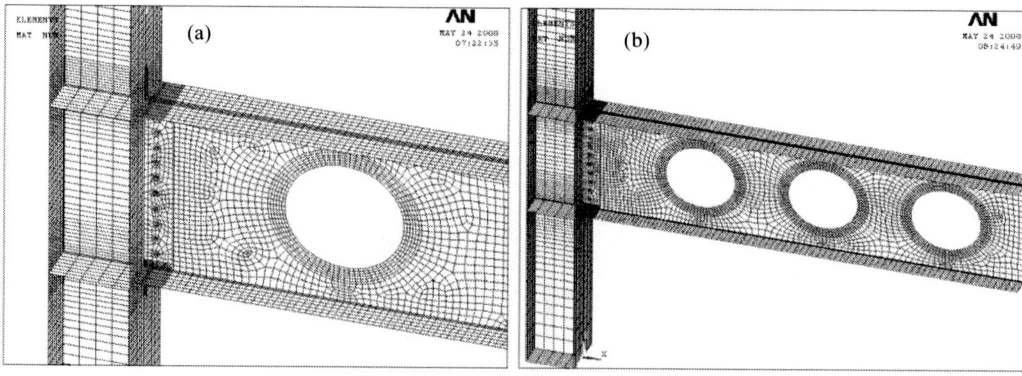

Figure 23. Finite element mesh for a typical RBW connection: (a) one void; (b) more than one void.

Both monotonic and cyclic analyses were conducted using full Newton-Raphson iterative scheme assisted by improved Arc-Length procedure presented by Forde and Stiemer [1987]. In order to verify the accuracy of modelling, the experimental results of specimens SAC3, SAC5 and SAC7 were compared with the analytical results obtained from the finite element models in terms of load and beam tip displacement. Figures 24.a to 24.c show this comparison for specimens SAC3, SAC5 and SAC7 under cyclic and monotonic loading respectively.

As Figure 24.a shows, generally, both analytical and experimental cyclic responses are close to each other, however, when loads become upward a minor mismatch can be seen between the two curves. This mismatch might be due to the use of different load history by Hedayat [2009c] and Lee et al. [2000]. Since this study is based on the chapter 3 of FEMA350 [2000], connection qualification, hence, the load history used was on that basis, whereas Lee et al. [2000] used the load history proposed by Clark et al. [1997] (Figure 25). Two 0.5% drift cycles were added after each cycle at drifts larger than 1% in the load history proposed by Clark et al. [1997]. It caused an increase in specimen strain hardening and consequently increased the load achieved at that particular displacement. However, the initial stiffness, maximum load achieved at the peak of the beam tip displacement and post-elastic envelopes under both cyclic and monotonic loadings for finite element model and test specimen were well matched. Similar behaviour can be observed for other SAC specimens, SAC5 and SAC3. Therefore, it might be concluded that the analytical results are in good agreement with experimental results.

7. Design Criterions

Three design criterions were investigated. The first criterion [Aschheim, 2000] is based on the shear yielding of the beam web. In this criterion, a circular opening (Figure 23.a) is provided in the beam web to cause the beam web to yield in shear at the void area when the maximum normal stress developed at the beam flanges near the column face is limited to a nominal target value, f_s, which is less than the yield strength of material ($f_s = \alpha . f_y$ where f_y is material yield stress and α is less than 1). Based on this definition, the void diameter (d') which is located at a distance equal to the beam depth (d) away from the column face is:

$$d' = d - 2\sqrt{3}.\alpha.S_b / (t_w . L_b) \tag{1}$$

where S_b, t_w and L_b are the beam section elastic modulus, the beam web thickness and the span length respectively. This criterion also considers the shear strength of the web at the location of voids in the longitudinal direction of the beam. In this case, n, the total number of voids along the entire beam span, is limited to L_b /d.

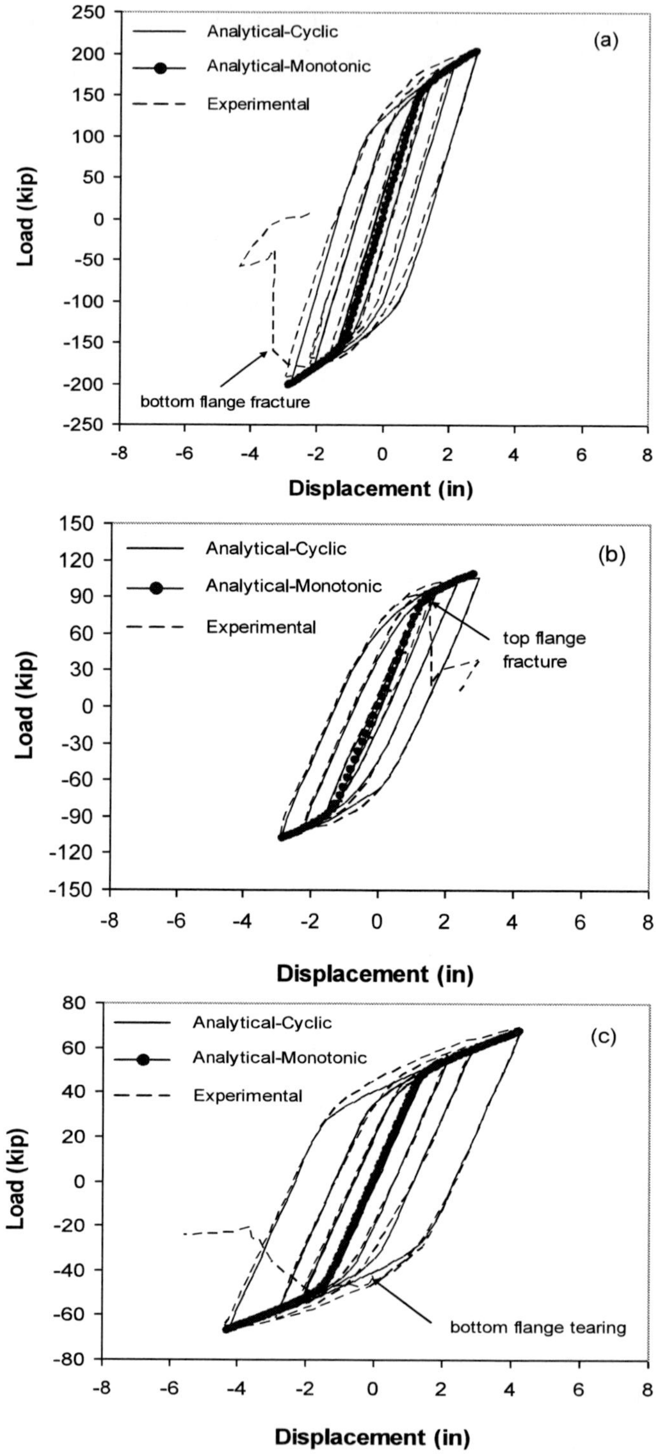

Figure 24. Comparison between analytical and experimental results in terms of beam tip load versus beam tip displacement for pre-tested specimens by Lee et al. [2000]: (a) SAC7; (b) SAC5; (c) SAC3.

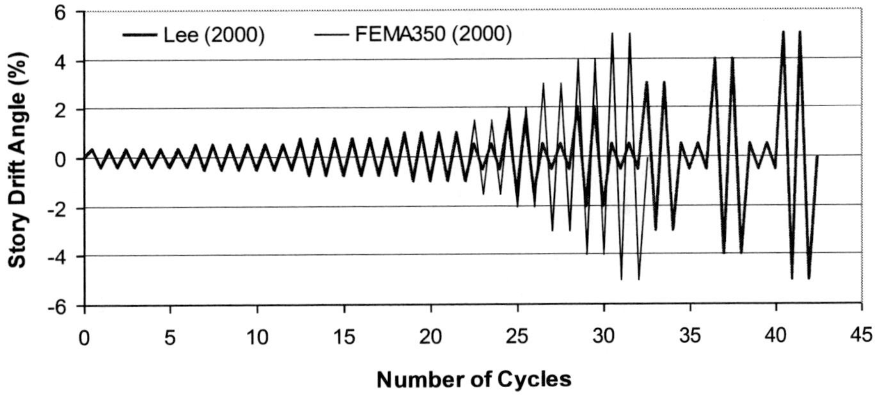

Figure 25. Comparison between load histories presented by FEMA350 [2000] and that of used by Lee et al. [2000]

The void diameter (d'), depends on the stress developed (f_s) in the column face. This stress was considered to be less than 1 by Aschheim [2000]. In this study, in order to determine the most suitable limit for f_s, this stress was chosen as proportion of yield stress, $f_y(f_s = \alpha.f_y)$. Therefore, a parametric study was done based on the parameter α for connections with one and more than one openings along the half beam span. Different values were used for parameter α and consequently the related void diameters for three specimens SAC3, SAC5 and SAC7 are presented in Table 1.

The second design criterion which was named moment yielding criterion was based on the moment yielding of the beam section. By considering the moment gradient, the diameter of the circular opening (d') was designed as such to cause the plasticization of the beam section at the column face and at the void location simultaneously (Equation 2). Parameter Z_b in Equation 2 is the beam section plastic modulus. The distance between the column face and the void center, L_d, was taken as the proportion of the beam depth, d (β=L_d / d =0.5, 0.75 and 1). Similar to the first criterion, the shear strength of the beam web at the location of voids in the longitudinal direction of the beam was used to determine the number of voids.

$$d' = \sqrt{\frac{4.Z_b}{t_w} - \frac{4.655S_b}{t_w}(1 - \frac{2L_d}{L_b})} \tag{2}$$

$$d' = d - 2\sqrt{3}.\beta.Z/(t_w.L_b) \tag{3}$$

$$L_d = L_b.d'^2.t_w/(8Z) \tag{4}$$

The number of voids was determined using the same procedure as the one used for the first and second criterion. The design parameters used for the second and the third design criterion of specimens SAC3 and SAC7 are presented in Table 2.

Table 1. Different values used for parameter α for specimens SAC3, SAC5 and SAC7 in the first criterion

Specimen	BEAM	α	d' (mm)	$L_d = d$ (mm)	Number of voids along the half beam length (n/2) & failure modes			
SAC3	W24x68	0.8	505	600	1	V	5	V
		0.9	495	600	1	V	5	V
		1	480	600	1	V	5	V
		1.1	470	600	1	V	5	V
		1.2	460	600	1	V	5	V
		1.3	445	600	1	W	5	W
		1.4	435	600	1	W	5	W
SAC5	W30x99	0.8	615	750	1	V	4	V
		0.9	600	750	1	V	4	V
		1	585	750	1	V	4	V
		1.1	565	750	1	V	4	V
		1.2	550	750	1	V	4	V
		1.3	530	750	1	W	4	W
		1.4	515	750	1	W	4	W
SAC7	W36x150	0.8	700	900	1	V	3	V
		0.9	670	900	1	V	3	V
		1	645	900	1	V	3	V
		1.1	620	900	1	W	3	W
		1.2	595	900	1	W	3	W
		1.3	565	900	1	W	3	W
		1.4	540	900	1	W	3	W

The third criterion was based on the yielding of the beam section in shear and moment at the same time and was named as shear-moment yielding criterion. The void diameter (d') was determined by using the Equation 3 where β is the over strengthening factor (1.2). However, the distance between the void center and the column face (L_d) was determined by using Equation 4 to cause the beam section plasticization in moment at void area and column face simultaneously.

Table 2. Design parameters used for the second and third design criterion of specimens SAC3 and SAC7

Specimen	BEAM	Second design criterion							Third design criterion					
		L_d / d	d' (mm)	L_d (mm)	n/2 & failure mode				d' (mm)	L_d (mm)	n/2 & failure mode			
SAC3	W24x68	0.5	285	300	1	W	8	W	435	625	1	W	5	W
		0.75	360	450	1	W	6	W						
		1	425	600	1	W	5	W						
SAC7	W36x150	0.5	555	450	1	W	3	W	540	450	1	W	3	W
		0.75	680	685	1	V	2	B						
		1	790	900	1	V	2	V						

8. Analytical Results

8.1. Failure Modes

Figures 23.a and 23.b show the finite element mesh for a typical RBW connection with a single and multiple circular opening at the beam web respectively. It was assumed that the material failure happens when the von-Mises strain reaches the ultimate strain of materials based on the stated material properties given by Lee [2000].

Based on this failure criterion, generally there are three different failure modes for RBW connections. These are:

- Beam flange fracture at void area (Figures 26.a and 26.b). This failure mode happened when a significant normal and shear strain concentration appeared at the beam flanges at the void area (Figure 27.a). It took place when the size of circular opening was relatively large which was corresponded to the small value of parameter α for the first design criterion and large value of parameter β for the second design criterion. The word "V" was used as abbreviation for this failure mode.
- Beam flange fracture at weld access hole (WAH) region (Figures 26.c and 26.d). This failure mode happened when the size of circular opening was relatively small. In this case, the normal and shear strains were mainly concentrated at beam flanges at WAH region (Figure 27.b). This was corresponded to large value of parameter α and small value of parameter β. The word "W" was used as abbreviation for this failure mode.
- Severe beam web buckling between the voids (Figure 26.e). This failure mode was rare. It took place when the distance between the openings became narrow. This happened only for specimen SAC7 in the second design criterion. The word "B" was used as abbreviation for this failure mode.

The failure modes of all RBW connections are listed in Tables 1 and 2. Table 1 shows that failure mode switching from the "V" state to the "W" state depends on the parameter α and the depth of beam section. The "W" failure mode started from a smaller value of parameter α (α=1.1) for deep beam specimens, SAC7, as opposed to the larger values of 1.3 for the other two specimens SAC5 and SAC3. Therefore, in order to prevent the WAH fracture for deep beams, the maximum developed normal stress at beam flanges at column face should be kept less than the yield stress of material. Whilst for shallower beams, the normal stress at column face can exceed the yield stress of material.

According to the proof-test protocol [Clark, 1997], a connection is acceptable if it sustains at least one complete displacement cycle with a plastic rotation of 0.03 rad (or total rotation of 0.04 rad), while maintaining resistance of 80% or more of beam plastic moment capacity during that displacement cycle. Investigation of the behaviour of a connection under cyclic loading is more accurate than it's behavior under monotonic loading. But cyclic loading analyses are time consuming and need a lot of hard disk space. Hence, all parametric studies were initially done using monotonic loading and then those connections which were acceptable based on monotonic loads were further checked using cyclic loading.

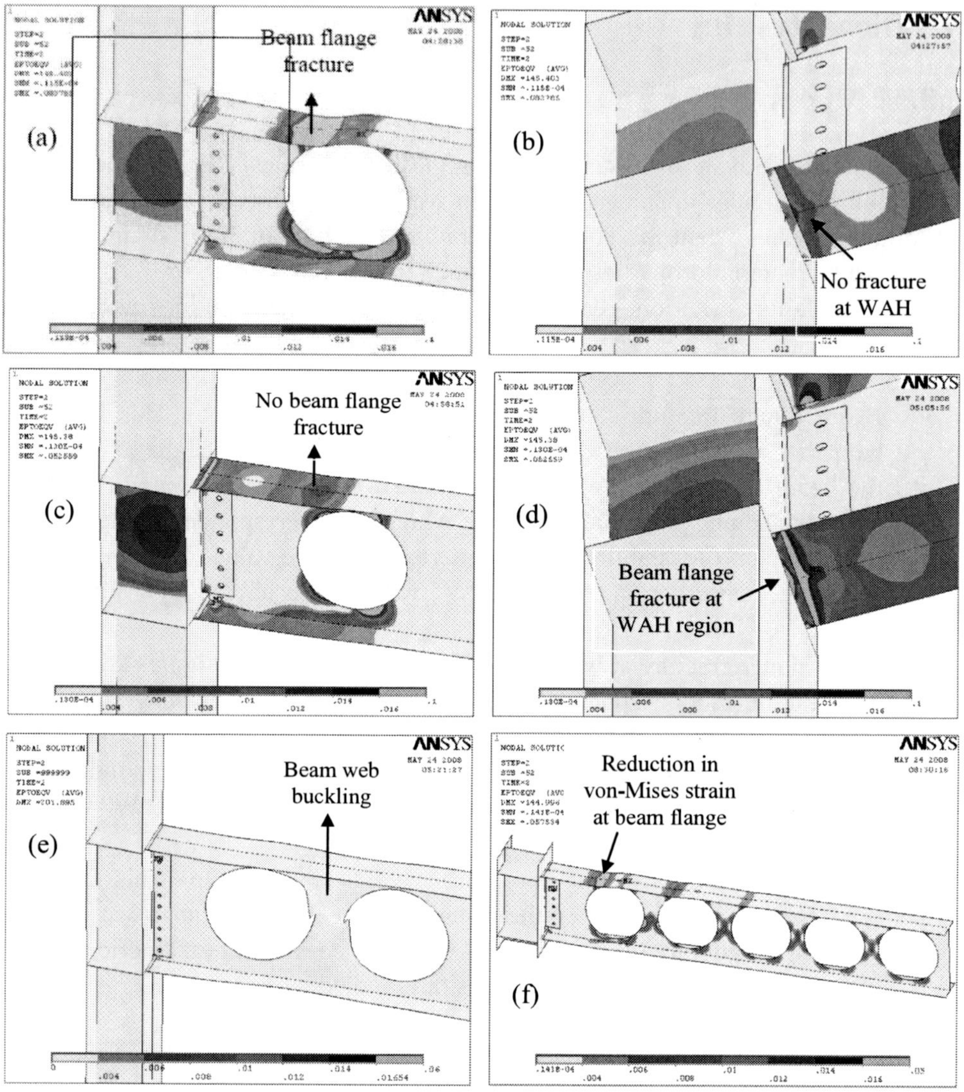

Figure 26. Different failure modes of a typical RBW connection.

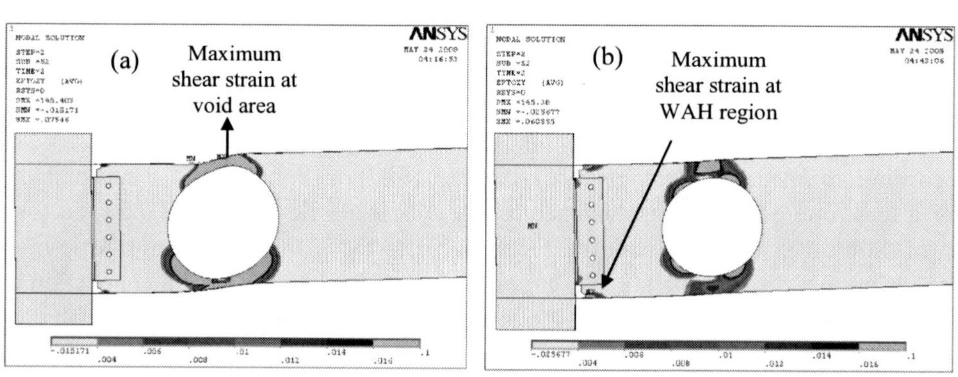

Figure 27. Location of maximum shear strain: (a) at void area for relatively large openings; (b) at WAH region for relatively small openings.

8.2. Behaviour of Rbw Connections Based on the Shear Yielding of the Beam Web

8.2.1. The Effects of Voids on the Moment-rotation Curves

Figure 28 shows the moment-rotation curves of specimen SAC3 in the presence of one and more than one void in the beam web for the first design criterion and using two different values of parameter α, 0.8 and 1.3. Connection total rotation was calculated based on chapter 3.9 of FEMA350 [2000], by dividing the beam tip deflection to the distance from the beam tip to the column web center.

Figure 28 shows that the presence of one void at the beam web does not always lead to increase in connection total rotation. For specimen of α equal to 0.8, which has relatively large opening size, the beam flange fracture happened at void area (Figure 26.a) earlier than the non-modified connection SAC3, while for specimen of α=1.3, which has smaller void size, there was a significant delay in the fracture time. However, this delay could not lead to the achievement of minimum required ductility (4 percent total rotation).

From the strength point of view for α=0.8, a significant reduction appeared in the connection strength while no remarkable change happened in the case of larger value of α, 1.3. When these two RBW connections are compared with the non-modified specimen SAC3, it is clear that the presence of opening causes a reduction in connection rotational stiffness. This reduction is more for smaller value of parameter α, 0.8. Increase in the number of voids along the beam length, reduced the von-Mises strain at beam flanges at critical locations, WAH region and void area (compare Figure 26.a and 26.f) which consequently led to increase in connection strength and ductility, such that specimen of α=1.3 reached to 4 percent total rotation. Increase in the number of voids increased the connection ductility, but it caused a significant reduction in the connection rotational stiffness, especially for α equal to 0.8. There is no information available in publications regarding the rotational stiffness of RBW connections. Therefore, investigation of the effect of voids on the initial rotational stiffness of connection was another aim of this study.

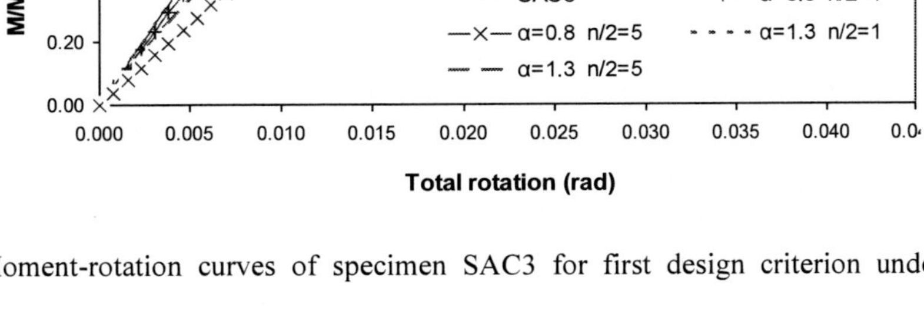

Figure 28. Moment-rotation curves of specimen SAC3 for first design criterion under monotonic loading.

8.2.2. Strength and Ductility

Figure 29 shows the maximum normalized moment-total rotation values for RBW specimens presented in Table 1. According to this figure, by increasing the parameter α, both total rotation and connection strength increased but after a specific value of parameter α, they started to decrease. In fact there was an optimum value for parameter α to achieve the highest connection strength and ductility. This optimum value depended on the beam depth and decreased by increasing the beam depth. These values were 1.3, 1.2 and 1 for specimens SAC3, SAC5 and SAC7 respectively.

These results are unlike those stated by Aschheim [2000] for shallower beams. The design criterion used by Aschheim [2000] was based on the fact that the openings should be designed as such to limit the maximum developed stress at the beam flanges (f_s) to a value less than the yield stress of material (α should be less than 1). The analytical results of deep beams showed that a small value of parameter α caused an early beam flange fracture at void area.

Among all the deep beam specimens only the shallowest one (SAC3 of α=1.2 and 1.3) with multiple voids reached the minimum required total rotation of 4 percent (Figure 29). For deep beams, the limitation of the design criterion is clear even when multiple same size openings are used. When connection is subjected to cyclic loading buckling of the beam web may cause reduction in the connection strength. Figure 30 shows normalized moment-rotation curve of a RBW connection with longitudinal voids (Figure 19) under cyclic loading. Furthermore, beam web buckling at the void area caused a reduction in the connection strength and create an unstable hysteretic curve.

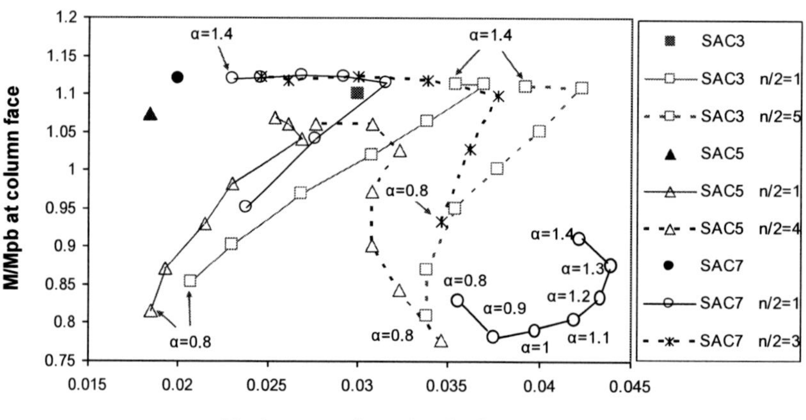

Figure 29. Maximum normalized moment-total rotation values for specimens SAC3, SAC5 and SAC7 for different values of parameter α for first design criterion.

For RBW connections with circular voids reduction in connection strength generally happens when voids are so close to each other. This creates a slender post between two adjacent circular voids. The moment-rotation curve of specimen SAC3 of 5 voids with α=1.3 and subject to cyclic loading is shown in Figure 31. This figure shows a stable hysteric curve for this specimen with no strength reduction under 4 percent total rotation. It also represents a suitable distribution of voids along the beam length. This connection satisfied both strength and ductility requirements based on proof-test protocol [Clark, 1997].

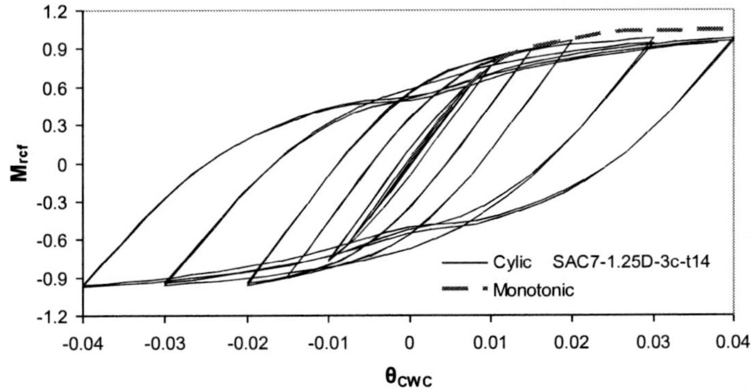

Figure 30. Normalized moment-rotation curve of specimen SAC7-1.25D-3c-t14 under cyclic loading. [Hedayat and Celikag, 2009b]

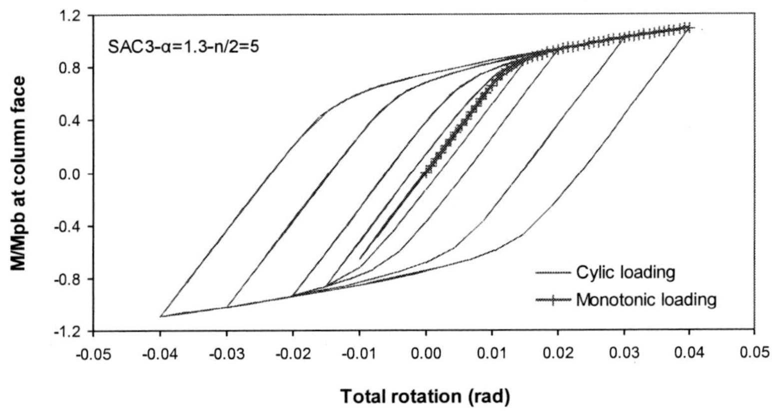

Figure 31. Moment-rotation curve of specimen SAC3 with 5 voids of α=1.3.

8.2.3. PZ Shear and Rotation

In welded connections there are only two sources to dissipate the released energy from the seismic loads, beam end and panel zone (PZ). One parameter which can represent the contribution of PZ in the energy dissipation is the ratio of the maximum PZ shear force at failure time to the PZ shear resistance (Vp). A statistical study on more than a hundred welded-flange connections [FEMA355D, 2000] indicated that specimens with large value of this ratio are those specimens that have large amounts of panel zone shear yielding and strain hardening and these specimens do not, on average, develop any more plastic deformation than specimens with less panel zone yielding.

This ratio versus the parameter α for different RBW connections is shown in Figure 32. This figure indicates a reduction in Vpz(failure)/Vp ratio, especially for small value of parameter α. It indicates a reduction in PZ yielding and consequently reduction in PEEQ index (plastic equivalent strain/yield strain of material) at WAH root and of course increase in PEEQ index at beam flanges at void area. It should be noted that a low value of Vpz(failure)/Vp ratio (e.g. 0.65 for specimen SAC5 n/2=4) is not also desirable since it represents a small contribution of PZ in the total plastic rotation of connection.

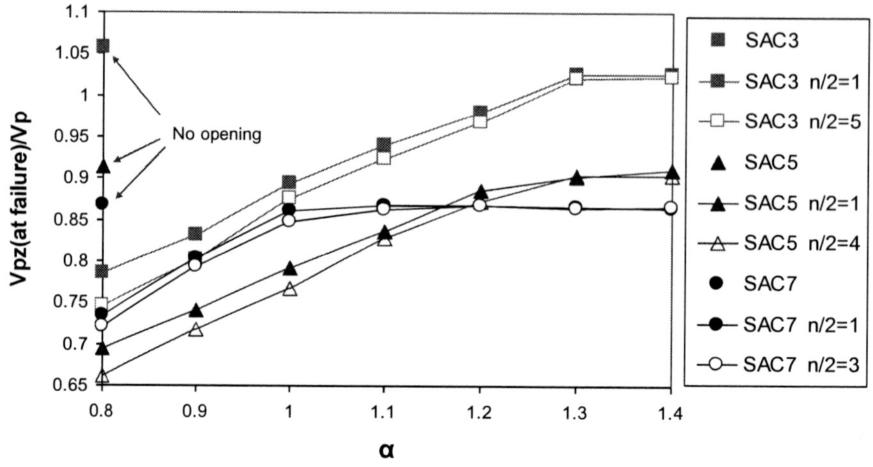

Figure 32. Vpz(failure)/Vp ratio versus the parameter α for RBW specimens SAC3, SAC5 and SAC7.

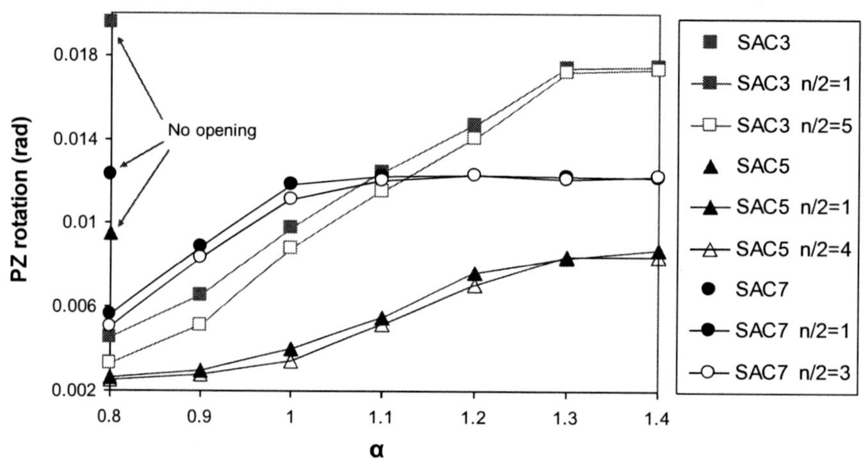

Figure 33. PZ rotation versus the parameter α for RBW specimens SAC3, SAC5 and SAC7.

Figure 32 shows that Vpz(failure)/Vp ratio increases by increase in the parameter α and becomes approximately constant for values of α greater than 1.3, 1.2 and 1 for specimens SAC3, SAC5 and SAC7 respectively. In fact, these values cause the maximum contribution of PZ in the plasticization while still keeping Vpz(failure)/Vp ratio less than that of the non-modified connections.

Figure 33 shows the PZ rotation versus the parameter α for RBW specimens presented in Table 1. The shear forces in PZ area causes the PZ rotation. By increasing the PZ rotation and consequently developing shear yielding in that area, the PZ contributes to the dissipation of seismic energies. However, excessive PZ rotation leads to strain concentration at WAH region and finally WAH fracture will be promoted.

It is also indicated that the presence of large openings at the beam web causes a significant reduction in PZ rotation and consequently reduces the contribution of PZ in the energy dissipation. The PZ rotation increases by increasing the parameter α and again becomes constant for α greater than 1.3, 1.2 and 1 for specimens SAC3, SAC5 and SAC7 respectively. These values can be considered as the best values for parameter α to achieve the highest contribution of PZ in the plasticization.

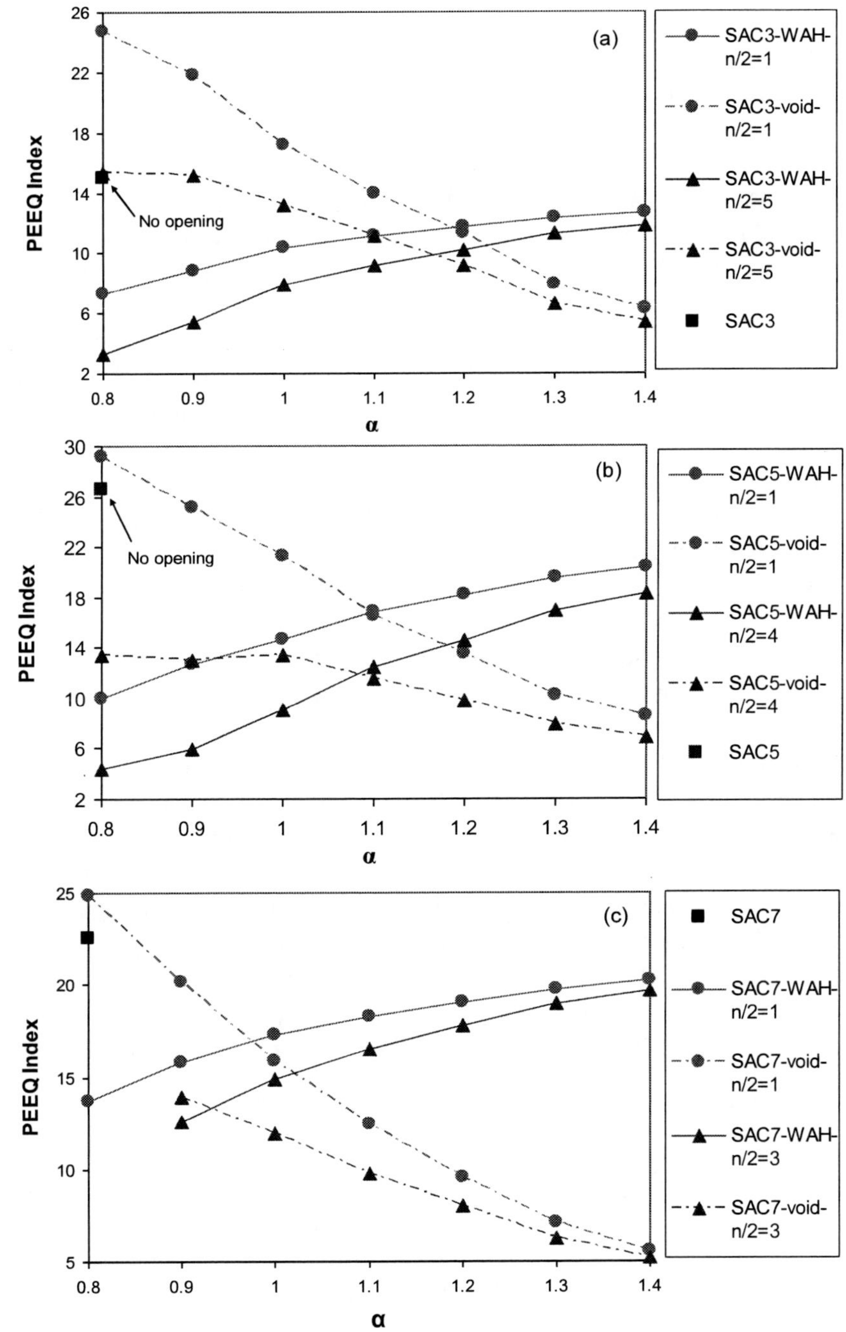

Figure 34. PEEQ index versus the parameter α at two critical locations, WAH root and void area for RBW specimens: (a) SAC3; (b) SAC5; (c) SAC7.

8.2.4. Normalized Equivalent Plastic Strain

The size of openings affects the fracture location of RBW connections. One of the parameters which can represent this behaviour is PEEQ index. The points with large value of this parameter are more likely to fracture. The PEEQ index versus the parameter α for specimens SAC3, SAC5 and SAC7 at beam flanges at two critical locations, WAH root and void area, are shown in Figures 34.a to 34.c.

Opening a relatively large void at the beam web (void with small value of α) caused a significant reduction in the PEEQ index at WAH root and significant increase at the beam flange near the void area. It indicates the early beam flange fracture at void area for specimens with large opening (Table 1). By increasing the parameter α, the PEEQ index at WAH root increased whilst decreasing at the void area, so that the failure mode of specimen SAC3 of α=1.4 was early beam flange fracture at WAH region (Table 1). Increasing the number of voids from one to five caused more reduction in PEEQ indexes at these two critical locations. This was the reason of achieving higher total rotation for specimens with multiple voids.

As was mentioned before, there is an optimum size for voids to achieve the highest connection strength and ductility. From the PEEQ index point of view, this point might be the intersection point of the PEEQ index lines at the WAH root and at the void area. These points were 1.2, 1.1 and 0.95 for RBW specimens SAC3, SAC5 and SAC7 respectively (Figures 34.a to 34.c). The optimum values for parameter α obtained from Figure 29 (1.3, 1.2 and 1) close to these intersection points (1.2, 1.1 and 0.95).

Therefore, it can be concluded that in the case of deep beams, for the first design criterion, the best value for parameter α is not always a constant value less than 1 (unlike what was proposed by Aschheim, 2000) but depends on the beam depth and reduces by increasing the beam depth. Figure 35 shows this relationship and using least square method a linear equation between the parameter α and the beam height was also proposed to predict the best value for parameter α for a specific beam height. In this figure, the ratio of the void diameter to the beam depth (d'/d) and the ratio of the beam section plastic moment at void area to the beam section plastic moment at column face (r ratio=Mpbv/Mpbf) for different beam height specimens are presented. The void diameter ratio (d'/d) and plastic moment ratio (r) for all specimens seem to be independent on the beam depth and are around 0.73 and 0.82 respectively.

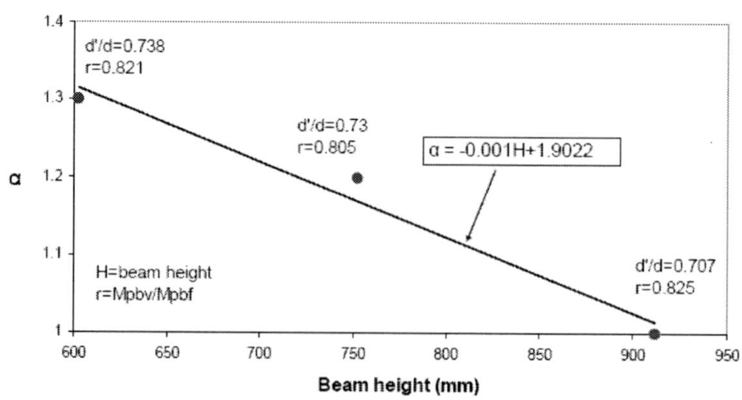

Figure 35. Relationship between the parameter α and the beam depth.

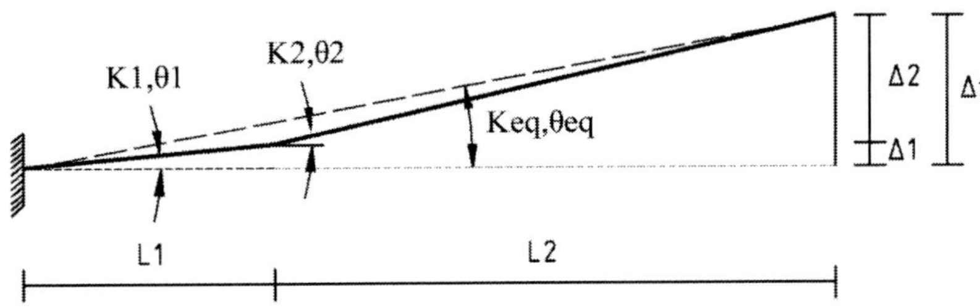

Figure 36. Equivalent initial rotational stiffness of a typical RBW connection.

8.2.5. Initial Rotational Stiffness

When connection rotational stiffness is considered, the presence of openings at the beam web divides the entire beam length into two regions. The first region, which starts from the column face to the center of the first void is a stiff region and has high initial rotational stiffness. The second region, which starts form the center of the first void to the beam tip is approximately a soft region and has lower initial rotational stiffness. Hence, the actual rotational stiffness of a RBW connection is a combination of the rotational stiffness of these two regions. Therefore, a method similar to the one defined by Mazzolani [2000] was used to calculate the equivalent initial rotational stiffness (Keq) for RBW connections. In this method, Keq represents the length and the initial rotational stiffness (Li and Ki) of each region (Figure 36).

Keq can be calculated as follow:

$$\Delta t = \Delta 1 + \Delta 2 \tag{5}$$

For each region $\theta i = Mi / Ki$ and $\theta i = \Delta i / Li$ where Mi and θi are moment and rotation of each region respectively. Substituting Δi in Equation 5, the connection equivalent initial rotational stiffness (Keq) will be as fallow:

$$\frac{1}{Keq} = \frac{L1}{L1+L2} \times \frac{1}{K1} + \left(\frac{L2}{L1+L2}\right)^2 \times \frac{1}{K2} \tag{6}$$

The Keq for all RBW connections were normalized with respect to the initial rotational stiffness of non-modified connections and are shown in Figure 37 versus the parameter α.

Figure 37 indicates that the presence of an opening at the beam web reduced the $\overline{Keq}$ and the reduction was higher for specimen with multiple voids. By increasing the parameter α and consequently reducing the opening sizes, the $\overline{Keq}$ increased. This figure clearly shows the detrimental effect of large openings on the connection rotational stiffness and finally on the lateral frame stiffness. The $\overline{Keq}$ for the optimum values of parameter α (1.3, 1.2 and 1) in the case of one and more than one openings at the beam web for specimens SAC3, SAC5 and SAC7 were 0.997, 0.97, 0.95 and 0.822, 0.77, 0.79 respectively. The stiffness reduction of multiple voids specimens is, on the average, 10 percent more than that reported for the dog bone connections (about 10%).

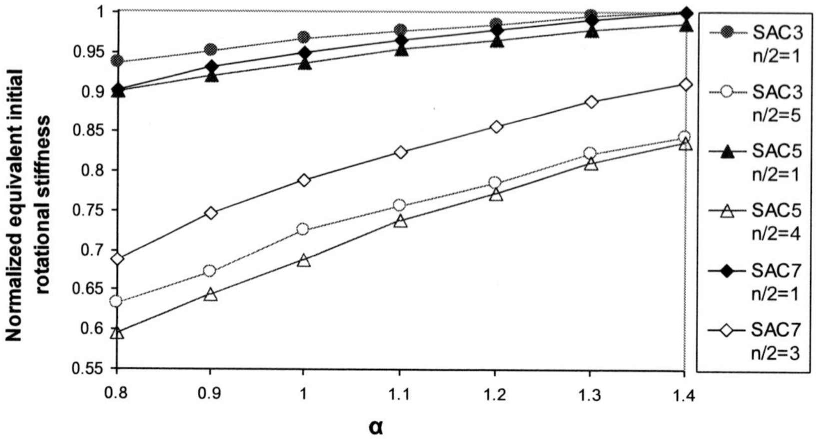

Figure 37. $\overline{Keq}$ for RBW connections versus the parameter α.

8.3. Behaviour of Rbw Connections Designed for the Second Design Criterion

8.3.1. Strength and Ductility

Figure 38 shows the maximum normalized moment versus total rotation values for the RBW specimens. Their design was based on the second design criterion (Table 2). First of all, the behaviour of SAC3 and SAC7 are not similar. For SAC3, by increasing the parameter β the total connection rotation increases while the connection strength remains approximately constant. The highest ductility was achieved for β=1 (3.7% rad) in the case of using multiple voids at the beam web. However, this rotation is less than the minimum required total rotation, 4% rad.

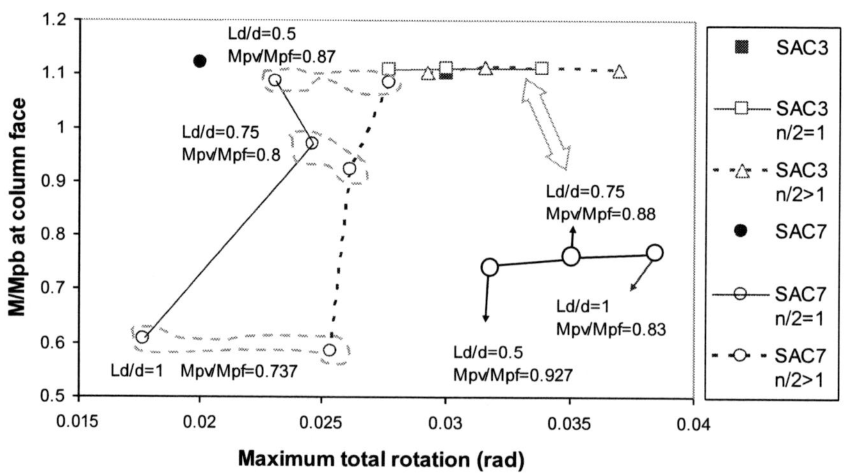

Figure 38. Maximum normalized moment-total rotation values for RBW specimens SAC3 and SAC7 for different values of parameter β for the second design criterion.

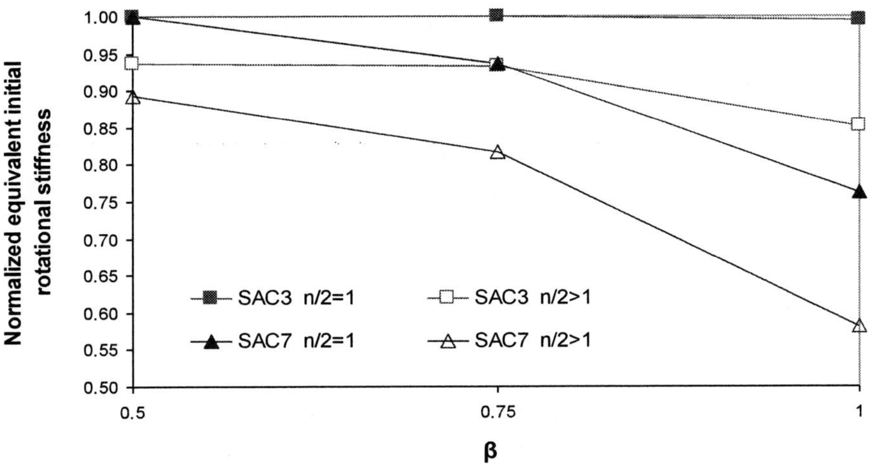

Figure 39. $\overline{Keq}$ for RBW connections versus the parameter β for the second design criterion.

For RBW specimens SAC7, as the parameter β increased the total rotation was also increased, however, finally reduced. Increase in parameter β always caused a significant reduction in the connection strength. For these specimens the maximum total rotation achieved could not even satisfy the minimum 3 percent total rotation required for ordinary moment frames.

For all specimens, the ratios of the beam plastic moment at void area to the beam plastic moment at the column face (r) are shown in Figure 38. The connection with the highest ductility (specimen SAC3, β=1, n/2=5) is corresponded to r=0.837 which is close to the r ratio of the specimen with the highest ductility in the first design criterion (SAC3, α=1.3, n/2=5, r=0.821). Both of these two specimens failed due to the beam flange fracture at WAH region. The reason for higher ductility of the specimen for the first design criterion originates from its larger void diameter size. This caused a decrease in von-Mises stresses at WAH region which finally led to achievement of 4% total rotation. Therefore, it might be concluded that the ratio r=0.83 can be considered as a design parameter that does not depend on any design criterion.

8.3.2. Initial Rotational Stiffness

Figure 39 shows the $\overline{Keq}$ for all RBW connections in the case of second design criterion. The presence of openings at the beam web area decreased the $\overline{Keq}$ and the reduction increased by increasing the beam depth and the number of voids. The $\overline{Keq}$ of the specimen with the highest ductility (specimen SAC3, β=1, n/2=5) is 85 percent which is 3 percent higher than the best specimen in first design criterion (specimen SAC3, α=1.3, n/2=5). This was due to the smaller void size in the case of second design criterion.

8.4. Behaviour of Rbw Connections Designed for the Third Design Criterion

8.4.1. Strength, Ductility and Rotational Stiffness

Figure 40 shows the maximum normalized moment-total rotation values for RBW specimens designed by considering the third design criterion (Table 2). The presence of voids at the beam web increased the connection total rotation for both specimens while no remarkable change happened in the connection strength. The shallower beam specimen achieved higher total rotation (3.92%) than the deep beam specimen SAC7 (2.77%). However, both of them were less than the minimum required total rotation (4%).

Specimen SAC3 achieved higher ductility than the most ductile specimen in the second design criterion and lower ductility than the most ductile specimen in the first design criterion (4.23>3.92>3.7). The following might be the reason:

Specimen SAC3 in the first design criterion has a void diameter of 445mm which is greater than 435mm and 425 mm in the third and second design criterion respectively. The larger void diameter caused a reasonable reduction in von-Mises stress at beam flanges at WAH region and consequently delays in the beam flange fracture which finally caused connection to achieve higher ductility. For specimen SAC7, the maximum connection total rotation in the second and the third design criterion were same. Figure 41 shows the $\overline{Keq}$ for all RBW connections for the third design criterion. The $\overline{Keq}$ of these specimens were similar to related specimens of the second design criterion.

8.5. Comparison of the Three Design Criterions

Table 3 compares those of specimens SAC3, SAC5 and SAC7 of three design criterions which achieved the highest ductility. This table shows that all design criterions achieved a connection strength between 0.97 and 1.113 which are more than the minimum required, M/Mpb=0.8. The highest connection ductility achieved for a r ratio is around 0.83 for all three design criterions where the first void is located at a distance equal to the beam depth away from the column face.

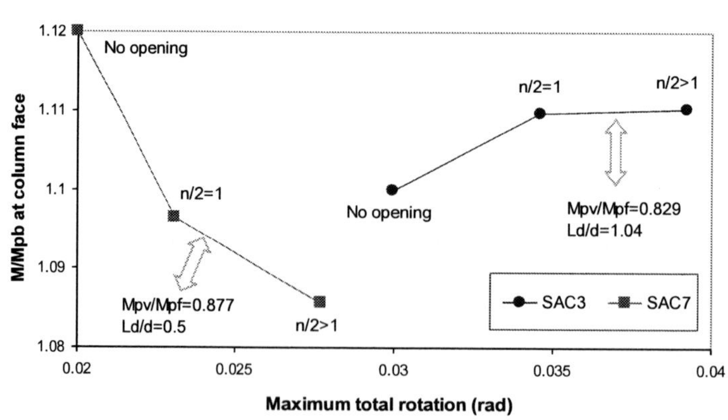

Figure 40. Maximum normalized moment-total rotation values for RBW specimens SAC3 and SAC7 for the third design criterion.

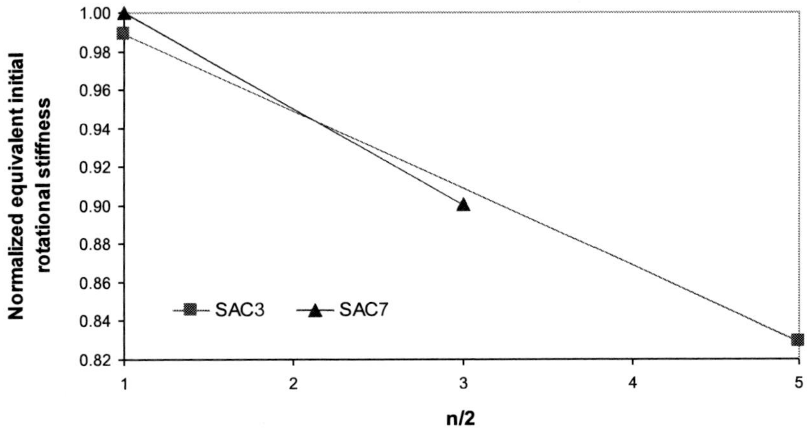

Figure 41. Normalized equivalent initial rotational stiffness for the RBW connections versus the number of voids for the third design criterion.

Table 3. The comparison of the best specimens in the three design criterions

Design Cri.	Design parameters	RBW Specimens					
		SAC3	SAC5	SAC7	SAC3	SAC5	SAC7
First design criterion	α	1.3	1.2	1	1.3	1.2	1
	d' (mm)	445	550	645	445	550	645
	L_d (mm)	600	750	900	600	750	900
	n/2	1	1	1	5	4	3
	r ratio	0.821	0.805	0.825	0.821	0.805	0.825
	total rotation (% rad)	3.7	2.7	3.16	4.23	3.23	3.77
	M/Mpb	1.113	1.042	1.115	1.108	1.026	1.098
	$\overline{Keq}$	0.99	0.97	0.95	0.823	0.77	0.79
Second design criterion	β	1	_	0.75	1	_	0.5
	d' (mm)	425	_	680	425	_	555
	L_d (mm)	600	_	685	600	_	450
	n/2	1	_	1	5	_	2
	r ratio	0.837	_	0.802	0.837	_	0.87
	total rotation (% rad)	3.38	_	2.46	3.69	_	2.77
	M/Mpb	1.111	_	0.97	1.109	_	1.084
	$\overline{Keq}$	0.99	_	0.94	0.85	_	0.89
Third design criterion	d' (mm)	435	_	540	435	_	540
	L_d (mm)	625	_	450	625	_	450
	n/2	1	_	1	5	_	3
	r ratio	0.829	_	0.877	0.829	_	0.877
	total rotation (% rad)	3.46	_	2.3	3.92	_	2.77
	M/Mpb	1.11	_	1.096	1.11	_	1.086
	$\overline{Keq}$	0.99	_	1	0.83	_	0.9

For all cases the first design criterion caused the highest connection total rotation. However, for this design criterion the use of multiple openings at the beam web area could only increase the total rotation of the shallowest beam specimen (SAC3, 600 mm) to the minimum required ductility (4 percent total rotation) and was failed to do so for the deeper beams. Therefore, in the following section, an attempt was made to increase the connection total rotation to the minimum required ductility for deeper beams, SAC5 and SAC7.

9. RBW Connection with Multiple and Different Size Voids

9.1. Specimen Sac7

In section 8.2, in order to increase the ductility of connection, multiple voids of same size were created at the beam web along the entire beam length. The void diameter was a function of parameter α (Equation 1), which was also a function of the beam depth (Figure 35).

For instance, unity was the best value for parameter α for multiple voids specimen SAC7 (Figure 29). Table 1 showed that for α less than 1 (e.g. 0.9), this connection failed due to the beam flange fracture at the first void area, while for values greater than unity (e.g. 1.1), early fracture at WAH region prevented the specimen from reaching 4% total rotation. Therefore, it can be concluded that a combination of n (n=Lb/d) voids of different sizes might lead to higher ductility. Similar work was also done by Aschheim [2000], but since this work is a patent then there is no information available in publications about the void sizes and their arrangements. However, the current study also aimed to develop the design criterion used by Aschheim [2000] for beams deeper than those used by Aschheim [2000].

Hence, for specimen SAC7 the diameter of the first void was taken as a variable (by considering α as a variable) and the diameter of the remaining voids was kept as a constant with the maximum size being 700 mm, α=0.8. The number of voids was kept constant as before, 3 (n=Lb/d). Figure 42 shows the configuration of voids along the half length of the beam for specimen SAC7. Analytical results showed that this void arrangement caused an early beam web buckling specially between the first and the second void. Therefore, in order to prevent this early failure mode, vertical stiffeners were used between the openings on both sides of the beam web to connect the top and the bottom beam flanges (Figure 42). The thickness of vertical stiffeners was equal to the thickness of the beam web.

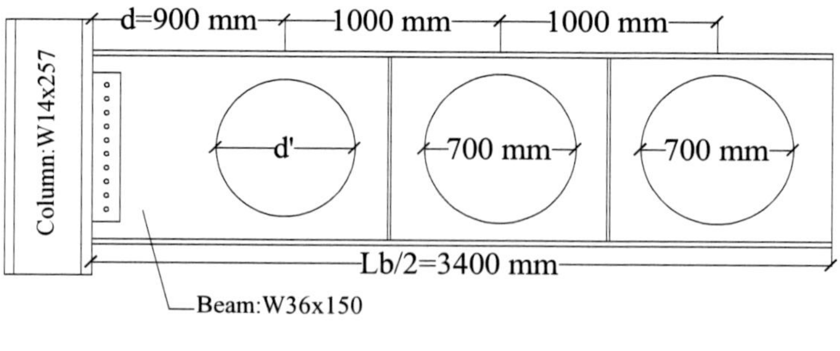

Figure 42. Voids arrangement along the half beam length for specimen SAC7 in the case of using different void sizes.

Figure 43 shows the maximum normalized moment-total rotation values of SAC7 specimens with one void and multiple voids of same size and different sizes at the beam web. This indicates a significant improvement in the connection total rotation when compared to multiple voids specimens of same sizes (compare the solid triangular curve with the solid circular curve). By increasing parameter α, the connection total rotation increased but after a value of α= 1.1 it started to decrease. From this figure it might be concluded that the best value of parameter α is 1 or 1.1. For both values of α a 5 percent total rotation was achieved.

Figure 44 shows the moment-rotation curve of RBW specimen SAC7 for α=1.1 under cyclic loading. This specimen has a stable hysteric curve with no strength reduction even at 5 percent total rotation. It is due to the efficiency of the vertical stiffeners used at the beam web between the voids to prevent undesirable web buckling. This connection satisfied both strength and ductility requirements for connections used in seismic regions.

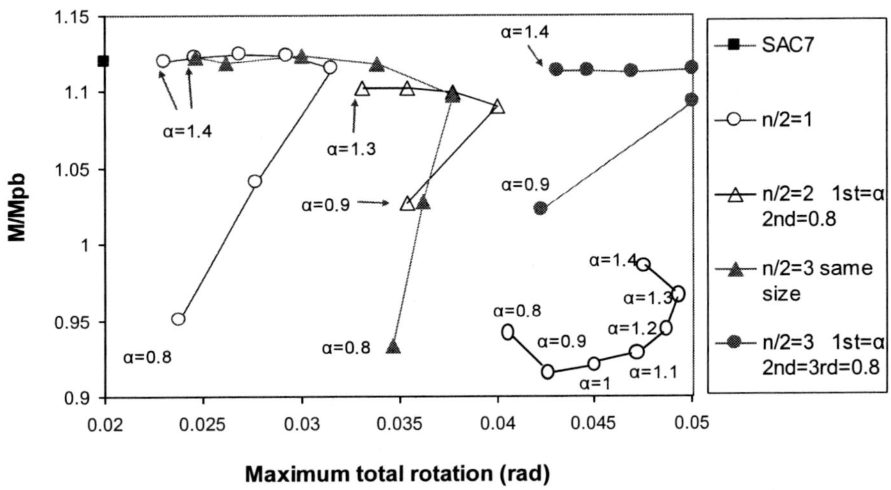

Figure 43. Maximum normalized moment-total rotation values for SAC7 specimens in the presence of different void arrangements at beam web area.

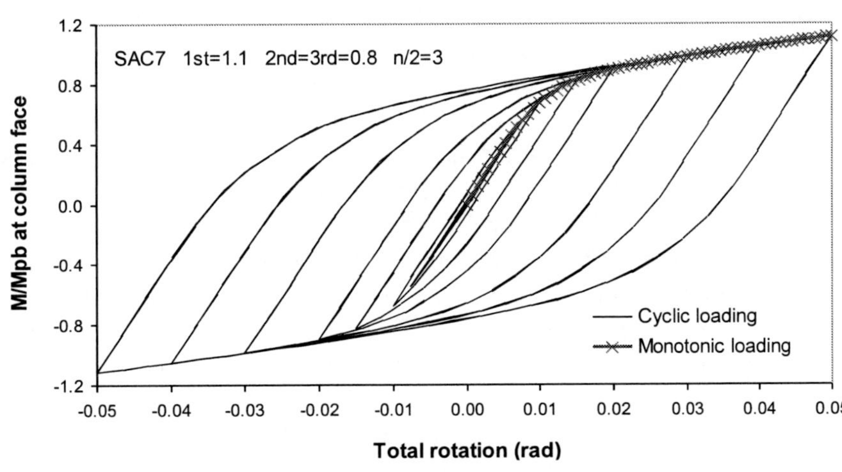

Figure 44. Moment-rotation curve of specimen SAC7 with voids of different size under cyclic loading where α is equal to 1.1 for first void.

Figure 45.a shows the von-Mises strain distribution at 4% total rotation for RBW specimen SAC7, having same diameter voids which corresponded to α=1. This void arrangement caused the highest connection ductility for specimen SAC7 in the first design criterion (3.77%, Figure 29). Figure 45.b shows strain distribution in the presence of different void diameter, where the first void diameter was corresponded to α=1 and the others were corresponded to α=0.8. Comparing these two figures it is clear that there is an increase in the contribution of the beam web at the second and the third void area to absorb von-Mises strain. This caused significant reduction in the concentration of von-Mises strains at the critical locations of the beam flanges, void area and WAH region. This caused a 20 percent reduction in the maximum value of von-Mises strain at the root of WAH which finally caused the specimen to reach 4 percent total rotation.

Finite element analyses showed that, from the rotational stiffness point of view, the entire beam length can be divided into three segments when the diameter of the first void is less than other voids. The first segment starts from the column face to the center of the first void, the second segment begins from the center of the first void to the center of the second void and the rest of beam length was taken as the third segment. Therefore, the equivalent initial rotational stiffness (Keq) can be obtained by the method mentioned in section 8.2.5 and as follows:

$$\frac{1}{Keq} = \frac{L1}{L} \times \frac{1}{K1} + \frac{L2(L2 + L3)}{L^2} \times \frac{1}{K2} + \frac{L3^2}{L^2} \times \frac{1}{K3} \tag{7}$$

Figure 46 compares the normalized equivalent initial rotational stiffness ($\overline{Keq}$) for specimens SAC7 with different void arrangements in the beam web area. This figure shows an average 10 percent reduction in $\overline{Keq}$ when the same size voids (n/2=3) compared to the one with different size of voids (n/2=3, 1st=α, 2nd=3rd=0.8).

For specimens with different void sizes and α equal to 1 and 1.1 the $\overline{Keq}$ were 0.72 and 0.73 respectively. Therefore, using different void sizes lead to achieve high connection strength and ductility but caused a significant reduction in connection rotational stiffness. This indicates the difficulty in achieving adequate connection strength, ductility and initial rotational stiffness simultaneously. In order to increase the connection rotational stiffness, it was proposed to decrease the number of voids. Hence, n/2-1 (3-1=2) voids were created at the half beam length. In this case the size of the first void was taken as a variable and the second void diameter was constant which was corresponded to α=0.8.

Figure 43 shows the maximum normalized moment-total rotation values for these specimens. The behaviour of these specimens were also same as other RBW connections, such that there is an optimum value for parameter α, in which connection shows highest ductility. Specimen with α =1 achieved the highest ductility which is equal to 4%. When the number of voids decreases, the total rotation decreases. Figure 47 shows the moment-rotation curve of this specimen (n/2=2, α =1) under cyclic loading. This is a stable hysteric curve for SAC7 with no strength reduction under 4 percent total rotation. This connection satisfied both strength and ductility requirements for connections used in seismic regions. $\overline{Keq}$ of these specimens (SAC7, n/2=2) are shown in Figure 46. $\overline{Keq}$ of the specimen with the highest ductility and two voids (n/2=2, α =1) was 0.82 which is 10 percent higher than the specimen with 3 voids.

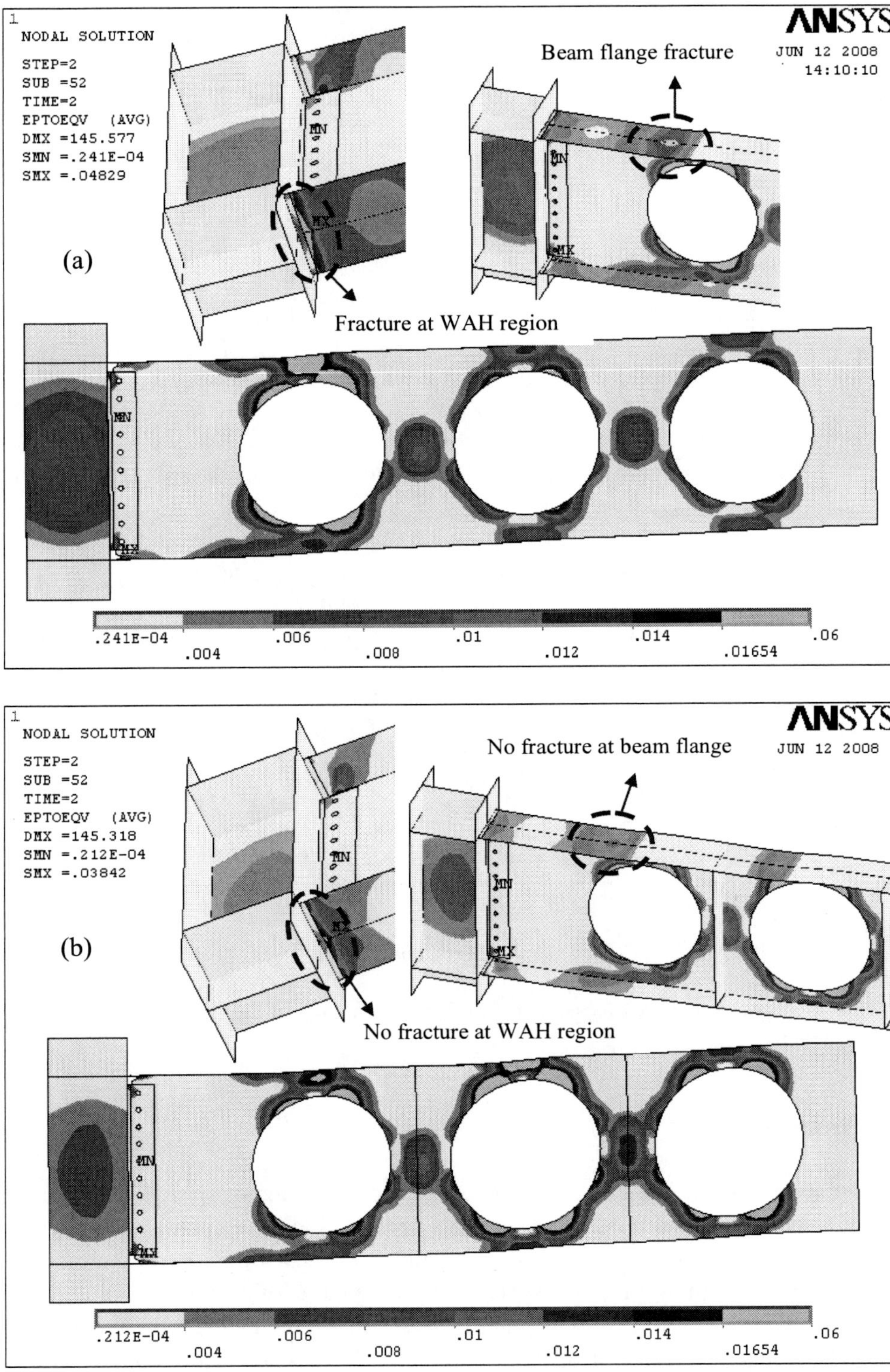

Figure 45. Von-Mises strain distribution at WAH region and at void areas for specimen SAC7, using voids of: (a) Same size, α=1; (b) Different size, α=1 & 0.8.

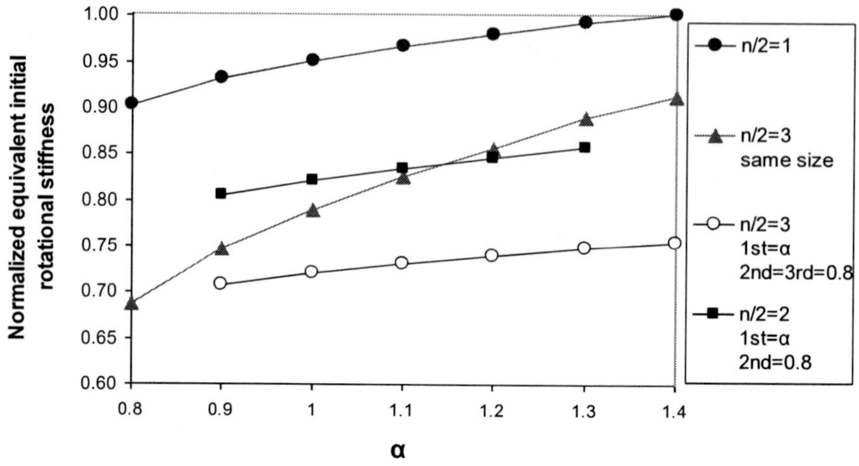

Figure 46. $\overline{Keq}$ for specimens SAC7 with different void arrangements.

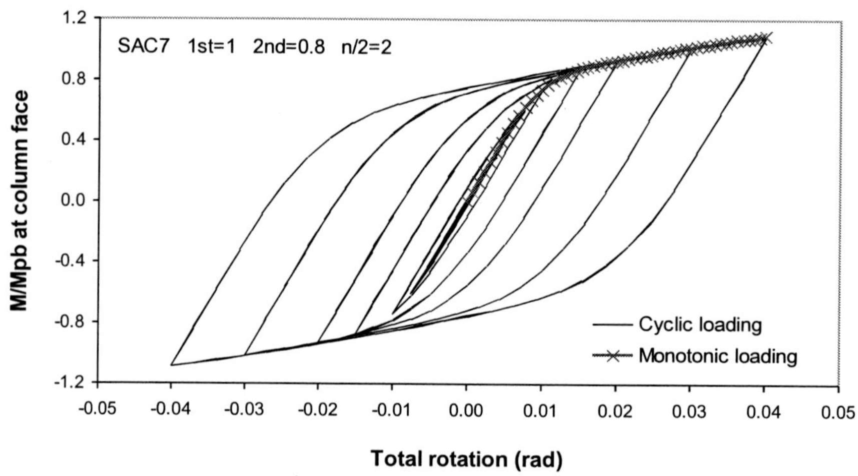

Figure 47. Moment-rotation curve of specimen SAC7 with two voids at beam web under cyclic loading where α is equal to 1.

9.2. Specimen Sac5

In order to investigate the efficiency of using voids of different sizes, a total number of voids equal to n/2, n/2-1 and n/2-2 along the half beam length were created at the beam web for specimens SAC5 and SAC3 (Table 4). Figure 48 shows the maximum normalized moment-total rotation values for SAC5 specimens presented in Table 4.

The presence of four voids of different sizes caused an increase in the connection ductility when compared to the specimens of four voids of same size (compare circle and triangle curves). All of these specimens, except specimen with α equal to 0.9, reached 4 percent total rotation. However, using different void sizes caused reduction in connection strength, but all specimens satisfied the minimum connection strength requirement (M/Mpb>0.8). For these specimens the highest connection strength and ductility was achieved for specimen of α=1.2. This indicates that 1.2 is the optimum value for parameter α to achieve the highest connection strength and ductility. Figure 49 shows the moment-rotation curve of this specimen (α=1.2) under cyclic loading. This figure shows a stable hysteric curve for SAC5 specimen with no strength reduction under 4 percent total rotation.

Figure 50 shows $\overline{Keq}$ for specimens SAC5 in the presence of different void arrangements at beam web area. As for specimen SAC7, the presence of voids of different size caused a remarkable reduction in the connection rotational stiffness (compare circle and triangle curves). $\overline{Keq}$ being 0.63 is very low for specimen with four voids of different size and α=1.2.

Table 4. RBW specimens of different void arrangements.

Spec.	Number of voids along the half beam length	Corresponded α for the			
		First void	Second void	Third void	Forth void
SAC5	n/2=4	variable	0.8	0.8	0.8
	n/2-1=3	1.2	variable	0.8	-
		1.2	1	1	-
		1.4	1.2	1	-
	n/2-2=2	1.2	variable	-	-
		1.4	1	-	-
SAC3	n/2-1=4	1.3	variable	variable-0.1≥1	1
	n/2-2=3	1.3	variable	1	-

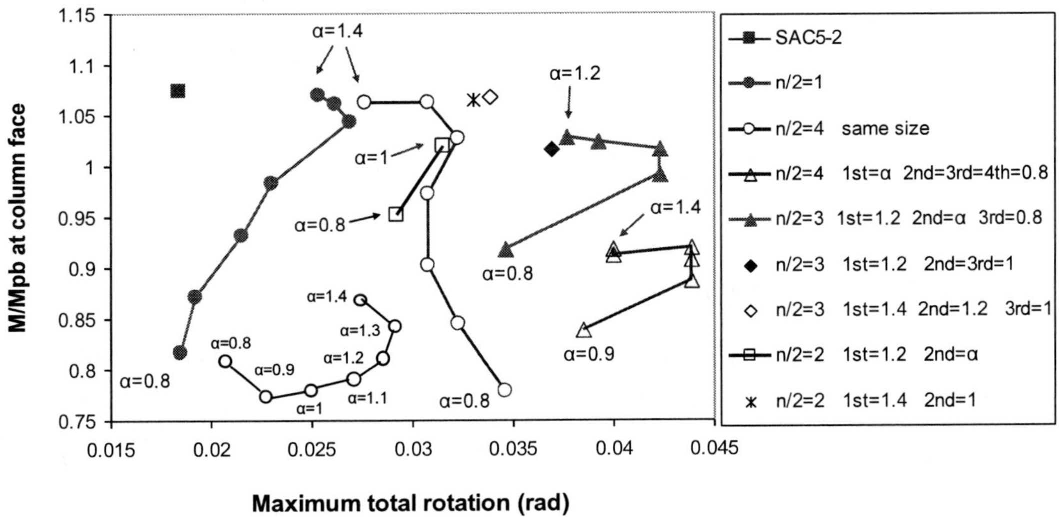

Figure 48. Maximum normalized moment-total rotation values for SAC5 specimens with different void arrangements at the beam web area.

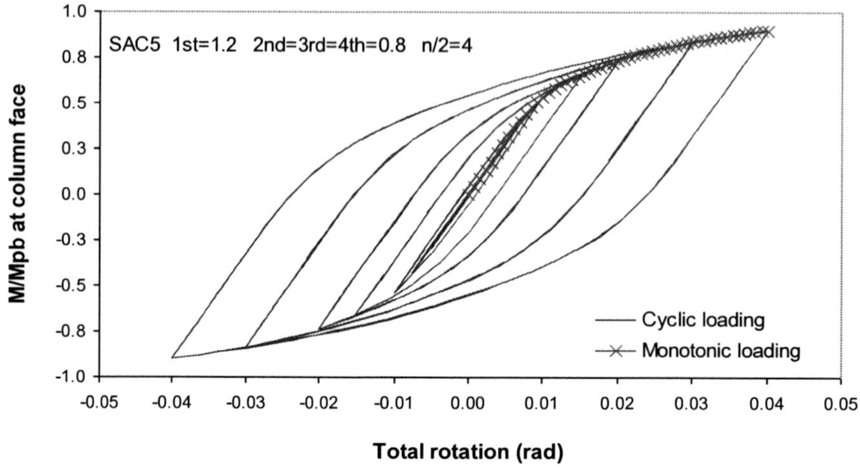

Figure 49. Moment-rotation curve of specimen SAC5 with four voids at beam web under cyclic loading where α for the first void is 1.2 and for others is 0.8.

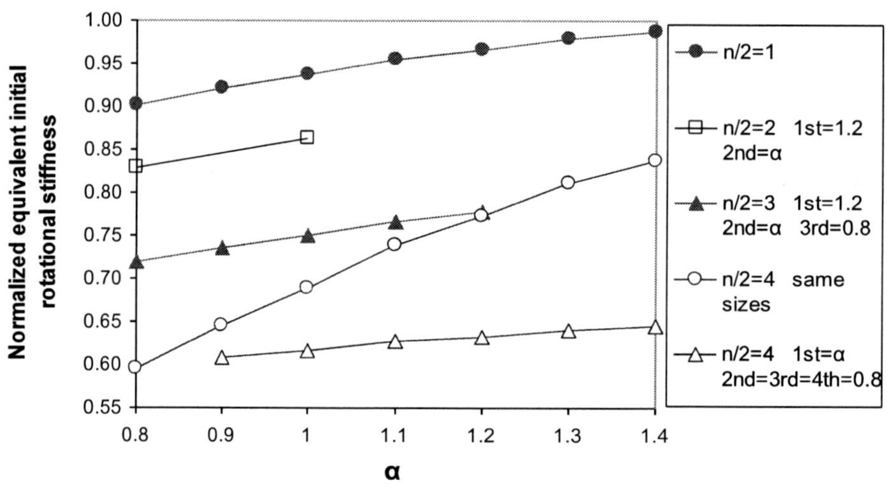

Figure 50. $\overline{Keq}$ for specimens SAC5 with different void arrangements at beam web area.

Therefore, in order to increase the connection rotational stiffness, the number of voids was reduced to n/2-1=3. In this case the diameter of the first and the third voids were kept as constant by setting α equal to 1.2 and 0.8 respectively, while the diameter of the second void was considered as a variable (Table 4). Both Figures 43 and 48 showed that for specimens of different void sizes the highest connection ductility can be achieved if the diameter of the first void is equal to the one corresponding to α presented in Figure 35. This was the reason of selecting of α=1.2 for the first void of specimen SAC5.

Figure 48 shows the normalized-moment rotation values for specimens SAC5 with three voids. By reducing the number of voids the connection strength increased and there was reduction in the connection rotational capacity (compare solid and non-solid triangle curves). For the specimens with three voids the connection with α=1 achieved the highest ductility and it's $\overline{Keq}$ increased to 0.75 (Figure 50) which is 20 percent higher than the specimen with four void. Figure 51 shows the moment-rotation curve of SAC5 specimen under cyclic loading in which the connection showed a stable hysteric curve with no strength reduction. This connection satisfied both strength and ductility requirements of connections used in seismic regions.

In order to increase $\overline{Keq}$, the number of voids was kept equal to 3 (n/2-1), the diameter of the third void was reduced by increasing α to 1 and α for the first and second voids were kept as 1.2 and 1 respectively (Table 4). Maximum total rotation of this specimen was 3.7 percent radian which is less than the minimum required and this specimen failed due to the beam flange fracture at first void area. Therefore, the diameter of the first void was decreased, such that the corresponding α for the first, the second and the third voids were selected as 1.4, 1.2 and 1 respectively. This specimen also could not achieve 4 percent total rotation and failed due to the fracture of beam flanges at WAH region. Therefore, it might be concluded that the best values for parameter α for the first, the second and the third void are 1.2, 1 and 0.8 respectively.

By decreasing the beam depth, the number of voids can be reduced. In this case, 2 (n/2-2) voids were created on the beam web area for specimen SAC5 (Table 4). Figure 50 shows however these specimens could achieve higher connection rotational stiffness but can not reach to 4 percent total rotation (Figure 48).

9.3. Specimen Sac3

Table 4 also shows the different void arrangements used for specimen SAC3. Since specimen SAC3 with five same size voids (α =1.3) could reach to a 4 percent total rotation (Figure 29) then, there was no need to use five voids of different sizes which could cause reduction in $\overline{Keq}$. Therefore, n/2-1=4 and n/2-2=3 voids were created at the beam web with different sizes. The values corresponding α for the first and the forth voids were kept as a constant and equal to 1.3 and 1 respectively. Since the depth of specimen SAC3 was less than the beam depth of specimens SAC5 and SAC7 (600mm<750mm<912mm), the size of the forth void was reduced by selecting α =1.

Figure 52 shows the maximum normalized moment-total rotation of these specimens. This figure clearly shows the efficiency of using different void sizes to increase the connection rotational capacity. For both specimens of 3 and 4 voids at the beam web, the highest ductility was achieved for α=1 and slight reduction was observed in connection strength.

Figures 53 and 54 show the moment-rotation curves of these two specimens under cyclic loading. The $\overline{Keq}$ for these two specimens (four and three voids) at α=1 are 0.78 and 0.84 respectively (Figure 55). It indicates that the shallower beam specimens can reach to 4 percent total rotation easier than deeper beams even by using smaller number of voids (n/2-2=3).

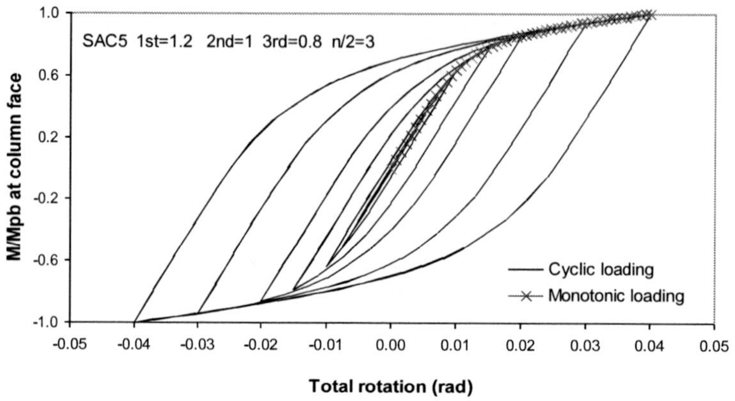

Figure 51. Moment-rotation curve of specimen SAC5 with three voids at beam web under cyclic loading where α for the second void is 1.

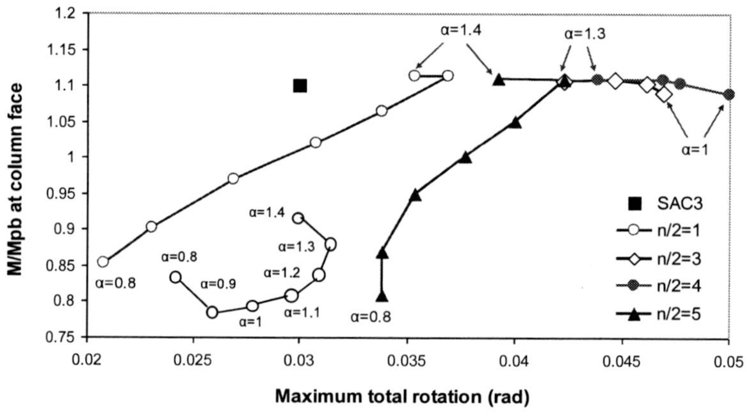

Figure 52. Maximum normalized moment-total rotation values for SAC5 specimens in the presence of different void arrangements at the beam web area.

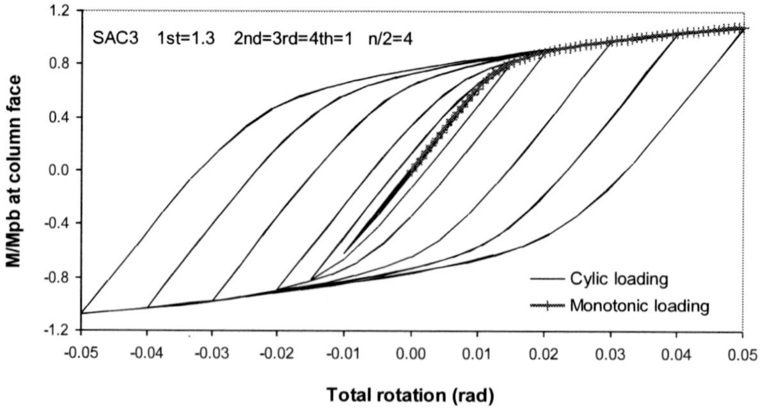

Figure 53. Moment-rotation curve of specimen SAC3 with four voids at beam web under cyclic loading where α for the second void is 1.

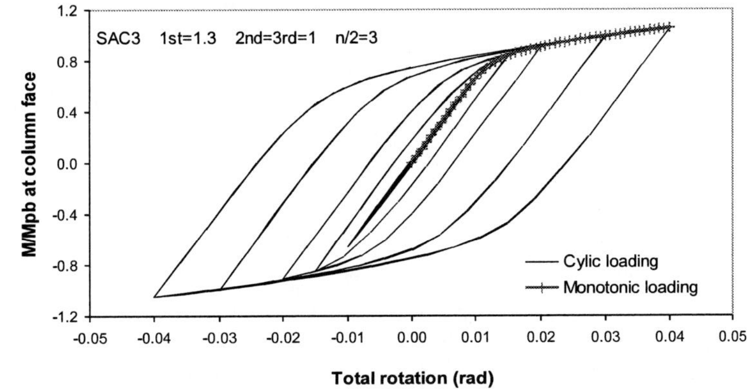

Figure 54. Moment-rotation curve of specimen SAC3 with three voids at beam web under cyclic loading where α for the second void is 1.

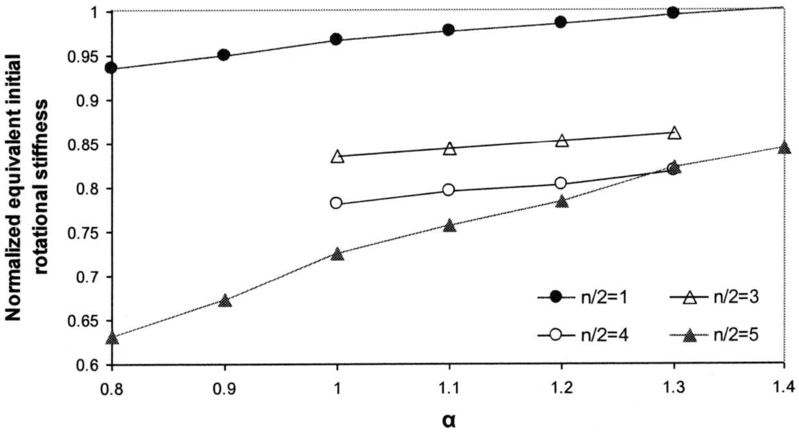

Figure 55. $\overline{Keq}$ for specimens SAC3 in presence of different void arrangements at beam web area.

9.4. Summary of Results

Table 5 summarizes the best void arrangements for three deep beam specimens SAC3, SAC5 and SAC7 which could achieve minimum required ductility. Based on this table, a void arrangement was proposed for deep beams in Table 6 which is expected to be effective to achieve adequate connection strength, ductility and rotational stiffness simultaneously. This proposed arrangement reduces the connection rotational stiffness by about 20 percent. Since in real structures, usually the number of deep beams is limited, this reduction in connection rotational stiffness may not significantly affect the lateral stiffness of the frames. On the other hand, a study was done by Faella [1994] to investigate the influence of joint behavior on the frame response. This study showed that the period of vibration of semi-rigid structure (T_K) increases by decreasing connection stiffness. From seismic point of view, generally it can be considered as an advantageous effect, since it corresponds to shifting in a period range of the elastic design response spectrum where the spectral acceleration is reduced [Faella, 1994].

Table 5. Summary of results of the proposed method for SAC specimens

Spec.	Number of voids along the half beam length	Corresponded α for					Total rotation (%rad)	M/Mpb	$\overline{Keq}$
		First void	Second void	Third void	Forth void	Fifth void			
SAC7	n/2=3	1	0.8	0.8	-	-	5	1.1	0.72
	n/2-1=2	1	0.8	-	-	-	4	1.1	0.82
SAC5	n/2=4	1.2	0.8	0.8	0.8	-	4.4	0.92	0.63
	n/2-1=3	1.2	1	0.8	-	-	4.23	1.02	0.75
SAC3	n/2=5	1.3	1.3	1.3	1.3	1.3	4.23	1.11	0.82
	n/2-1=4	1.3	1	1	1	-	5	1.09	0.78
	n/2-2=3	1.3	1	1	-	-	4.7	1.09	0.84

Table 6. Proposed void arrangement for deep beam RBW connections

Connection void arrangement		Beam height (mm)		
		600	750	900
Number of voids along the half beam length		Lb/2d-1	Lb/2d-1	Lb/2d
α	1st void	Figure 7.11	Figure 7.11	Figure 7.11
	2nd void	1	1	0.8
	others	1	0.8	0.8

10. Conclusion

The 1994 Northridge earthquake demonstrated that connections in existing WSMFs are vulnerable to premature fracture. The ability of a beam-to-column connection to withstand earthquake demands has commonly been evaluated by the connection's ductility (plastic or total rotation capacity). As part of the SAC Phase 1 research, inelastic time-history analyses were conducted on 10 WSMF buildings that experienced varying degrees of connection damage in the Northridge earthquake [SAC 1995]. Analyses of these buildings, which ranged from 2 to 17 stories in height, indicated that the plastic (total) rotation demands resulting from the Northridge Earthquake ground motions were in the range of 0.01 (0.02) radian to 0.015 (0.025) radian at the most severely loaded connections. This same SAC analytical study also examined the response of the ten buildings subject to a variety of other, potentially more damaging ground motions. It was found that maximum plastic hinge rotations in the order of 0.015 to 0.025 radians were obtained when the buildings were subjected to a suite of actual ground motion records roughly consistent with the response spectrum that has a 10 percent probability of exceedance in 50 years. While ongoing research suggests that this range may not be conservative for all conditions, it appears to be reasonable over a wide range of practical design situations. Furthermore, in the judgment of Engelhardt et al. [2003], modified connections are capable of developing at least 0.02 radian of plastic rotation (or 0.03 total rotation) provide a reasonable basis for the seismic rehabilitation of many buildings constructed with WSMFs. However, under some conditions a higher level of plastic rotation capacity may be needed and may be appropriate in the rehabilitation of a WSMF.

The objective of the connection modifications described in this chapter was to improve the performance of post-Northridge connections to a desirable level for possible future earthquakes. For this purpose a study was done to investigate the effects of presence of two and more than two circular voids along the entire beam length on the ductility, strength and initial rotational stiffness of RBW connections with deep beams. Three different design criterions, shear yielding, moment yielding and shear-moment yielding were used to determine the void sizes. Analytical results showed that:

1. There are generally three different failure modes, beam flange fracture at WAH region or at void area or severe beam web buckling.
2. Among the three design criterions for the RBW specimens, the shear yielding criterion caused the highest ductility.
3. When deep beams are used the highest connection ductility may not always be achieved for a constant value of α less than one. This is unlike what was proposed in previous studies for RBW specimens designed by using the shear yielding criterion. But α depends on the beam depth and reduces by increasing the beam depth. Consequently a linear equation was proposed to predict the best value for parameter α in terms of the beam depth.
4. In the case of using the shear yielding criterion, the use of multiple voids of same size at the beam web area could only increase the total rotation of the shallowest beam specimen (SAC3, 600 mm beam depth) to the minimum required ductility (4% total rotation) and was failed to do so for deeper beam specimens.
5. The use of multiple voids of different sizes along the beam length was effective in increasing the ductility of deep beam specimens to achieve the minimum required ductility. Therefore, an arrangement of different void sizes was proposed for deep beams in terms of the beam depth.
6. The proposed void arrangement caused an average 20 percent reduction in connection rotational stiffness.

As stated by Engelhardt [2003], the use of the modified connections should reduce potential economic losses and mitigate safety concerns for existing and new WSMFs in future earthquakes. But it should be recognized that regardless of the detail chosen for a connection modification, some damage should still be expected in a very strong earthquake. Thus, using modified connections in a new or existing WSMF does not preclude damage in future earthquakes. However, modified connections should be capable of sustaining larger earthquakes with less damage.

Among the connection modification methods described in this chapter, choosing the most suitable method for a particular project will depend on a number of project specific factors. Consequently, no general recommendation can be provided herein on a preferred method. Nonetheless, the designer should consider the potential advantages and disadvantages of each method prior to making a choice. Some of the issues that may affect the choice of a connection modification method include total rotational requirements, cost, constructability issues, the ability to satisfy strong column-weak beam requirements, and other factors.

References

AISC, (1993), Load and Resistance Factor Design Specification for Structural Steel Buildings. *Amer. Inst. Steel Const.*, Chicago, IL.

Allen, J., Richard, R. M. & Partridge, J. (1998). "Seismic connection designs for new and existing steel moment frame structures", Proc., 2nd World Conf. on Steel in Constr, *Elsevier Science Ltd.*, Oxford.

Anderson, J. C. & Duan, X. (1998). Repair/upgrade procedures for welded beam to column connections. Report no. PEER 98/03, *Pacific Earthquake Engineering Research Center*, Richmond, CA.

ANSYS User Manual, (2007). *ANSYS Multi Physics Version*. **10**, Inc.

ASCE, (1998). Specification for structural steel beams with web openings, *ASCE standard*, SEI/ASCE 23-97.

Aschheim, M. A. (2000). Moment-resistant structure, sustainer and method of resisting episodic loads. United State Patent, Patent number, **6**, 012, 256.

Azuma, K., Kurobane, Y. & Makino, Y. (2000). Cyclic testing of beam-to-column connections with weld defects and assessment of safety of numerically modelled connections from brittle fracture. *Engineering Structures*, **22**, 1596-608.

Beedle, L. S., Ln, L. W. & Ozer, E. (1973). "Recent developments in steel building design." *AISC Engrg. J*, **10**(4), 98-111.

Burdekin, F. M. & Hamour, W. (2000). Derivation of partial safety factors for BS 7910:1998. Tech Rept. CIV/SAG/98/2, UMIST, 1998. Also in: Partial safety factors for SINTAP procedure. *Tech Rept.* OTO 2000/020, HSE, www.hsebooks.co.uk, 2002.

Carpenter L. D. & Lu, L. W. (1973). "Reversed and repeated load tests of full-scale steel frames." Tech. Rep. Steel Res, for Constr. Bull No. 24, *American Iron and Steel Institute*, Washington, D.C.

Chen, C. C., Chen, S. W., Chung, M. D. & Lin, M. C. (2005). Cyclic behaviour of unreinforced and rib-reinforced moment connections. *Journal of Constructional Steel Research*, **61**(1), 1-21.

Chen, C. C., Lee, J. M. & Lin, M. C. (2003a). Behaviour of steel moment connections with a single flange rib. *Engineering Structures*, **25**, 1419-28.

Chen, C. C., Lin, C. C. & Tsai, C. L. (2004). Evaluation of reinforced connections between steel beams and box columns. *Engineering Structures*, **26**, 1889-1904.

Chen, C. C. & Lu, C. A. (2003b). Cyclic behaviour of steel moment connections strengthened by flange rib. In: *Proceedings of the 5th Korea–Taiwan–Japan joint seminar on earthquake engineering for building structures*, 71-9.

Chen, S. h., Yeh, C. & Chu, J. (1996). Ductile steel beam-to-column connections for seismic resistance. *Journal of Structural engineering*, 1292-1299.

Chen, S. J. & Chao, Y. (2001). Effect of composite action on seismic performance of steel moment connections with reduced beam sections. *Journal of Constructional Steel Research*, **57**, 417-434.

Chen, W. E. & Patel, K. V. (1981). "Static behaviour of beam-to-column moment connections." *J. Struct. Div., ASCE*, **107**(9), 1815-1838.

Chi, W. M., Deierlein, G. G. & Ingraffea, A. R. (1997). *Finite Element Fracture Mechanics Investigation of Welded Beam-Column Connections*. Report No. SAC/BD-97/05, SAC Joint Venture.

Chi, W. M., Deierlein, G. G. & Ingraffea, A. (2000). Finite-element fracture analyses of welded beam-column connections. Fatigue and fracture mechanics, *American Society for Testing Materials, Philadelphia*, vol. 30, ASTM STP 1360, 439-455.

Chi, W. M., Deierlein, G. G. & Ingraffea, A. (2000b). Fracture toughness demands in welded beam-column connections. *Journal of Structural Engineering (ASCE)*, **126**(1), 88-97.

Chia, B., Uang, C. M. & Chen, A. (2006). Seismic rehabilitation of pre-Northridge steel moment connections: A case Study. *Journal of Constructional Steel Research*, **62**, 783-792.

Chung, K. F., Liu, T. C. H. & Ko, A. C. H. (2003). Steel beams with large web openings of various shapes and sizes: An empirical design method using a generalized moment–shear interaction curve. *J Construct Steel Res.*, doi:10.1016/S0143-974X(03)00029-4.

Clark, P., Frank, K., Krawinkler, H. & Shaw, R. (1997). *Protocol for Fabrication, Inspection Testing, and Documentation of Beam-Column Connection Tests and Other Experimental Siemens*, SAC Report SAC/BD-97/02, SAC Joint Venture.

Coons, R. G. (1999). Seismic Design and Database of End-Plate and T-Stub Connections, *A thesis submitted in partial fulfillment of the Master of Science in Civil Engineering Degree*, University of Washington, Seattle, Washington.

Crawford, J., Houghton, D., & Karns, J. (2002). "Multi-Hazard Mitigation in Steel Frame Air Traffic Control Towers as a Function of Redundancy, Lateral Configuration and Connection Geometry", *All-Regional FAA Structural Seminar on Seismic and Progressive Collapse Resistant Structures for Air Traffic Control Towers*, Renton, Washington.

El-Tawil, S., Mikesell, T., Vidarsson, E. & Kunnath, S. K. (1998). *Strength and Ductility of FR Welded-Bolted Connections*. Report No. SAC/BD-98/01, SAC Joint Venture.

Engelhardt, M. D., Uang, C. M., Gross, J. L., Kasai, K. & Iwankiw, N. (2003). *Modification of existing welded steel moment frame connections for seismic resistance*, AISC.

Engelhardt, M. D., Sabol, T. A., Aboutaha, R. S. & Frank, K. H. (1995). Testing connections. *Modern Steel Construction, AISC*, **35**(5), 36-44.

Engelhardt, M. D. & Sabol, T. A. (1998). Reinforcing of steel moment connections with cover plates: benefits and limitations. *Engineering Structures*, **20**(4-6), 510-20.

Engelhardt, M. D. & Popov, E. P (1989). 'Behaviour of long links in eccentrically braced frames', Report no. UCB/EERC-89/OI, *Earthquake Engineering Research Center*, University of California, Berkeley, CA.

Engelhardt, M. D., Winneberger, T., Zekany, A. & Potyraj, T (1996). 'The dogbone connection - Part 11', *Modern Steel Construction*.

Faella, C., Piluso, V. & Rizzano, G. (1994). Connection influence on the elastic and inelastic behaviour of steel frames. International workshop and seminar on behaviour of steel structures in seismic areas. *STESSA 94*, Timisoara, Romania, 26 June-1 July 1994.

Federal Emergency Management Agency, (2000). FEMA 350 -, *Recommended Seismic Design Criteria for New Steel Moment-Frame Buildings*.

FEMA, (2000). *State of the art report on connection performance*. Report No. FEMA 355D, Washington, September 2000.

FEMA350, (2000). *Recommended Seismic Design Criteria for New Steel Moment-Frame Buildings*, June 2000.

Forde, B. W. R. & Stiemer, S. F. (1987). Improved arc-length orthogonality methods for nonlinear finite element analysis. *Computer and Structures*, **27**(5), 625-630.

Gilton, C. S. & Uang, C. M. (2002). Cyclic response and design recommendations of weak-axis reduced beam section moment connections, *Journal of Structural Engineering*, Vol. 128, No. 4.

Hedayat, A. A. & Celikag, M. (2009a). Review: Fracture Moment and Ductility of Welded Connection", *Proceedings of ICE, Structures and Buildings*, Vol.162, Issue SB6, pp.405-418. Hedayat, A. A. & Celikag, M. (2009b). Post-Northridge connection with modified beam end configuration to enhance strength and ductility. *Journal of Constructional Steel Research*, Vol.65, Issue 7, 1413-1430.

Hedayat, A. A. (2009c). *Behaviour of newly proposed reduced beam web connections. Ph.D. thesis*, Eastern Mediterranean University.

Hedayat, A. A. & Celikag, M. (2010). Wedge design: A reduced beam web (RBW) connection for seismic regions", *Advances in Structural Engineering*, Vol.13, No:2Hennessey, J., Dinehart, D., Hoffman, R., Gross, S. & Yost, J. (2004). Effect of Cope Geometry on the Strength and Failure Behaviour of Non-Composite Open Web Expanded Beams: *Research Report 7 to SMI Steel Products*. Villanova University.

Houghton, D. & Karns, J. (2001). "Effective Mitigation of Progressive Collapse in Steel Frame Buildings using Ductile High-Capacity Girder-to-Column Moment Connections Exhibiting Discrete Structural Continuity Across a Failed Column", *Proceedings of the Society of American Military Engineers (SAME) National Symposium on Comprehensive Force Protection*, The Citadel, Charleston, South Carolina.

ICBO, (1994). Uniform Building Code. *International Conference of Building Officials (ICBO)*, Whittier, CA.

Kasai, K., Hodgon, I. & Mao, C. (1997). "Bolted Repair Methods for Fractured Welded Moment Connections," *Proceedings*, Behaviour of Steel Structures in Seismic Areas (Stessa '97), 939-946, Kyoto, Japan, 3-8 August.

Kasai, K., Hodgon, I. & Bleiman, D. (1998). "Rigid-Bolted Repair Methods for Damaged Moment Connections," *Engineering Structures*, Vol. 20, No. 4-6, 521-532.

Kaufmann, E. J., Fisher, J. W., Di Julio, Jr, R. M. & Gross, J. L. (1995). Failure analysis of welded steel moment resisting frames damaged in the Northridge earthquake. *Tech Rept*, **5625**, NIST.

Kiral, B. G. & Erim, S. (2005). Prediction of fracture behaviour of steel beam-to-column connections with weld defect using the SINTAP. *Engineering Structures*, **27**, 760-768.

Lee, C. H., Jung, J. H., Oh, M. H. & Koo, E. S. (2003). "Cycling seismic testing of steel moment connections reinforced with welded straight haunch". *Engineering Structures*, **25**, 1743-1753.

Lee, C. H. & Uang, C. M. (2001). Analytical modelling and seismic design of steel moment connections with welded straight haunch. *J Struct Eng*, **127**(9), 1028-35.

Lee, C. H. & Yoon, T. H. (1999). Analytical re-examination of shear transfer in welded steel moment connection. In: *Proceeding of the 1st Japan-Korea joint seminar on earthquake engineering for building structures*.

Lee, C. H. (2006). Review of force transfer mechanism of welded steel moment connections. *Journal of constructional steel research*, **62**, 695-705.

Lee, H., Stojadinovic, B., Goel, S. C., Margarian, A. G., Choi, J. & Wongkaew, A. at al. (2000). *Parametric Tests on Unreinforced Connections*. SAC/BD-00/01, Volume I – Final Report.

Load and Resistance Factor Design Specification for Structural Steel Buildings, *AISC*, December, 27, 1999.

Lu, L., Ricles, J., Mao, C. & Fisher, J. (2000). Critical issues in achieving ductile behaviour of welded moment connections. *Journal of Constructional Steel Research*, **55**, 325-341.

Mao, C., Ricles, J., Lu, L. & Fisher, J. (2001). Effect of local details on ductility of welded moment connections. *Journal of Structural engineering*, 1036-1044.

Matos, C. G. & Dodds, R. H. Jr. (2000). Modelling the effects of residual stresses on defects in welds of steel frame connections. *Eng Structure*, **22**, 1103-20.

Mazzolani, F. M. (2000). Moment resistant connections of steel frames in seismic areas, *Design and reliability*. London, E & FN Spon.

Miller, D. K. (1998). Lessons learned from Northridge earthquake. *Engineering Structures*, **20**, 249-260.

Moslehi Tabar, A. & Deylami, A. (2005). Instability of beams with reduced beam section moment connections emphasizing the effect of column panel zone ductility. *Journal of Constructional Steel Research*, **61**, 1475-1491.

Naeim, F. (2001). *The seismic design handbook*. 2nd ed. Kluwer Academic Publishers, 418.

Noel, S. & Uang, C. M. (1996). "Cyclic Testing of Steel Moment Connections for the San Francisco Civic Center Complex," Report No. TR-96/07, *Division of Structural Engineering*, University of California, San Diego.

Parfitt, J. & Chen, W. F. (1976). "Tests of welded steel beam-to-column moment connections." *J. Struct. Div., ASCE*, **102**(1), 189-202.

Popov, E P. & Tsai, K. C. (1989). Performance of large seismic steel moment connections under cyclic loads. *Engineering Journal, AISC*, **26**(2), 51-60.

Popov, E. P., Yang, T. & Chang, S. (1998). Design of steel MRF connections before and after 1994 Northridge earthquake. *Engineering Structures*, **20**(12), 1030-1038.

Popov, E. P. & Bertero, V. V. (1973). "Cyclic loading of steel beams and connections." *J. Struct. Div., ASCE*, **99**(6), 1189-1204.

Popov, E. P. & Pinkney, R. B. (1969). "Cyclic yield reversal in steel building connections." *J. Struct. Div., ASCE*, **95**(3), 327-353.

Popov, E. P. & Stephen, R. M. (1970). "Cyclic loading of full-size steel connections." Tech. Rep. UCB/EERC-70/3, *Earthquake Engrg, Res. Ctr.*, University of California, Berkeley, Calif.

Popov, E. P. & Stephen, R. M. (1972). "Cyclic loading of full-size steel connections." Tech. Rep. Steel Res. for Constr. Bull. No. 21, *American Iron and Steel Institute*, Washington, D.C.

Ricles, J. M., Fisher, J. W., Lu, L. W. & Kaufmann, E. J. (2002). Development of improved welded moment connections for earthquake-resistant design. *Journal of Constructional Steel Research*, **58**, 565-604.

Ricles, J. M., Mao, C., Lu, L. W. & Fisher, J. W. (2003). Ductile details for welded unreinforced moment connections subject to inelastic cyclic loading. *Engineering Structures*, **25**, 667-680.

Righiniotis, T. D., Omer, E. & Elghazouli, A. Y. (2002). A simplified crack model for weld fracture in steel moment connections. *Engineering Structures*, **24**(9), 1133-40.

SAC, (1995). "Analytical and Field Investigations of Buildings Affected by the Northridge Earthquake of January 17, 1994," Technical Report SAC 95-04, *SAC Joint Venture*, Sacramento, CA.

SAC, (1996). "Experimental Investigations of Beam-Column Sub assemblages," *Technical Report SAC-96-01*, Parts 1 and 2, SAC Joint Venture, Sacramento, CA.

Satish Kumar, S. H. & Prasada Rao, D. V. (2006). RHS beam-to-column connection with web opening—experimental study and finite element modelling. *Journal of Constructional Steel Research*, **62**, 739-746.

SEAOC, (1990). *Recommended Lateral Force Requirements and Commentary*, Structural Engineers Association of California (SEAOC), Sacramento, 1990.

Tsai, K. C. & Popov, E. P. (1993). "Seismic design of steel beam-to-box column connections", *in Proc. Structures Congr, '93*, ASCE.

Uang, C. & Bondad, D. (1998). Cyclic performance of haunch repaired steel moment connections: experimental testing and analytical modelling. *Engineering Structures*, **(20)**, 552-561.

Uang, C. M., Yu, Q. S., Noel, S. & Gross, J. (2000). Cyclic testing of steel moment connections rehabilitated with RBS or welded haunch. *J Struct Eng, ASCE*, **126**(1), 57-68.

Uang, C. M., Bondad, D. & Noel, S. (1996). Cyclic Testing of the MNH-SMRF™ Dual Strong Axes Moment Connection with Cruciform Column (Report No. TR-96/04), *University of California, San Diego*, La Jolla, California, U.S.A.

Uang, C. M. & Latha,m, C. (1995). *Cyclic Testing of Full-scale MNH-SMRF™ Moment Connections* (Report No. TR-95/01), University of California, San Diego, La Jolla, California, U.S.A.

Wilkinsona, S., Hurdmanb, G. & Crowtherb, A. (2006). A moment resisting connection for earthquake resistant structures. *Journal of Constructional Steel Research*, **62**, 295-302.

Zahrai, S. M. & Moslehi, Tabar, A. (2006). Cyclic behaviour of steel braced frames having shear panel system, *Asian journal of civil engineering (building and housing)*, **7**(1), 13-26.

Zekioglu, A., Mozaffarian, H., Chang, K. L., Uang, C. M. & Noel, S. (1997). Designing after Northridge. *Modern Steel Construction, AISC*, **37**(3), 36-42.

Zhang, J. & Dong, P. (2000). Residual stresses in welded moment frames and implications for structural performance. *Journal of Structural engineering*, 306-315.

In: Structural Steel and Castings
Editor: Lena M. Becker, pp. 57-106

ISBN: 978-1-61728-104-4

Chapter 2

RESEARCH ON STEEL PLATE SHEAR WALL: PAST, PRESENT AND FUTURE

***Siddhartha Ghosh*[1] *and Swapnil B. Kharmale*[2]**
Department of Civil Engineering, IIT Bombay, Mumbai, India

Abstract

Due to its robust post-buckling strength, substantial ductility, stable hysteretic characteristics and high initial stiffness, the steel plate shear wall (SPSW) system is now considered to be an appealing alternative to conventional lateral load resisting systems used for earthquake resistant design of structures. SPSW systems, when compared to a traditional reinforced concrete shear wall, offer lighter structures, increased floor area, faster speed of construction, considerable economy and better quality control. Although research work on SPSW has been going on since the early 1980s, the satisfactory structural performance of SPSW systems in the Northridge, USA (1994) and Kobe, Japan (1995) earthquakes led researchers and practicing engineers to study and implement the SPSW system to a greater extent. Experimental and analytical research works, conducted primarily in Canadian, US and UK universities on SPSW considered various aspects of the seismic behavior of SPSW and developed the fundamental guidelines for its implementation as an effective lateral load resisting system. This chapter presents a comprehensive review of the stages of the development in SPSW research and implementation over the past three decades. The primary focus of this review is on the recent experimental and analytical research works on unstiffened thin steel plate shear walls carried out all over the world. The existing design guidelines for the SPSW system in current US and Canadian codes are also discussed. An overview of the ongoing research activities across the globe is presented with an emphasis to the development of design methodologies for SPSW systems conforming to the performance-based seismic design (PBSD) philosophy. The final sections of this chapter deal with the research areas needing development in the immediate future, roadblocks in the process of implementation, standardization of design guidelines and a roadmap for the future aiming at a wide acceptance of this system in the field of earthquake engineering.

[1] E-mail address: sghosh@civil.iitb.ac.in. (Assistant Professor and Corresponding Author)
[2] E-mail address: swapnil.kharmale@iitb.ac.in. (Doctoral Candidate)

1. Introduction

The engineering use of steel plate shear walls originated in aerospace applications where steel panels were used in stiffened and unstiffened forms. The use of steel plate shear wall (SPSW) as a primary lateral load resisting system in new building constructions, and also for upgrading the lateral load resistance of existing structures, began in the early 1980s in the USA and Japan. In its typical form, the SPSW system consists of a steel shear panel added as an infill to the building structural frame composed of beams (horizontal boundary elements or HBE) and columns (vertical boundary elements or VBE), as shown in Figure 1. The structural frame may use either simple or moment resisting connections between beams and columns and the steel panel is either bolted or welded to these boundary elements (usually through a fish plate). Depending on the design philosophy, the steel plate panels are either stiffened or unstiffened.

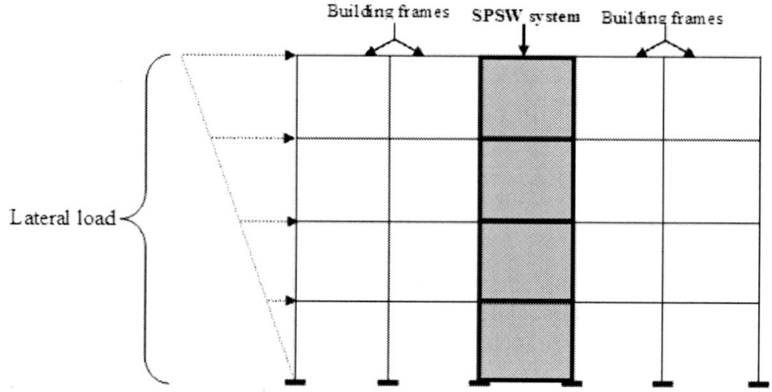

(a) SPSW system as a lateral load resisting system in a building frame

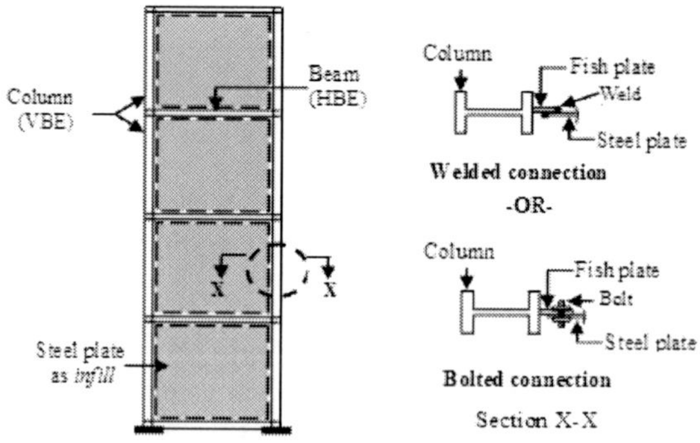

(b) Components of typical SPSW system

Figure 1. Typical steel plate shear wall (SPSW) system.

Initially, SPSW were constructed either with thick steel plates or with stiffened steel plates. Those designs were based on the concept of avoiding the (elastic) out-of-plane buckling of the steel plates. In Japan (for example, in the Nippon Steel Building in Tokyo), this was achieved by using heavily stiffened plates while in the USA (for example, in the Sylmar Hospital in Los Angeles), moderately thick plates were used. Using the thick or stiffened SPSW is an unattractive option because of its higher cost in comparison to reinforced or prestressed concrete shear walls. This has led to a gradual and general shift towards the use of thin unstiffened steel plate shear walls.

Research work on thin unstiffened SPSW systems started in the late 1970s. Several analytical and experimental studies carried on these SPSW systems in various universities around the world have shown that the thin unstiffened SPSW systems have sufficient post-buckling reserve strength, which makes it a more economical alternative to various other traditional lateral load resisting systems. The primary advantages of the thin unstiffened SPSW are high initial stiffness, substantial ductility, stable hysteretic characteristics and a large capacity for plastic energy absorption. Moreover, this system offers a light-weight structure, increased floor area, higher speed of construction, and considerable economy and better quality control when compared to a conventional reinforced concrete shear wall system (Sabouri-Ghomi et al. 2005). The thin infill panels which are the main lateral resisting component in a SPSW system are allowed to buckle out-of-plane under relatively small shears and the shear resistance of the panel is dominated by the tension field action.

Considering the extensive analytical as well as experimental work on the post-buckling behavior of SPSW during the 1980s and early 1990s by various researchers (Thorburn et al. 1983, Timler and Kulak 1983, Tromposch and Kulak 1987, Elgaaly 1998, Driver et al. 1998a), primarily from University of Alberta, the design standards/codes started incorporating design guidelines for SPSW systems. In 1994, the Canadian steel design standard, CAN/CSA-S16-94 (CSA 1994) included design requirements for unstiffened thin SPSW, although as an appendix to the main code. In the last decade, the plastic analysis and design methods for SPSW, developed primarily in University of Buffalo, resulted in the inclusion of capacity design provisions in the AISC Seismic Provisions (AISC 2005a). The current AISC Design Guide 20 (Sabelli and Bruneau 2007) provides a detailed design guideline for steel plate shear wall systems considering different seismicity conditions.

In the recent years, steel plate shear wall systems have been used as a part of the lateral load resisting system in a number of buildings mainly in highly seismic areas in Japan and North America. The 56-story, LA Live Hotel and Residence in Los Angeles, USA is the latest example of high-rise structures where SPSW systems are used. Nabih Youssef Associates, the structural consultant of this project, decided to replace heavy 30 in (762 mm) thick concrete shear walls with light 1/4-3/8 in (6.13-9.53 mm) steel plate shear walls. This resulted in availability free valuable real estate space, reduced seismic design forces and foundation sizes by eliminating 35% of the weight of the structure, compressed construction schedule and budget, and allowance for simplified and more efficient construction (Youssef et al, 2009).

The 6-story Sylmar hospital building in Los Angeles, USA and the 35-story high-rise Kobe City Hall building in Kobe, Japan are typical examples of buildings with SPSW as part of the lateral load resisting system that had experienced real strong earthquakes. The Sylmar Hospital in the Los Angeles area was built as a replacement for the Olive View Hospital which had been so severely damaged during the 1971 San Fernando earthquake (magnitude 6.5 on the Richter scale). The new structure, consists of a steel structure with reinforced concrete shear walls in the lower two stories and steel plate shear walls in the perimeter walls of the upper four floors. The structure experienced the 1987 Whittier earthquake (magnitude 5.9 on the Richter scale) and seven years later the 1994 Northridge earthquake (magnitude 6.7 on the Richter scale). As the California Strong Motion Instrumentation Program (CSMIP) data indicate, the accelerations at roof level were more than 2.3*g* while the ground floor acceleration was about 0.66*g*. The investigation on seismic damage to this building in the aftermath of the 1994 Northridge earthquake indicated that there was severe damage to some non-structural elements. The non-structural damage was clearly an indicator of the very high elastic stiffness of this structure, which was also the cause of relatively large amplification of accelerations from the ground to the roof level.

The 35-story high-rise in Kobe was subjected to the 1995 Kobe (Hyogoken-Nanbu) earthquake. Researchers in Japan (Fujitani et al. 1996) studied the seismic performance of this building during this earthquake, which indicated that the damage was minor and consisted of local buckling of stiffened steel plate shear walls on the 26th story and a permanent roof drift of 225 mm and 35 mm in the two orthogonal horizontal directions. The results of inelastic analysis of this structure reported in Fujitani et al. (1996) indicates that soft stories may have formed at floors between 24th and 28th level of the building. A visual inspection of the structure two weeks after the earthquake did not show any sign of visual damage (Astaneh-Asl 2001).

Kulak et al. (2001) and Bruneau et al. (2007) had presented brief reviews on SPSW covering specific aspects of this still emerging lateral load resisting system. A more comprehensive review of past research works on SPSW, including applications and future trends projections, is presented in this chapter. The following sections of this chapter provide: i) A brief review of fundamental works on the post-buckling strength of shear panel; ii) A detailed review of analytical and experimental research works on unstiffened thin steel plate shear walls carried out all over the world; iii) A review of the plastic design and analysis methods along with existing design guidelines for the SPSW system in current US and Canadian codes; iv) Introduction to and a review of performance based seismic design (PBSD) methodology for SPSW systems; and v) An overview of research needs in this area and a projection for the next few years.

2. Post-Buckling Strength of Shear Panel: "Diagonal Tension Field Theory"

This section deals with the pre-history of steel plate shear walls as we know them today. Today's SPSW systems rely on their post-buckling force and deformation capacities to withstand seismic shaking. The original research work that inspired a shift from the use of thick or stiffened SPSW to the thin unstiffened SPSW was on the evaluation of post-buckling strength of metal shear panels through the *diagonal tension field theory*. The theory of semi-tension fields was originally developed by Wagner (1931) based on an observation of a phenomenon that occurred in thin webs under a shear load. Thin webs are generally used in the design of aircraft wings, which typically consist of an upper flange and a lower flange fastened by thin webs.

Wagner (1931) demonstrated that when a thin shear web with transverse stiffeners buckles between the stiffeners, it does not "fail"; it merely forms diagonal folds and remains functional as a series of tension diagonals, while the stiffeners act as compression posts. The action of the web may be understood qualitatively by considering a frame as shown in Figure 2. Under a small applied load *P*, the two diagonals share the shear load equally. If the load *P* is increased to, say, $P+\Delta P$, a stage will be reached at which the compression diagonal buckles. The shear in the frame resulting from a further increase of the load will be carried chiefly by the tension diagonal, because the buckled compression diagonal is not capable of taking a significant amount of additional load. Consider now the same frame braced by a solid sheet. The shear in the sheet is equivalent to numerically equal tensile and compressive stresses on faces obtained at ± 45° as long as the sheet has not buckled. If the shear load is increased sufficiently, the compressive stresses will begin to form buckles or folds in the sheet, and a further increase in the shear load results primarily in an increase only of the tensile stresses, since the diagonal along which compression is acting has already buckled. If it is possible to increase the load even further without rupturing the sheet, the compressive stresses will gradually become negligible compared to the tensile stresses. This asymptotic limiting condition is referred to as "pure diagonal tension". Because the webs used in aeronautical design applications are so thin, the force required to produce the theoretical web buckling load is relatively small; it is customary to ignore this contribution when calculating the total shear capacity of a web. Furthermore, since the surrounding framing members are much stiffer than the thin web, a complete tension field develops throughout the metal shear panel. Wagner proposed a unique strip model representation encompassing both of these concepts for the strength prediction of thin webbed aircraft membranes.

Following Wagner's (1931) study, designers considered the ultimate strength of the shear web as being in either of two categories: webs could be deemed a) as "shear resistant", wherein instability was not permitted prior to yield, or b) as "pure diagonal tension webs", in which case the shear carried by the web before buckling was disregarded. Kuhn (1956) elaborated on the theory brought forth by Wagner by introducing the idea of "incomplete" diagonal tension and proposed a method of interpolating between the two extremes postulated by Wagner. The proposed solution involved a trial and error procedure if the flanges bounding the shear panel were not infinitely stiff, thus limiting the application. The first civil engineering application of this theory was suggested by Basler, following the research of Wagner (1931) and Kuhn (1956) made with particular reference to the aircraft industry. Basler (1961) judged that for a plate girder, because of the relatively low bending strength in its flanges, the tension field in the web would develop only partially. He also considered the ultimate strength of the web plate as the contribution of two parts: the shear capacity of the web due to a beam action and the additional resistance of the web from the formation of the tension diagonals following out-of-plane buckling.

Although many variations of tension field theory have since been introduced for plate girders, their differences lie largely in the configuration of the assumed tension band and the type of failure mechanism used to define it. Most provide solutions which are marginally better than Basler's (1961) but since they tend to be more complicated, Basler's (1961) approach is still favored, particularly in the application of this theory to steel plate shear walls with boundary elements.

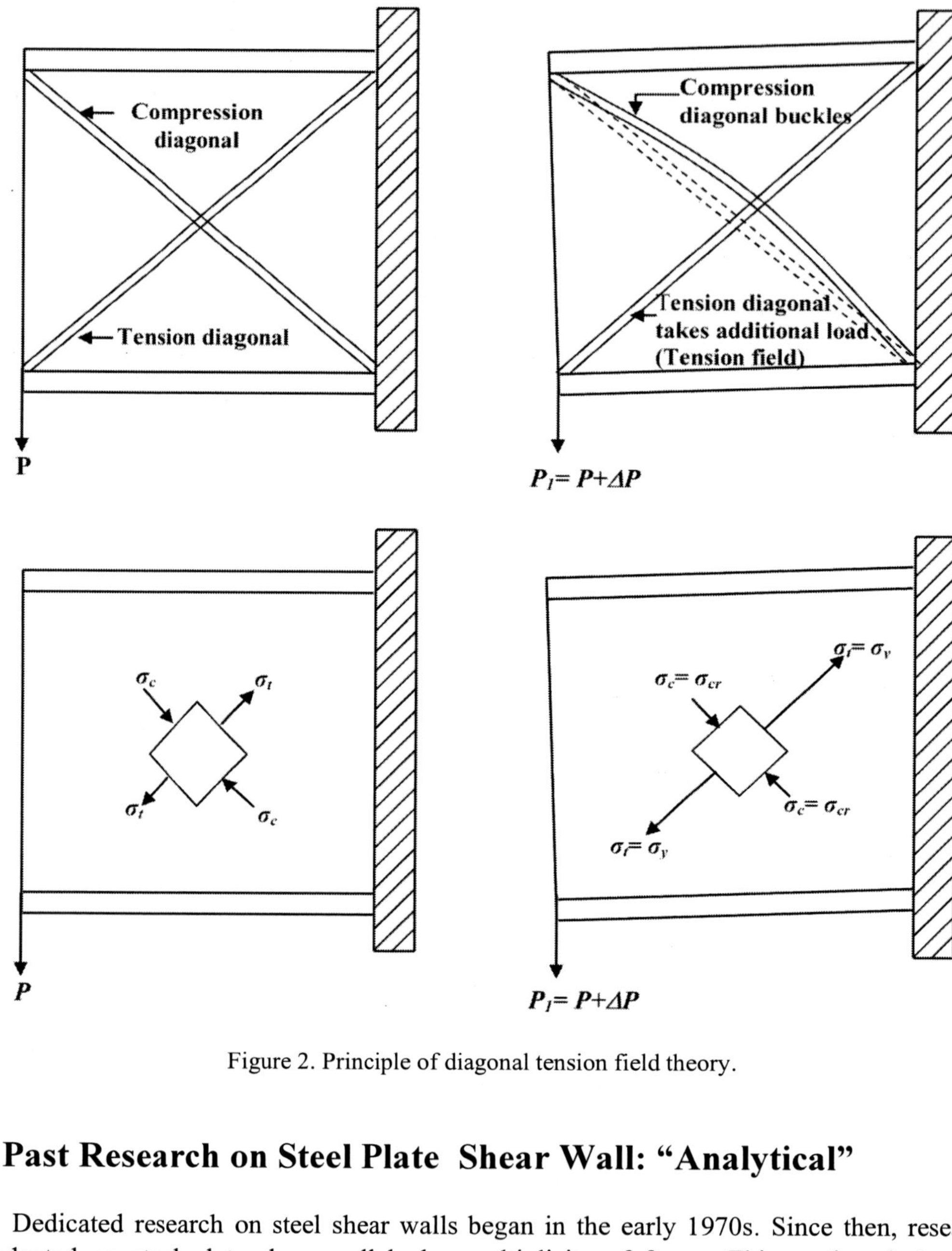

Figure 2. Principle of diagonal tension field theory.

3. Past Research on Steel Plate Shear Wall: "Analytical"

Dedicated research on steel shear walls began in the early 1970s. Since then, research conducted on steel plate shear wall had a multiplicity of forms. This section deals with research works that can be loosely termed as "analytical", signifying that these studies can be categorized under the non-laboratory theoretical works primarily focusing on the behavioral aspects of SPSW. The fundamental objective of these analytical works was to facilitate the analysis and design of SPSW system without introducing much complexity. These works were dedicated primarily to the various modeling and analysis aspects of the thin unstiffened SPSW in order to represent the real pre- and post-buckling behavior of the SPSW.

3.1. Hysteresis Model

Mimura and Akiyama (1977) developed a general method for predicting the monotonic and cyclic behavior of unstiffened steel plate shear panels through a series of experimental and analytical studies. The monotonic behavior of a shear wall panel was obtained by superimposing the behavior of the infill plate and the frame. Classical plate theory was used to predict the infill plate buckling capacity and a diagonal tension field action was assumed in the post-buckling range. The hysteresis model developed by Mimura and Akiyama (1977) to describe the hysteretic behaviour of a steel plate shear wall panel is shown in Figure 3.a. It was assumed that the deformation required for forming the tension field when loading in the opposite direction is equal to one-half of the plastic deformation of the previous load cycle. In Figure 3.b, Q is the lateral load applied to the panel and δ is the resulting lateral deflection. Other notable assumptions included setting the plastic Poisson's ratio of the plate to 0.5 and a constant angle of inclination of the tension field that was set to 45°. The path OAB describes the initial positive loading of the steel plate shear wall. The unloading of the steel plate shear wall, as described by BC', was assumed to be parallel to the initial loading path, OA. C'C describes the loading of the wall in the opposite direction, or negative loading. Shear buckling of the infill plate was assumed to have occurred at point C and the tension field to have re-formed in the plate at point D. The point where the tension field re-formed was located on a line parallel with OA and starting at point D', which was set at the halfway point between O and C', a direct result of setting the Poisson's ratio of the plate to 0.5. Assuming a negative monotonic curve OA' E, the hysteresis model continued down the path DA'E. The removal of the negative load from the wall, as described by EF', was assumed to be parallel to OA. In order to validate the proposed model, they had conducted tests on small-scale simply supported stiffened plate girders subjected to a single cyclic point load at mid-span. The test results were in good agreement with their proposed model except in the redevelopment phase of the tension field where stiffness of the frame was neglected.

3.2. Multi-Strip Model

The development of the multi-strip modeling technique for thin unstiffened SPSW was the first major breakthrough towards seismic design of buildings incorporating steel plate shear walls. Based on the diagonal tension field theory proposed by Wagner (1931) and modified later by several others, Thorburn et al. (1983) developed a simple analytical model to study the shear behavior of thin unstiffened SPSW systems. In this model, referred to as the *multi-strip model* (Figure 4.a), the action of the tension field was modeled by a series of pin-ended inclined tension-only members. These strips were oriented parallel to the direction of the tension field. Each strip was assigned an area equal to the width of the strip times the plate thickness. The angle of the tension field was obtained using the principle of least work. The boundary beams were assumed to be (flexurally) infinitely rigid and are pin-connected to columns, whereas the boundary columns were assigned to actual stiffnesses. By assuming a thin SPSW infill panel buckling under compressive diagonal load, they derived the inclination angle (α) for the tension field:

$$\tan^4\alpha = \sqrt{\frac{1+\dfrac{Lt}{2A_c}}{1+\dfrac{ht}{A_b}}} \tag{1}$$

where t is the thickness of the infill plate, A_c and A_b are the cross-sectional areas of the column and beam, respectively, L and h are bay width and story height, respectively. In the derivation of the above expression using the principle of least work, only the energy of the tension field and the axial energy in the beams and columns were considered.

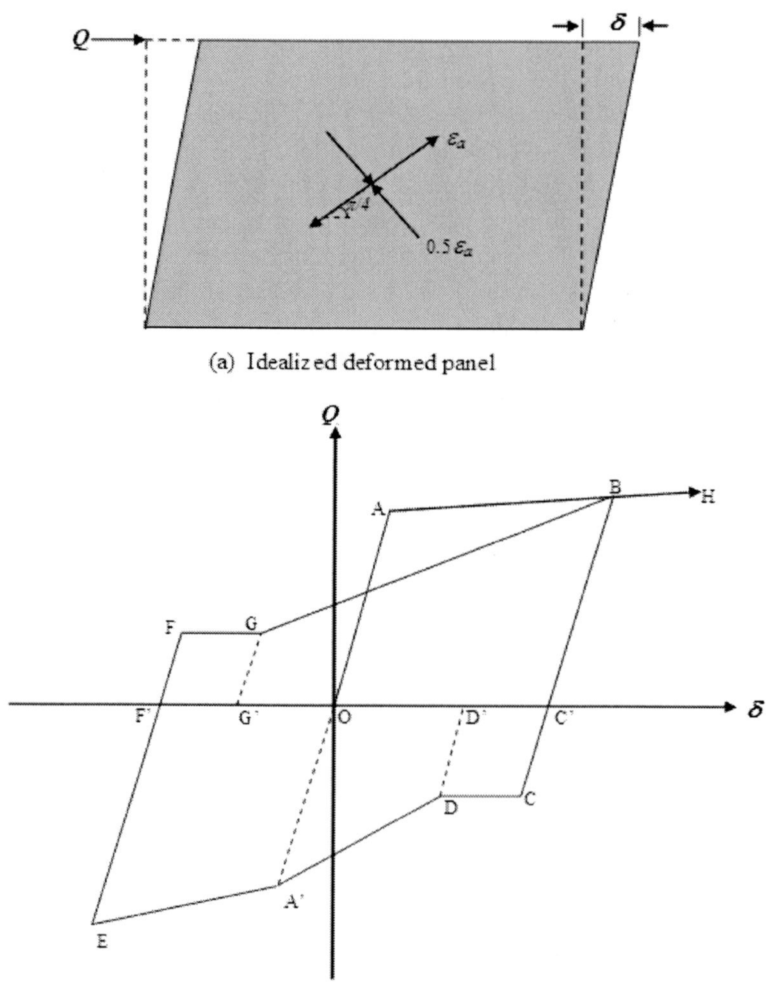

(a) Idealized deformed panel

(b) Hysteresis model

Figure 3. Hysteresis model proposed as per Mimura and Akiyama. (1977)

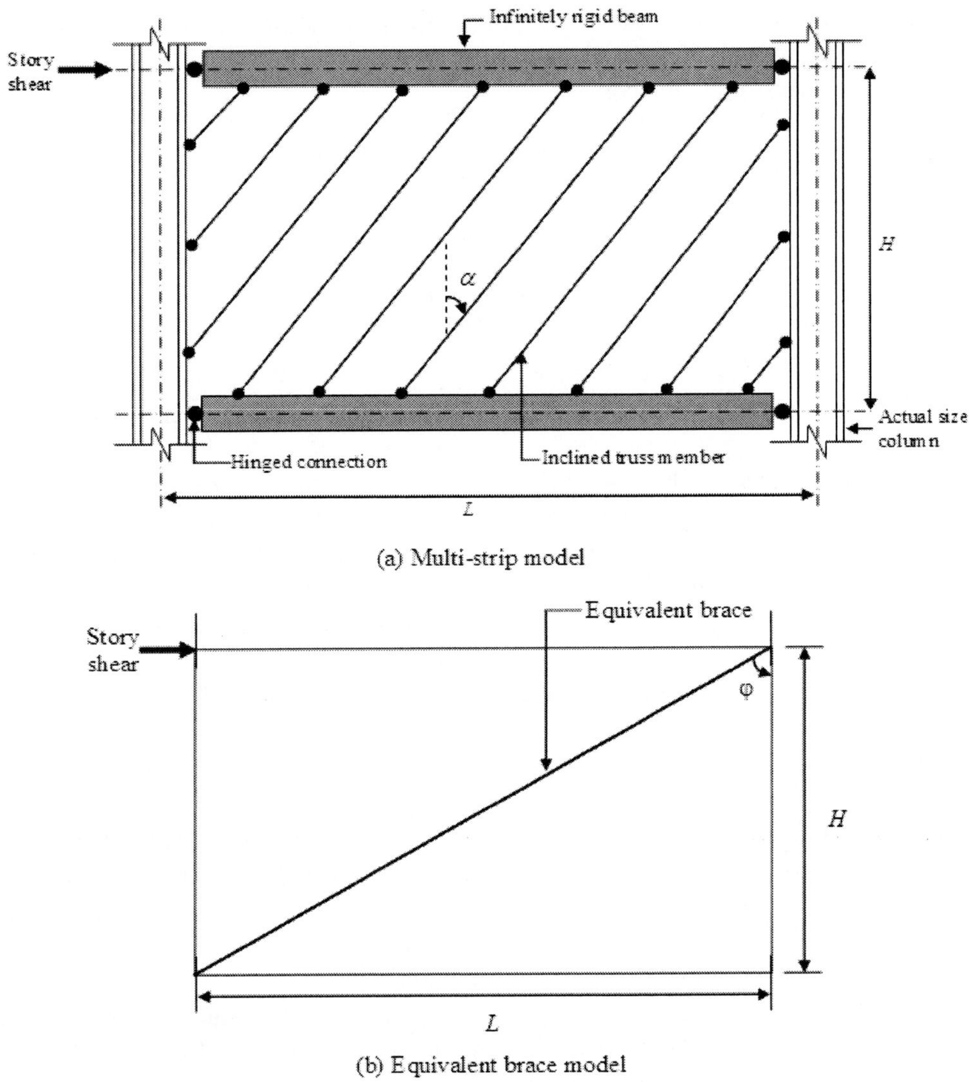

Figure 4. Analytical models for SPSW as per Thorburn et al. (1983)

Timler and Kulak (1983) conducted cyclic load tests on a large-scale, single-story steel plate shear wall specimen in order to validate the multi-strip model proposed by Thorburn et al. (1983). The researchers modified the angle of inclination of the principal tensile direction (α), proposed by Thorburn et al. (1983), by considering the bending strain energy of the boundary columns in the derivation:

$$tan^4\alpha = \sqrt{\frac{1+\frac{tL}{2A_c}}{1+th\left[\frac{1}{A_b}+\frac{h^3}{360I_cL}\right]}} \tag{2}$$

where I_c is the moment of inertia of the boundary column taken perpendicular to the plane of the web plate. Thorburn (1983) also derived an expression of the angle of tension field for SPSW with rigid beam-to-column connections:

$$tan^4 \alpha = \sqrt{\frac{1+Lt\left[\frac{1}{A_c}+\frac{L^3}{120I_b h}\right]}{1+th\left[\frac{1}{A_b}+\frac{h^3}{360I_c L}\right]}} \tag{3}$$

where I_b is the moment of inertia of the boundary beam taken perpendicular to the plane of the web plate.

The strip model proposed by Thorburn et al. (1983) and the expression for the angle of inclination of tension field proposed by Timler and Kulak (1983) not only constituted a major breakthrough in research on steel plate shear walls, but was also the beginning of a long line of experimental and analytical studies on various aspects of SPSW behavior in the University of Alberta, Canada. These research works finally resulted in the first incorporation of the thin unstiffened SPSW system in a design standard, when the Canadian Standard CSA-S16-1 1994 (CSA 1994) included the multi-strip modeling as a simple approach for seismic analysis of the unstiffened SPSW system.

Elgaaly et al. (1998) performed sensitivity studies to investigate the influence of the number of strips (truss members) to be used in a multi-strip model and their angle of inclinations. It was found from this analysis that the number of strips to be used depends on the slenderness of shear panels and stiffness of boundary elements. It was also found that a small variation in angle of inclination has a negligible effect on the initial stiffness. Rezai (1999) also conducted sensitivity analyses to assess the effect of various structural properties on the angle of inclination of the tension field. The Canadian Standard (CSA 2001) provisions and commentary of to AISC Seismic Provisions (AISC 2005a) recommends that a minimum of ten strips be used to model the web plate in order to approximate the effects of a distributed load on the boundary elements of the frame. The multi-strip modeling technique for steel plate shear walls, with minor variations, remains the most commonly used idealization for analysis and design of SPSW systems till date.

3.3. Equivalent (Story) Brace Model

In addition to the multi-strip model described in the previous section, Thorburn et al. (1983) also developed a Pratt truss model for the analysis of thin SPSW, known as the *equivalent brace model* or the *equivalent story brace model* (Figure 4.b). In this technique, the infill plate is represented by a single equivalent diagonal brace in such a way that the stiffness of the infill panel is equal to that derived from the multi-strip model of the panel. The area of the brace (A) is obtained as

$$A = \frac{tLsin^2 2\alpha}{2sin\varphi sin2\varphi} \tag{4}$$

where φ is the angle of the brace with respect to the column and all other parameters are as defined earlier for the multi-strip modeling. CAN/CSA-S16-01 (CSA 2001) recommended the equivalent brace model as a preliminary design tool for steel plate shear walls. The main advantage of the equivalent brace model over the multi-strip model is in the reduced computation. The multi-strip modeling becomes cumbersome for the geometry of node locations and for the large number of elements and degrees of freedom to handle. However, the equivalent brace model suffers from the fact that it does not represent the distributed forces applied by the plate on the boundary beams and columns, in any way. The multi-strip modeling technique is preferred by most users because of the level of accuracy in the results are higher than those obtained using the single equivalent brace model.

3.4. Multi-Angle Strip Model

Research conducted at the University of British Columbia, Canada by Rezai (1999) showed that the multi-strip model is significantly incompatible and inaccurate for a wide range of SPSW arrangements. The experimental studies conducted in the University of British Columbia indicated that the angle of the tension strips was closer to vertical at the corners and more horizontal around the mid-point of the plate (Rezai 1999). This was primarily related to the interaction of the infill plate and boundary elements at the corners. In order to overcome this deficiency of the multi-strip model, Rezai (1999) proposed a *multi-angle strip model* (Figure 5) for steel plate shear walls. In this model, all the strips are not placed parallel to the tension field. Instead, the multi-angle strip model has five truss members connecting a beam-column joint corner to a) middle of beam not meeting at this corner, b) middle of the corner not meeting at the corner, and c) the diagonally opposite corner. The diagonally opposite corner has the same connections through strips. These strips, oriented at various angles, account for the variation in the angle of tension field across the panel. Rezai (1999) also gave the equations for calculating the cross-sectional area to be assigned to each strip. In addition, a concept of effective width was employed so as to account for the incomplete tension field action.

Using a nonlinear analysis program and the multi-angle strip model, researchers were able to produce analytical predictions that are reasonably close to results. However the model was found to be conservative in predicting the ultimate capacity, besides being a complex one to handle without any significant gain in accuracy when compared to the multi-strip model.

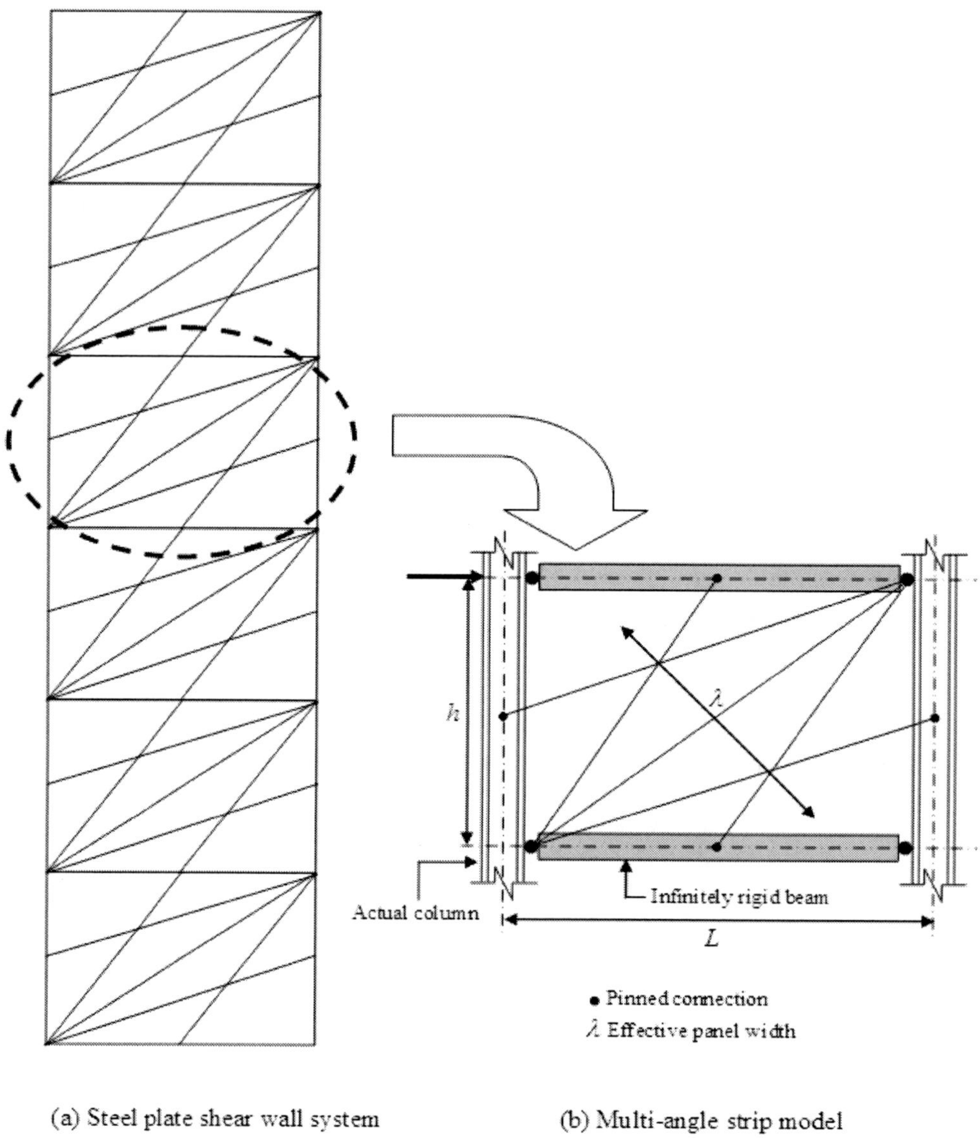

Figure 5. Multi-angle strip model as per Rezai et al. (1999)

3.5. Cross Strip Model

To predict the hysteretic behavior of a SPSW, a symmetric *cross-strip model* (Figure 6), that uses hysteretic stress-strain relationship, was developed by Elgaaly et al. (1993). The cross truss members were used to model the tension field action in opposite directions during cyclic load reversals and a hysteretic stress-strain relationship for these truss members was developed based on test results. Although this modeling technique is a little more computation intensive compared to the multi-strip technique, for obvious advantages it has gained acceptance over the years, specifically for nonlinear response/time-history analyses of SPSW systems.

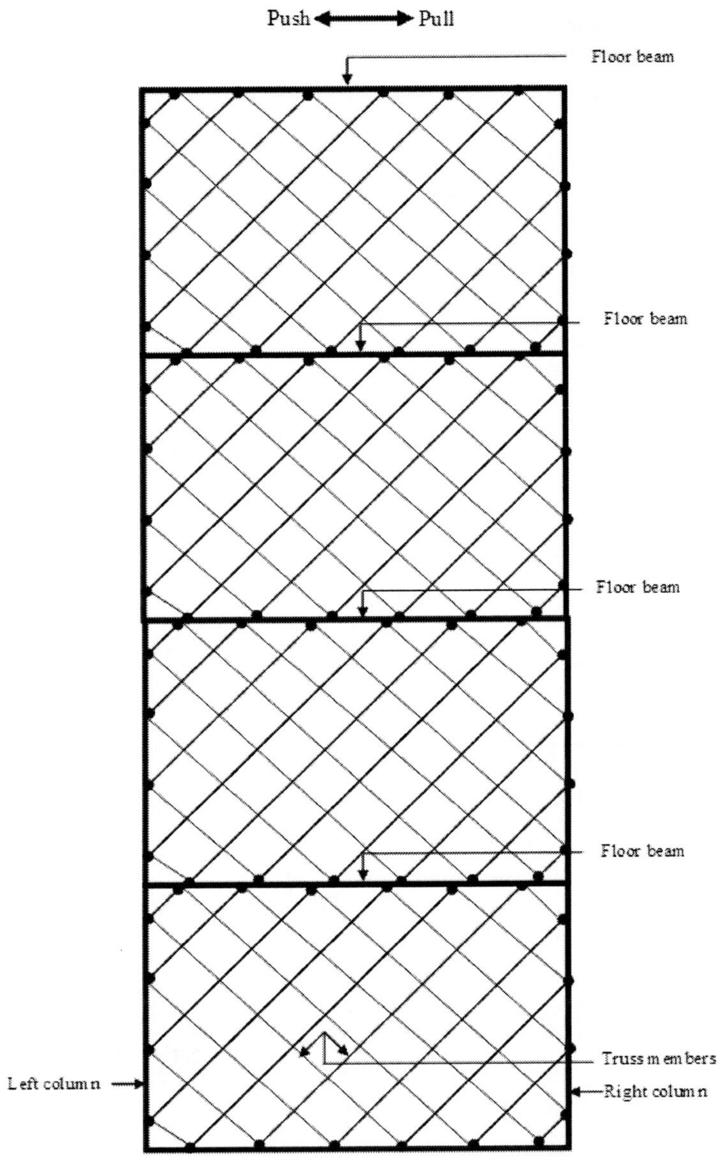

Figure 6. Cross-strip model for cyclic loading as per Elgaaly. (1993)

3.6. Plate-Frame Interaction Model

Sabouri-Ghomi et al. (2005) developed a general modeling technique for analysis and design of SPSW systems with different configurations. This method considers the behavior of the steel plate shear wall and the frame separately, and accounts for the interaction of these two structural elements. Thus, it was named the *plate frame interaction* (*PFI*) *model* (Figure 7). Sabouri-Ghomi and Roberts tested the effectiveness of the PFI model by comparing the analytical results to the results from various tests conducted earlier by Timler and Kulak (1983), Tromposch and Kulak (1987), Driver et al. (1997), etc. Kharrazi et al. (2008) later enhanced the PFI model as the *modified plate-frame interaction* (*M-PFI*) *model* by modifying the load-displacement diagram to include the effect of overturning moments on the SPSW response. Kharrazi et al. (2008) demonstrated the implementation of the M-PFI modeling technique in the design of a steel plate shear wall system considering different heights of the system. Evaluation of the M-PFI design methodology was performed using finite element analysis using the commercial general purpose nonlinear finite element program Abaqus. Good agreement was observed for stiffness and strength of the steel plate shear wall models obtained from both the M-PFI and FE methods. However, one disadvantage of the M-PFI method reported that it can not consider the material strain hardening effects.

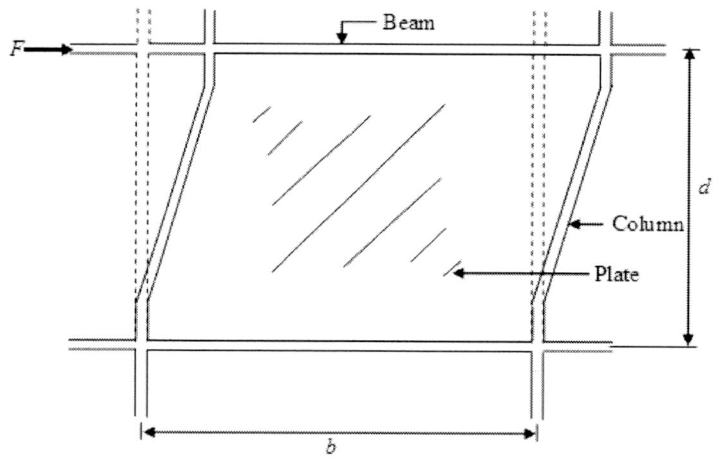

(a) Plate-frame interaction model

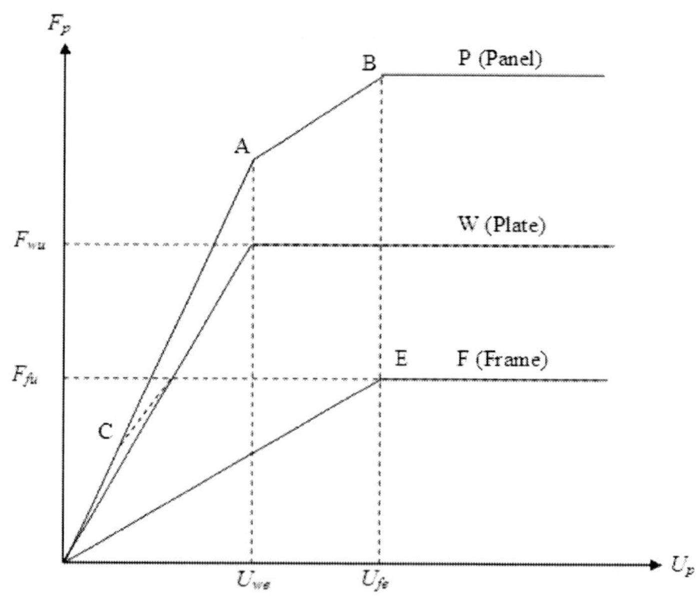

(b) Components of plate-frame interaction model shear load-displacement

Figure 7. (a) Plate-frame interaction model, (b) Components of plate-frame interaction model as per Sabouri-Ghomi and Roberts. (2002)

3.7. Finite Element Model

Over the past 20 years, various researchers adopted the finite element (FE) approach to study the post-buckling behavior of thin steel plate shear walls. The primary advantage of using a finite element model is that the elastic out-of-plane buckling of the thin plate can be explicitly modeled in a 3D finite element analysis. This formulation also has various other advantages in terms simulating the actual physical behavior of a structural system, albeit at the cost of being highly computation intensive. Various researchers (Elgaaly et al. 1993, Driver et al. 1997, Rezai 1999, Behbahanifard and Grondin 2001) used the finite element formulation including an explicit modeling of the out-of-plane buckling of the steel panel. These finite element model formulations were validated against various experimental observations carried out in different universities and research institutes.

Elgaaly et al. (1993) performed a nonlinear finite element analysis including both material and geometric nonlinearities. The results from experiments conducted in University of Maine were used for validating the FE model. Three-dimensional isoparametric doubly-curved shell elements and isoparametric three-noded beam elements were used to model the panel and boundary elements, respectively. The FE model comprised of a 6×6 mesh to represent the steel plate in each story, and six beam elements for each frame member. A monotonically increasing lateral load was applied until a loss of stability developed due to column plastic hinge formation and flange local buckling. The NONSAP software program which used the Newton-Raphson iteration method for nonlinear systems was used for the analysis. It was found that the wall with thicker plates was not significantly stronger because column yielding was the governing factor for both cases. The finite-element models using shell elements significantly over-predicted both the capacity and stiffness compared to the experimental results. These discrepancies were attributed to the difficulty in modeling initial imperfections in the plates and the inability to model out-of-plane deformations of the frame members. However, the finite element analysis using shell elements gave important information regarding the elastic buckling and the post-buckling behavior of thin panels.

Xue and Lu (1994) performed analytical studies on four twelve-story three-bay steel plate shear wall configurations. The objective of the study was to investigate the effect of beam-to-column and plate connections on the behavior of SPSW. Four different configurations were considered: (i) moment-resisting beam-to-column connections and infill plates fully connected to the surrounding frame, (ii) moment-resisting beam-to-column connections and the infill plates attached to only the beams, (iii) simple beam-to-column connections and fully-connected infill plates, and (iv) simple beam-to-column connections with infill plates connected only to the beams. Single bay, 12-story SPSW were modeled using elastic line elements for boundary members and four-noded shell elements with large-deformation capacity for infill plates. A 6×6 mesh was used for all panels, with the exception of the bottom panel where a 6×8 mesh was used. The structures were loaded monotonically with lateral forces at each story. Based on this study, Xue and Lu (1994) concluded that the beam-to-column connection type had a very small effect on the lateral stiffness of the frame. Connecting the infill plates only to the beams and using simple beam-to-column connections in the interior bay was found to be the optimal configuration because this drastically reduced the shear forces in the interior columns and helped avoiding a premature column failure.

Driver et al. (1998b) had developed the numerical model of four-story SPSW test specimens using the commercial general purpose nonlinear FE package Abaqus (Abaqus

1994). The shear wall infill plates were modeled using eight-noded quadratic shell elements (Abaqus element S8R5) and boundary elements were modeled using three-noded quadratic beam elements (Abaqus element B32). The actual connection between plate and boundary element was not modeled in FEM model. The material stress-strain behavior determined from coupon tests was used with bilinear representation in the FE model. The finite element simulation predicted the ultimate strength and initial stiffness well for all stories. However, at displacements larger than the yield displacement; the simulation overestimated the stiffness of the steel plate shear wall. It was concluded that this discrepancy was due to the inability to include the second-order geometric effects.

Rezai (1999) developed an orthotropic FE model (Figure 8.b) for SPSW in which the shell elements representing the infill panel were assigned orthotropic material properties in order to simulate the buckling of the compression diagonal in the plates. An orthotropic material allows different moduli of elasticity and shear moduli for three principal directions of the plate. This permits a modeling of the compression diagonal with much less stiffness than the tension diagonal, and thus ensures that it will attract much less shear in proportion to the tension diagonal. The in-plane local axis of the infill plates was oriented in 45° to the horizontal. The orthotropic material properties were assigned to the plates, taking into account the full modulus of elasticity of the plates along the tension diagonal axis and only a 2 to 5% elastic modulus along the compression diagonal axis. The shear modulus used in the orthotropic material is set to zero. The FE models were validated with the results from the tests conducted at University of British Columbia.

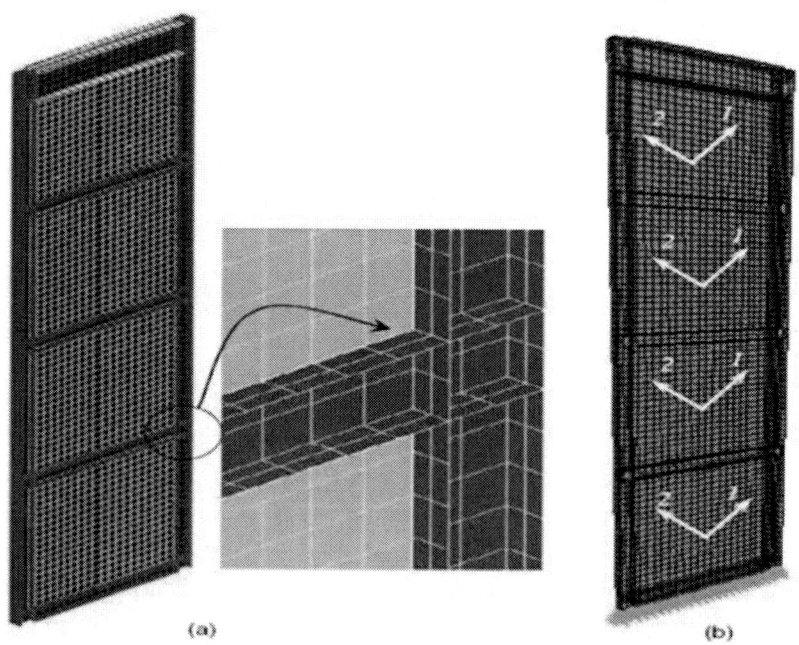

Figure 8. (a) 3-Dimensional FEM model, (b) 3-Dimensional orthotropic model of 4-story SPSW tested at UBC as per Rezai et al. (2000)

Later, the numerical models of SPSW specimens at University of British Columbia were developed by Rezai et al. (2000) using LSDYNA, a general purpose nonlinear finite element program (Livermore Software, 2003). The boundary elements and infill panels were all modeled by three- or four-noded shell elements considering both geometric and material nonlinearity (Figure 8.a). Trilinear stress-strain relationship obtained from coupon tests was used for modeling the material. The SPSW FE models were loaded with vertical and horizontal loads at each floor level. The horizontal load was increased linearly from zero to the ultimate capacity of the frame. A small load perpendicular to the plane of the plates was applied to simulate the effect of plate imperfection. The comparison of FE model results with UBC test result showed varying degrees of accuracy. Both the detailed and the orthotropic finite element models over-predicted the elastic stiffness.

A finite element model based on nonlinear dynamic explicit formulation was developed in Abaqus by Behbahanifard and Grondin (2001). This explicit formulation included the kinematic hardening material model to simulate the Bauschinger effect. After validating this model using experimental test results, it was used for the parametric study to identify parameters affecting the stiffness and strength of SPSW systems. The simulations were conducted for different steel panel aspect ratios (story height to bay width ratio) under monotonically increasing lateral and constant gravity forces. It was found that changing the aspect ratio within the range of 1 to 2 had negligible effect on the behavior of a shear wall panel. For aspect ratios less than 1, normalized shear capacity of panel, which is the ratio of shear load to shear yield capacity, increases. It was also concluded that initial imperfection in infill panel can have significant influence on the stiffness of steel plate shear walls, especially when subjected to low amplitude cyclic loading, but have low effect on shear capacity. The parametric study showed that panel out-of-plane imperfections were found to be of insignificant consequences, provided they were limited to 1 percent of $\sqrt{(Lh)}$ (that is, within normal fabrication tolerances). It was also found that increase in gravity loads and overturning moments on SPSW reduces the elastic stiffness and strength of the shear wall panel, as well as the drift at which the peak strength is reached.

In the last few years, 3D FE formulations including both material and geometric nonlinearities have become quite common for studying the behavior of SPSW systems. With the great advancement in the computational capabilities of modern computers, FE approaches are adopted to study collapse behavior of SPSW systems under lateral and other loads, such as blast, and for simulating experiments realistically. Although a finite element modeling gives valuable information about the buckling and post-buckling behavior of steel plate shear walls, it is much more complex compared to the other modeling schemes such as the multi-strip modeling or the equivalent brace modeling. It needs an experienced researcher to use this tool. Also, going for a finer mesh in order to obtain convergence results in increased computational time, which may not be a feasible option for a practicing engineer. Finally, there is not yet a standardized method of FE analysis for steel plate shear walls and often variations from experimental observations are significant. At present, the FE formulation is not considered to be a tool for design calculations but a great tool for research.

4. Past Research on Steel Plate Shear Wall: "Experimental"

Similar to the "analytical" research on SPSW, the "experimental" studies on steel plate shear walls, over the last forty years, have had multitude of forms. Major research works covered both stiffened and unstiffened SPSW systems, from single-story to four-story specimens of various configurations, under both monotonic and cyclic quasi-static shear loads, and even some dynamic tests on shake tables. Often, these experimental studies were conducted in conjunction with related analytical research works. Some of the significant experimental research works are reviewed in this section.

4.1. Research on Stiffened SPSW

The first extensive research program on the behavior of steel plate shear walls was conducted by Takahashi et al. (1973). The objective of this experimental research was to study both unstiffened and stiffened SPSW with various configurations of stiffeners for shear panels under inelastic cyclic loading in order to determine their stability for use as a lateral load resisting system for building. The experimental program was carried out in two phases. The first phase consisted of cyclic test on 12 SPSW panels with varying plate thicknesses (2.30 mm to 4.50 mm) and stiffeners arrangements. The stiffeners were cut, in various widths, from flat plates and welded to panels on one or both sides. All test panels were 1200 mm in length and 900 mm in height. Each panel was bounded by very stiff rectangular pin-jointed frames using high strength bolts. Four to six cycles of shear loading were applied with increasing deformation in each cycle. The important findings from this experimental research were: (a) stiffened panels dissipated significantly more energy than unstiffened panels, (b) both stiffened and unstiffened panels behaved in stable and ductile manner; (c) the panels with stiffeners on both sides tended to show more stable behaviour than those with single-sided stiffeners. Based on these results of test, Takahashi et al. recommended that stiffened plate shear wall be designed so that the shear panel dose not buckles elastically.

The second phase of the experimental program consisted of two cyclic full-scale tests on a representative portion of a stiffened shear wall taken from the design of a 32-story building. The test specimens were one bay wide and two story high, one is stiffened (4 mm thick) with door opening and the other without any opening (6 mm thick). Horizontal in-plane loads were applied at the top of the specimen and loaded and unloaded in one direction, with a few fully reversed loading cycles interspersed. Both specimens showed good ductility and energy dissipation characteristics. An analytical study accompanied these tests. The test specimens were modeled using finite element method, considering an elasto-plastic material response together with the von Mises yield criterion, and only planar behavior of the plate. This signifies that the plate buckling was not explicitly modeled in the analysis. The finite element analyses used monotonic loading, but the results achieved a good agreement with the envelope of the experimental load versus deflection curves.

4.2. Quasi-Static Cyclic Tests

In order to validate the multi-strip idealization of thin unstiffened SPSW under lateral load, developed by Thorburn et al. (1983), Timler and Kulak (1983) tested two full-scale specimens that represented single-storey, single-bay steel plate shear wall elements. Figure 9 provides a schematic representation of the test set-up, which consisted of vertically oriented beams and horizontally oriented column connected by pin joints at four extreme corners. The member sizes were so chosen as to represent typical building constructions. The specimen was loaded statically with three complete cycles of loading to a deflection limit of $h_s/400$ (h_s = height of one story) or 6.25 mm (under service load). It was observed that: (i) during the cyclic loading the test specimen behaved elastically; (ii) the infill plate material used in the test specimen showed a continuously curved stress vs. strain relationship of cold-formed steel which was approximated in analysis (multi-strip model) by a elastic-perfectly plastic curve.

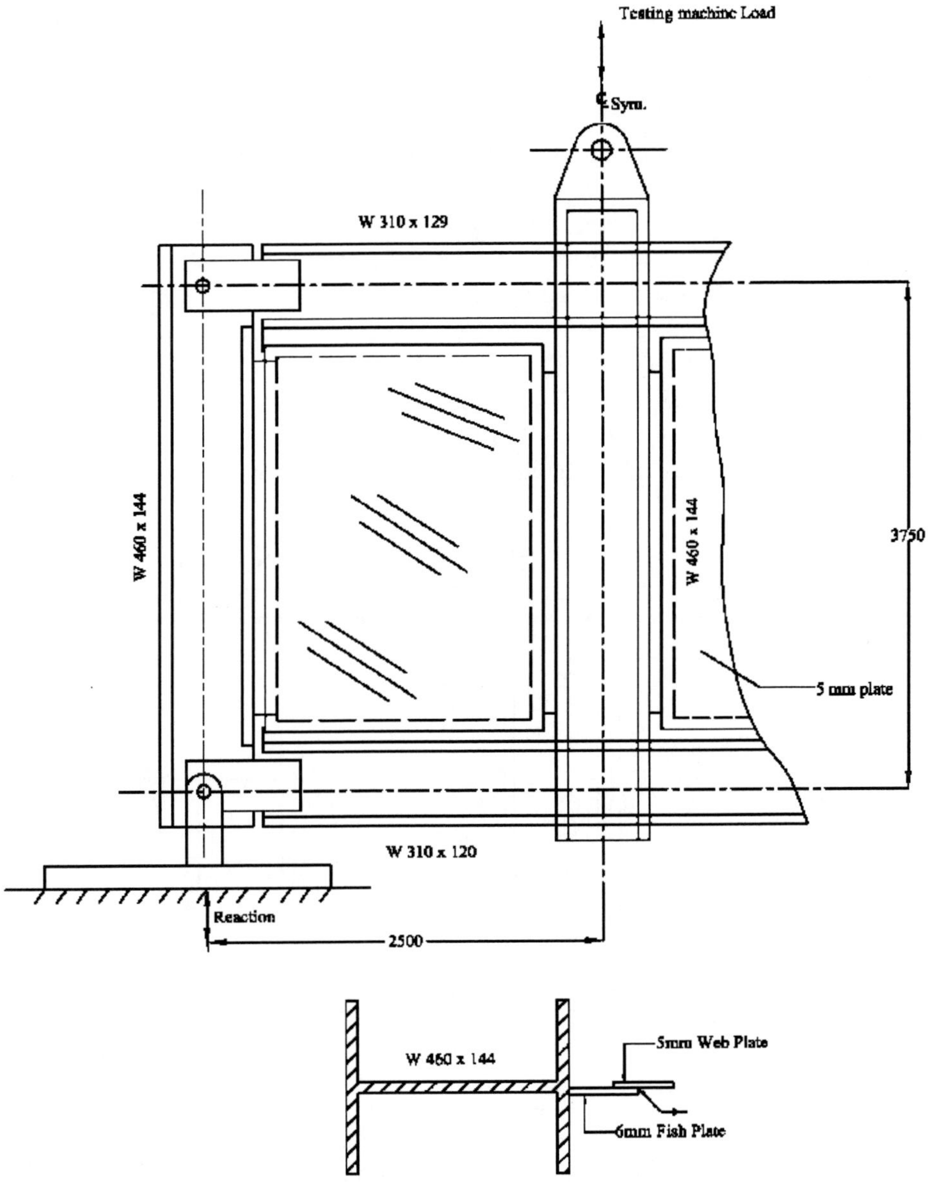

Figure 9. Single-story test specimen as per Timler and Kulak. (1983)

Timler and Kulak (1983) analyzed their test specimen using the then recently developed multi-strip model. Since the analysis program could not account for inelastic behavior, it was simulated in the boundary members by successive reductions in the cross-sectional properties of the entire length of the members and in the strips by limiting the stress to the static yield stress measured from tension coupons. Good correlation was observed between predicted and actual values of the infill plate stresses, axial strains, and in the load vs. deflection curves. It was concluded that the flexural stiffness of the columns affected the value of the angle of inclination of the tension field (α). This led to a modification of the original expression for α (Equation 1) proposed by Thorburn et al. (1983).

Tromposch and Kulak (1987) tested a one-storey, two-panel specimen similar to that tested by Timler and Kulak (1983), except that bolted beam-to-column connections were used that would typically represent the connection with stiffer columns as practiced in the field. The new specimen had a thinner infill plate (3.25 mm). Also, the columns were pre-stressed before testing in order to simulate gravity loads on the structure as shown in Figure 10. The main objectives of the tests were to examine the hysteretic behaviour of the specimen and to verify the analytical multi-strip model proposed by Thorburn et al. (1983) for load reversals. Fully reversed cyclic lateral loads were applied on the specimen. The column prestressing rods were removed prior to the final loading in order to eliminate the possibility of restraints to the specimen occurring at large deformations. The primary findings from this study were: (i) the cyclic lateral loads were gradually increased and these reached up to a maximum of 67% of the ultimate load calculated using multi-strip idealization with a corresponding maximum lateral deflection of $h_s/129$ and (ii) response of test specimen during the cyclic loading phase indicated very ductile behaviour but hysteresis curves are severely pinched.

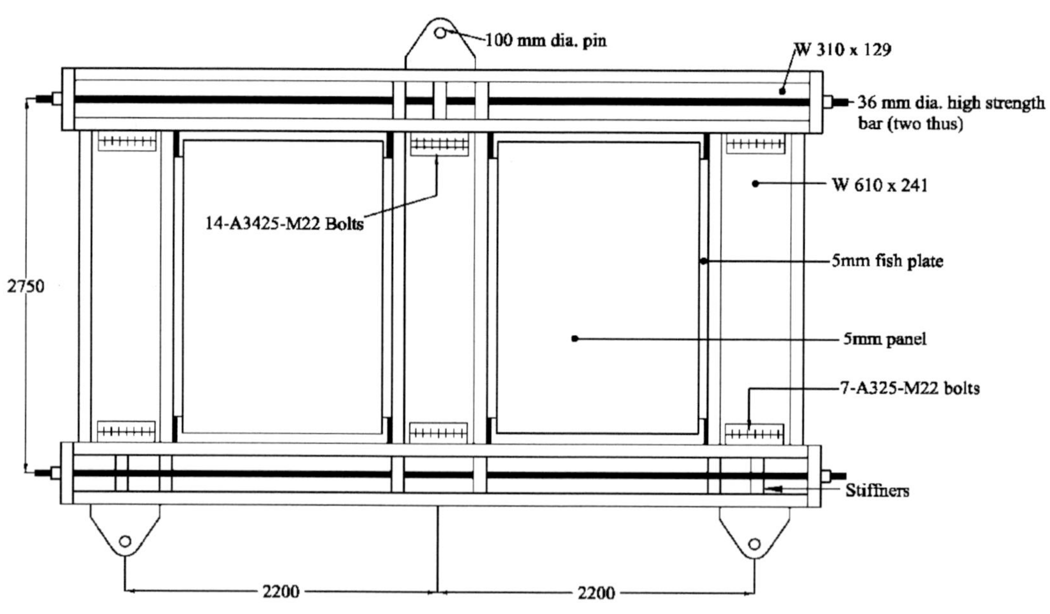

Figure 10. Test specimen as per Tromposch and Kulak. (1987)

Sabouri-Ghomi and Roberts (1992) conducted a series of quasi-static cyclic loading tests on small-scale SPSW models. The specimens were composed of single-panel unstiffened plates with a stiff, pin-ended boundary frame. The panels were either 300 mm × 300 mm or 300 mm × 450 mm in size and having a thickness ranging from 0.54 mm to 1.23 mm. The shear panel was loaded at two opposite corners in the direction of panel diagonals. Initially, the tensile load was applied until a significant inelastic behavior was observed. This was followed by a similar compressive load. At least four complete cycles of loading with gradually increasing peak displacement were applied to each specimen in this manner. The observations from this model led to the development of a general method of dynamic analysis of thin-panel steel plate shear walls (Sabouri-Ghomi and Roberts 1992). This method uses a time-stepping finite-difference technique to solve the governing differential equation of motion. The SPSW structure is idealized as a vertical cantilever beam with masses lumped at each floor level. The time-dependent loading is also assumed to act discretely at each floor. In order to include the nonlinear material behavior, an approximate elasto-plastic hysteresis model was proposed that included the influence of shear buckling and yielding of the web plate and the surrounding frame.

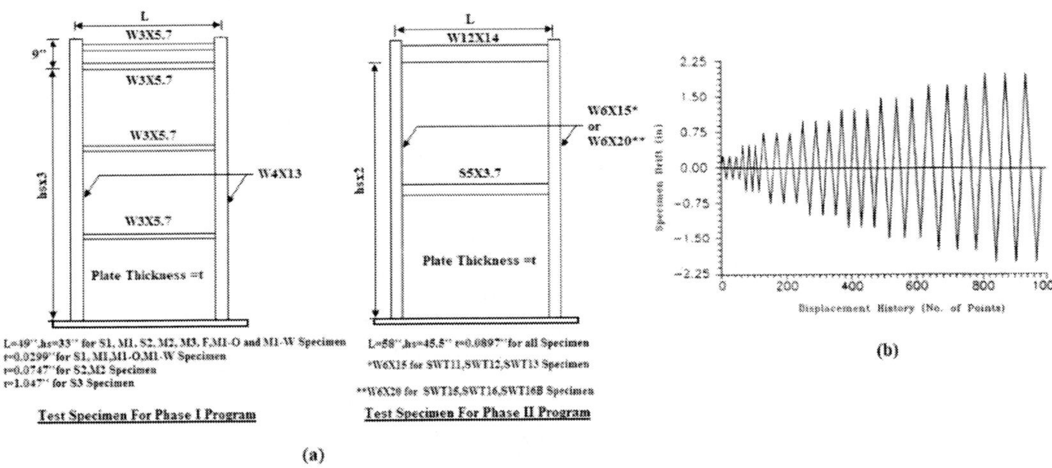

Figure 11. (a) Test specimens tested at University of Maine and (b) Cyclic load history used for test as per Elgaaly et al. (1993)

Elgaaly (1998), Elgaaly et al. (1993), Caccese et al. (1993) conducted a number of cyclic tests of small scale steel frames infilled with steel plate shear walls. The objective of these tests was to study the post-buckling behavior of unstiffened thin SPSW under cyclic loading. The capacity of SPSW with different connections, such as bolted and welded connections, were also investigated. This experimental research program conducted at the University of Maine consisted of two phases – eight quarter scale specimens were tested in phase I and seven one-third scale specimens were tested in phase II. The specimens tested are shown in Figure 11. The significant outcomes from this study were: (i) the effect of the presence of the column axial compressive loads (upto a magnitude equal to 50% of the column's nominal axial force capacity) was almost negligible; (ii) the bolt spacing did affect the mode of failure and the specimen with the bigger bolt spacing failed due to a combination of plate rupture and shearing of the bolts; and iii) the specimen with the welded plates exhibited higher stiffness compared to the bolted specimens.

To assess the effectiveness of using thin plate shear wall systems in seismic zones, Cassese et al. (1993) conducted quasi-static cyclic load tests on six quarter scale single-bay three-storey unstiffened SPSW specimens. The goal of these tests was to observe the effect of the beam-to-column connection type and the panel thickness on the overall behavior of the

SPSW system. They concluded that the beam-column connection type has minor effect on the SPSW behavior. It was later argued by Kulak et al. (1994) that greater energy dissipation can be achieved with the use of moment connections as obtained by Tromposch and Kulak (1987). Regarding the plate thickness, Cassese et al. (1993) concluded that the plate has an optimum thickness which if exceeded, produces no increase in strength and that the wall fails by either column yielding or buckling.

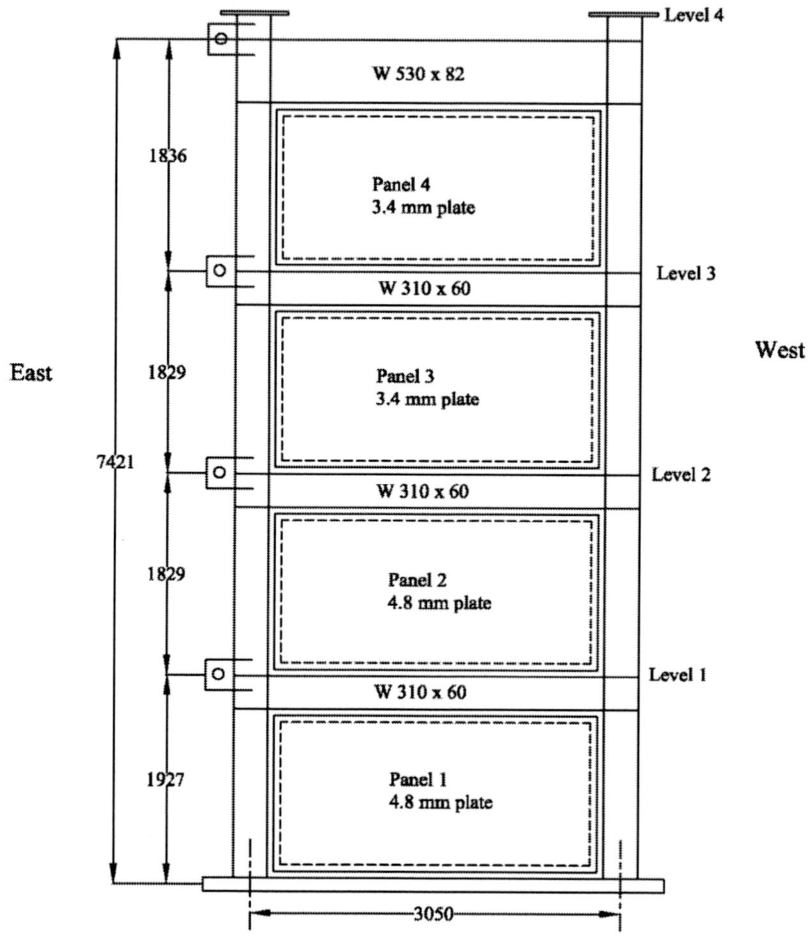

Figure 12. Test specimens of 4-story SPSW tested at University of Alberta as per Driver et al. (1998)

Driver et al. (1998b) performed tests on a large-scale four-storey single-bay SPSW specimen (Figure 12). This test on multi-story SPSW was some sort of a pioneer and the results from this study have been used by various researchers to validate various finite element as well as simplified analytical models of SPSW systems. The specimen had moment resisting beam-to-column connections and the infill plates were welded to the boundary members using fish plates. Gravity loads were applied at the top of each column and cyclic lateral loads of equal magnitude were applied at each floor level, as per the requirements of ATC-24 (ATC 1992). The primary findings of this research work were: (i) the specimen was able to resist increasingly higher loads at each successive cycle until a deflection of five times the yield deflection was reached; (ii) after the ultimate strength was attained, the deterioration of the load-carrying capacity was gradual and stable; (iii) the maximum deflection attained by the lowest storey before the failure occurred was nine times the yield deflection; (iv) the amount of energy dissipated during the loading cycles was significantly greater than that shown by similar specimens, but with shear-type beam-to-column connections as in the tests carried by Tromposch and Kulak (1987); and (v) overall, the test results confirmed that a properly designed steel plate shear wall system is an excellent lateral load-resisting system for seismic loading.

Lubell et al. (2000) conducted experiment consisting of two one-storey steel plate shear wall specimens (SPSW1 and SPSW2) and one four-storey specimen (SPSW4), as shown in Figure 13. In all these specimens, the beams were connected to the columns using moment-resisting connections. A relatively stiffer beam was used at the roof level for the specimen SPSW2 in order to develop a full tension field. Steel masses were placed at each storey of specimen SPSW4 to simulate gravity loading. Quasi-static cyclic load was applied as per ATC-24 (ATC 1992) requirements. These experiments were also simulated analytically through a series of numerical studies to asses the ability of the simplified analysis technique presented in Canadian steel design standard, CAN/CSA-S16-1 (CSA 1994). There was a varying degree of consistency in capacities predicted by test results and by analytical models. Furthermore, it was found that the use of a stiffer roof beam in the specimen SPSW2 lead to significant increase in the ultimate strength and stiffness of the SPSW system.

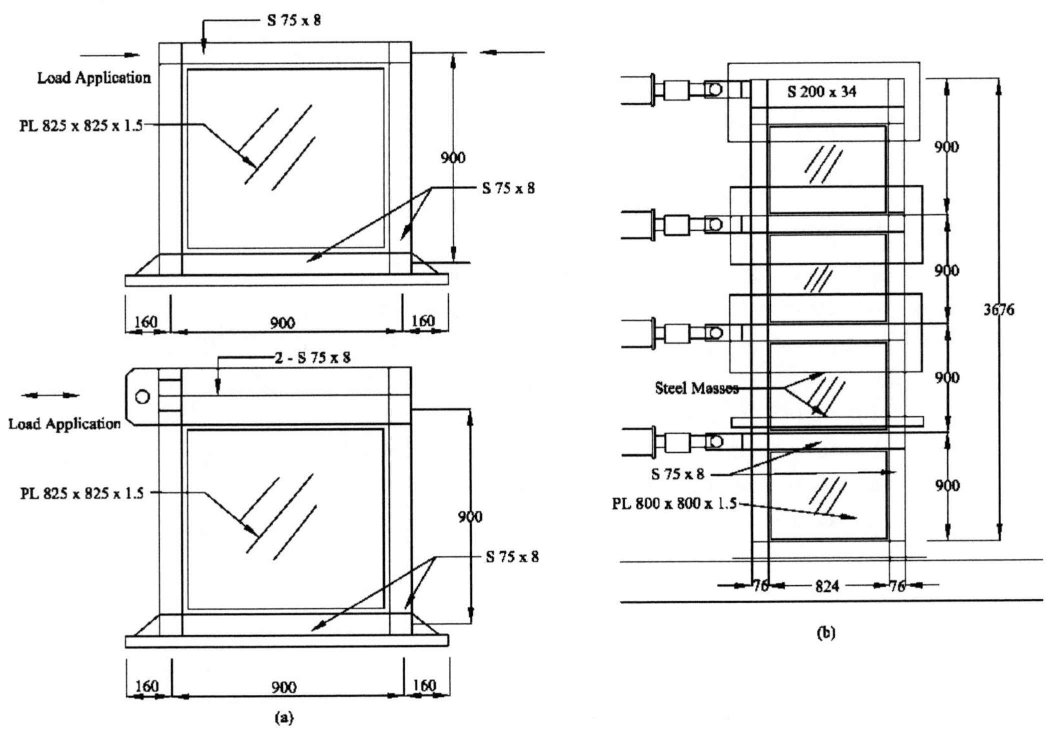

Figure 13. Test specimens of (a) Single story SPSW (SPSW1), (b) Single story SPSW with stiffer beam (SPSW1), (c) Four story SPSW (SPSW4) tested at University of British Columbia as per Lubell et al. (2000)

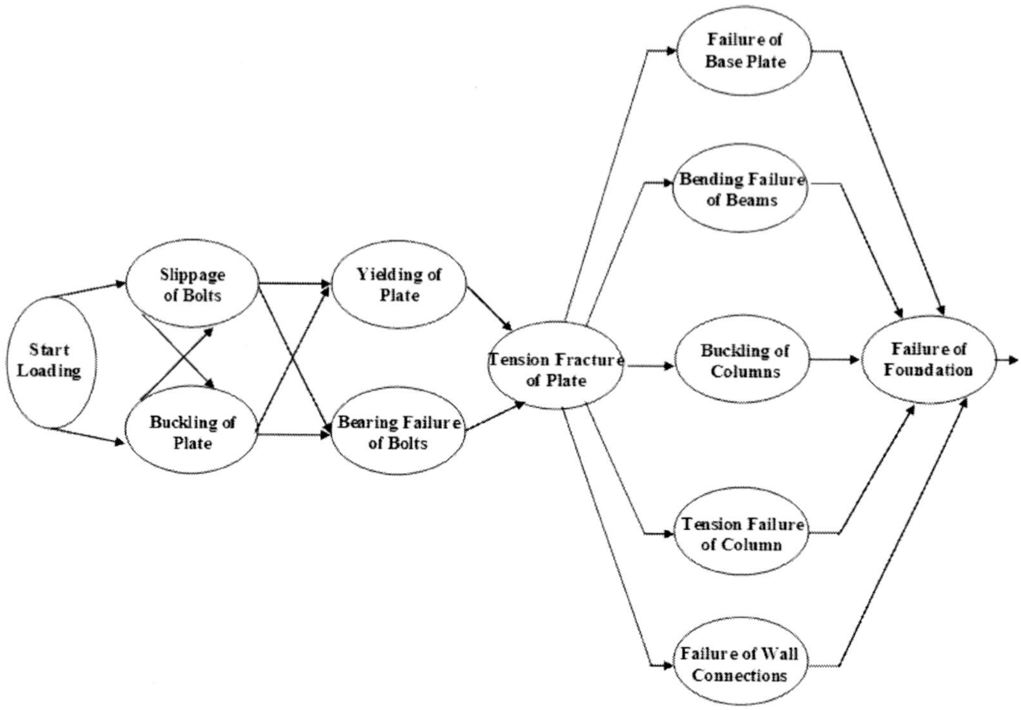

Figure 14. Major failure mode of a typical steel plate shear wall as per Astaneh-Asl. (2000)

Astaneh-Asl and Zhao (2002) conducted tests on two half-scale three-story SPSW specimens to investigate the cyclic behavior of a steel plate shear wall system. The specimens used were derived from the subassembly of a prototype building. Each specimen included one or two full stories in the middle and two half stories at the top and bottom. The structural component consists of concrete filled tubes or steel pipes (CFT) as gravity columns, wide flange (WF) steel beams and columns and steel wall panel. Both specimens were subjected to a large number of elastic and inelastic load-deformation cycles. The failure modes for these test specimens were the local buckling of the wide flange column (not a gravity column) for one specimen and the fracture at the upper floor beam-column junction for the other. Both the test specimens demonstrated large ductility capacities before failure. Astaneh-Asl (2001) also compiled a comprehensive document detailing the behavior and design of SPSW. Figure 14 shows a list of possible failure mechanisms that is organized into a hierarchical order of failure modes. The ductile failure modes are ranked as more desirable than brittle failure modes and are arranged first. This chart can be a very effective design guideline for checking individual members in a steel plate shear wall system.

4.3. Shake Table Tests on SPSW

Unlike quasi-static tests on steel plate shear walls of various forms and configurations, dynamic tests on SPSW specimens on a shake table are few and relatively recent. One of the first shake table tests was conducted by Rezai (1999), on a 25% scale model of four-story single-bay unstiffened steel plate shear wall specimen (Figure 13.c). The main objective of this shaking table test was to provide more information regarding seismic performance of multi-story steel plate shear walls under the effect of intense input excitations. The four-story test specimen was subjected to three recorded and one synthetically generated ground motions. A similar specimen was also tested earlier under quasi-static loading by Lubell (1997). The records selected for the shake table tests were such that they represent different excitation intensities, frequency contents and epicentral distances. The floor mass of 1700 kg per floor was applied through stacks of steel plates of size 1.5 m × 0.6 m × varying thickness. Rezai (1999) reported that the limited capacity of the shake table used prevented the attainment of any significant inelastic response of. It was found that the majority of the input energy was dissipated by the first floor shear panel while the upper floor panels behaved as a single rigid body rotating about the first floor. The first natural frequency of the specimen was found to decrease with the increase in shaking intensity. This was related to the severity of the out-of-plane buckling behavior of the infill panel, which in turn influenced the overall stiffness of the SPSW frame. The uniaxial strains at the top and bottom flanges of first story beams were quite small which indicated that the flexure generated in beams due to the infill panel forces was negligible. The hysteresis loops were pinched because of the dominance of the frame action during unloading and redevelopment of tension field in opposite direction.

5. Design and Analysis Methods

In early literatures on steel plate shear wall, the behaviour of SPSW was considered to be analogous to the vertical cantilevered plate girder where columns act like “flanges”, steel plates act like the “web”, and intermediate beams act like “stiffeners” in a plat girder (Timler et al. 1998, Sabouri-Ghomi et al. 2005). Astaneh-Asl (2001) recommended the use of equations that describe the behaviour of plate girders (AISC 1999) for the design of unstiffened steel plate shear walls. Later, Berman and Bruneau (2004), through an analytical study, showed that this plate girder analogy is just qualitative and not quantitative. As per Berman and Bruneau, the angle of inclination of the tension field, in case of plate girders, is not much influenced by flanges or web stiffeners, but it heavily depends on the stiffnesses of boundary elements for a SPSW. They suggested that designing a SPSW following the standard plate girder design requirements (AISC 1999) can lead to conservative and uneconomical designs. The plate girder analogy for SPSW behaviour also leads to ineffective capacity based design provision.

The subsequent text in this section deals with plastic analysis and plastic design concepts for steel plate shear wall systems (Berman and Bruneau 2003a), capacity design provisions, and a performance-based design method for SPSW (Ghosh et al. 2009)

5.1. Plastic Analysis

Using a plastic mechanism based analysis and the idealization of the infill plate as discrete parallel strips, Berman and Bruneau (2003a) derived an equation to calculate the ultimate shear strength of single-story SPSW with pinned beam-to-column connections (Figure 16.a) as follows:

$$V = \frac{1}{2} F_y t L sin2\alpha \tag{5}$$

where V is the maximum shear taken by the steel panel, F_y is the yield stress of the infill panel material, L is the (bay) width of panel and α is the angle of inclination of the tension field of infill panel with respect to the vertical. This equation was modified further (Berman and Bruneau, 2003a) for single-story SPSW with rigid beam-to-column connections by accounting for the inelastic work done through the rotation of plastic hinges formed at beam ends and in columns:

$$V = \frac{1}{2} F_y t L sin2\alpha + \frac{4M_p}{h} \tag{6}$$

where M_p is the smaller of the beam and column plastic moment capacities, and h is the height of the infill panel

For multi-storey shear walls, design equations were developed based on two types of failure mechanisms that provide a rough range of ultimate strengths of SPSW: soft-story failure (Figure 15.b) and uniform yielding of the infill plates in all storeys simultaneously (Figure 15.b). Thus, for a soft-story collapse mechanism at the ith story, the ultimate shear strength of the shear panel can be obtained as

$$\sum_{j=1}^{n_s} V_j = \frac{1}{2} F_y t_i L sin2\alpha + \frac{4M_{pci}}{h_{si}} \tag{7}$$

where V_j = applied lateral forces above the ith soft story, t_i = plate thickness at i^{th} story, M_{pci} = the plastic moment capacity of the column at soft story, h_{si} = height of the ith soft story and n_s = total number of stories.

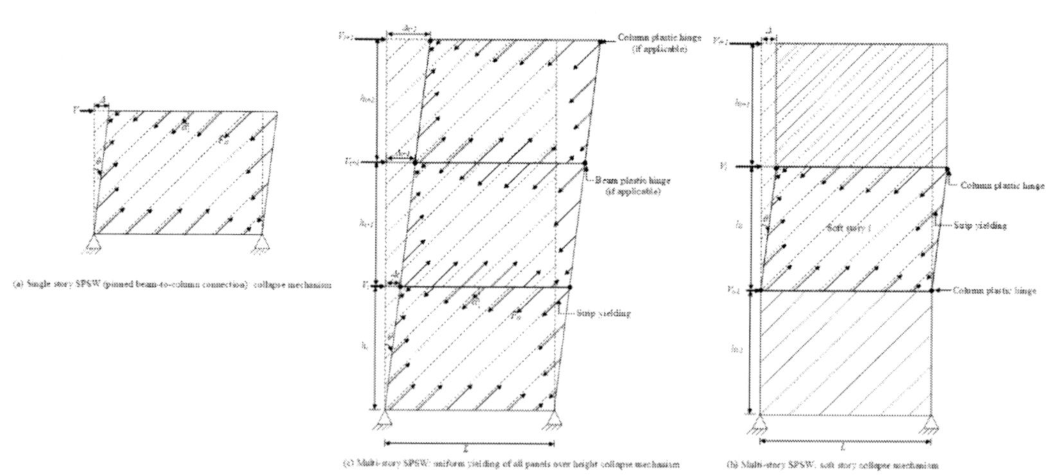

Figure 15. Single-story and multi-story SPSW collapse mechanisms as per Berman and Bruneau. (2003a)

Similarly, considering a uniform yielding of the plates over every story (which is the most desirable collapse mechanism) and having plastic hinges formed at the beam ends, the ultimate shear strength of multi-story SPSW can be calculated as

$$\sum_{i=1}^{n_s} V_i h_i = 2M_{pc1} + 2M_{pcn} + 2\sum_{i=1}^{n_s-1} M_{pbi} + \sum_{i=1}^{n_s-1} \frac{1}{2} F_y L h_i (t_i - t_{i+1}) sin2\alpha \tag{8}$$

where M_{pc1} and M_{pcn} are the first and top story plastic moment capacities, M_{pbi} is the plastic moment capacity of ith story beam. It should be noted that depending upon the aspect ratio the terms M_{pc1} and M_{pcn} may require to be replaced by M_{pb1} and M_{pbn} and the angle of tension field of all panels is assumed uniform through out.

Berman and Bruneau (2003a) validated their proposed equations (Eqn. 5 and Eqn. 6) for ultimate shear strength of a single-story SPSW having simple and rigid beam-to-column connections with the experimental test conducted by various researchers, such as, Timler and Kulak (1983), Roberts and Sabouri-Ghomi (1992), Caccese et al. (1993), Driver et al. (1997), Elgaaly (1998), Lubell et al. (2000), etc. It was found that the average error between the ultimate shear predicted by equations and that obtained form the experiment was –5.9% in case of pinned beam-to-column connections and was 17.0% in case of rigid beam-to-column connections. Hence, it was reported that the equations derived from the plastic analysis of a strip model were generally conservative for calculating the expected ultimate shear strength of the steel plate shear wall.

5.2. Plastic Design of Steel Plate Shear Walls

Berman and Bruneau (2003a) proposed a revised on the design procedure for steel plate shear walls. In this, the design base shear and its distribution over the height of a building was obtained as per the relevant building code, and then the minimum plate thickness required for each story was calculated using equations that were derived from the plastic analysis of the strip model of SPSW:

$$t_i = \frac{2V_i \Omega_s}{F_y L sin2\alpha} \tag{9}$$

where Ω_s is the system overstrength factor and V_i is the design story shear. After fixing the thickness of the infill steel panel at each story, the boundary elements were selected as per the minimum stiffness requirement and the strip model was developed for analyses. The angle of inclination of the tension field was recalculated and beams and columns were designed according to capacity design provisions. This procedure of designing the SPSW is purely an iterative one, which starts with an assumption of the angle of inclination of the tension field of the shear panel. It is then followed by the calculation of the plate thicknesses as per Eqn.9, and the selection of the boundary elements as per the minimum stiffness requirement. These preliminary sizes are used to develop the multi-strip model for analysis and the sizes of plate and boundary elements are later revised as per analysis results/design requirements.

Recently, Berman and Bruneau (2008), Vian et al., (2009), Vian (2005) proposed capacity design provisions for VBE and HBE with reduced beam sections that are discussed in detail in Section 7.1.

5.3. Performance Based Design of Steel Plate Shear Walls

As discussed earlier, the SPSW has significant inelastic deformation capacity which cannot be utilized properly using the elastic force-based or even the capacity design provisions, available in current seismic design codes and guidelines that only implicitly incorporate this deformation capacity. The performance-based seismic design (PBSD) methodology is a more general, reliable, and efficient method and it explicitly considers the inelastic behaviour of a lateral load resisting system. Thus a shift towards this methodology for SPSW system was felt very necessary by Ghosh et al. (2009). Considering this, they have recently proposed a displacement/ductility-based design methodology of steel plate shear wall systems with pin-connected boundary beams. The method proposed by Ghosh et al. (2009) considers the target displacement ductility ratio (μ_t) as the design criterion. Thus it can utilize the ductility capacity of SPSW systems efficiently. In addition, their proposed design method is also based on a pre-selected failure mechanism, thus satisfying the PBSD requirement of controlling both the quantity and the quality of seismic structural damage. The performance-based design method as proposed by Ghosh et al., (2009) is based on equating the inelastic energy demand on a structural system with the inelastic work done through the plastic deformations (for a selected yield mechanism) subjected to a monotonic loading up to the target drift.

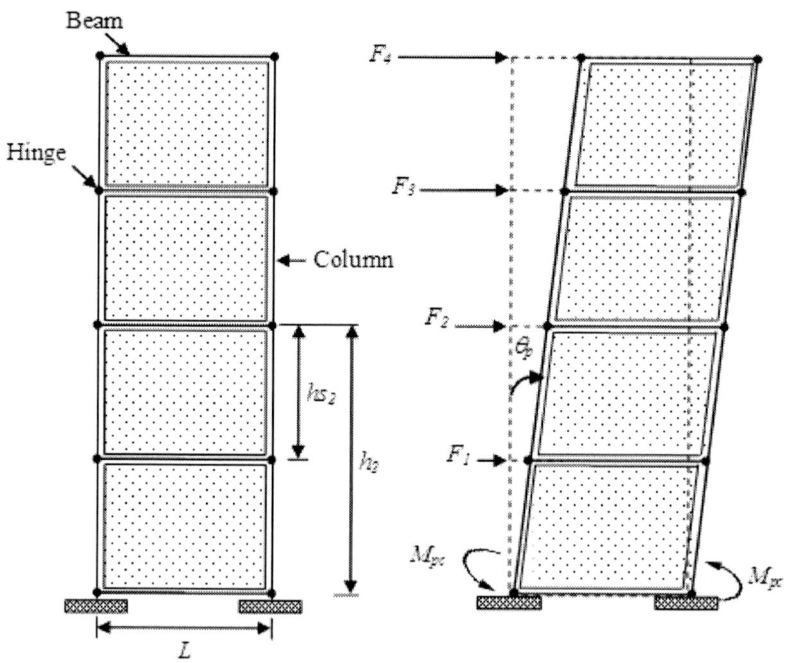

Figure 16. (a) Schematic of the SPSW with pin-connected beams, (b) Selected yield mechanism as per Ghosh et al. (2009)

Following is a very brief overview of the design formulation presented by Ghosh et al. (2009). A simple SPSW system is considered for this where the beams are pin-connected at their ends to the columns, while the columns are fixed at their bases and are continuous along the height of the system, as shown in Figure 16.a. The total strain energy (elastic and plastic) which is imparted to an inelastic system, is estimated as

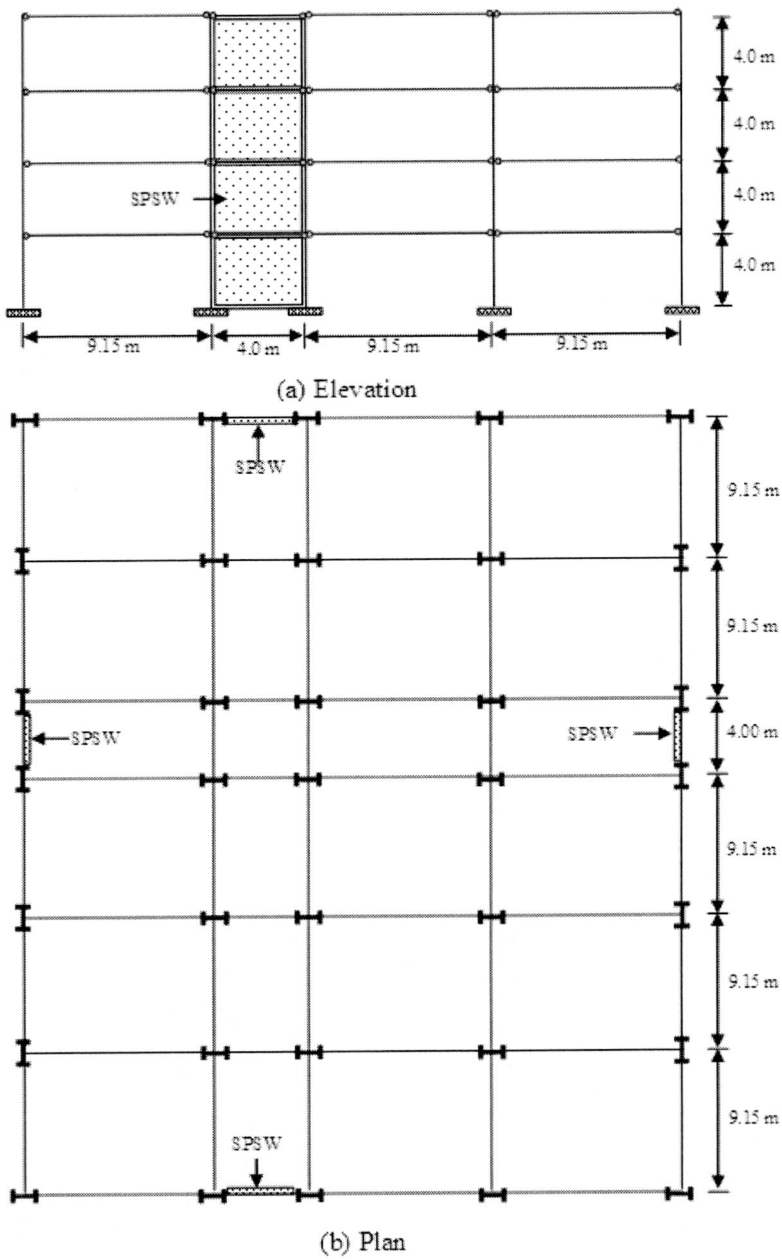

Figure 17. Four-story steel building with pin-connected beams and one SPSW bay as per Ghosh et al. (2009)

$$E_e + E_p = \gamma\left(\frac{1}{2}MS_v^{\ 2}\right) = \frac{1}{2}\gamma M\left(\frac{T}{2\pi}C_e g\right)^2 \tag{10}$$

where, E_e = elastic strain energy demand, E_p = plastic strain energy demand, γ = energy modification factor, M = total mass of the structure, S_v = pseudo velocity corresponding to T, T = fundamental period, C_e = elastic force coefficient, and g = gravitational acceleration. The energy modification factor is calculated based on the target ductility ratio of the system (μ_t) and ductility reduction factor (R):

$$\gamma = \frac{2\mu_t - 1}{R^2} \tag{11}$$

The structure is idealized as an inelastic equivalent single degree system by selecting a typical yield mechanism for the peak monotonic demand, where the mechanism is composed of the yielding of all shear plates and plastic hinge formation at the bases of the boundary columns (Figure 17.b). The elastic strain energy demand (E_e) during this monotonic push is calculated based on the yield base shear, V_y, and substituting this in Equation (10), the plastic energy demand (E_p) is obtained:

$$E_p = \frac{WT^2 g}{8\pi^2}\left[\gamma C_e^2 - \left(\frac{V_y}{W}\right)^2\right] \tag{12}$$

This E_p is equated with the inelastic work done (W_p) through all the plastic deformations in the SPSW system:

$$W_p = \sum_{i=1}^{n} P_i h_{si} \theta_p + 2M_{pc}\theta_p \tag{13}$$

where n = number of stories, P_i = plastic shear capacity of the ith story steel plate, h_{si} = i^{th} inter-story height, and M_{pc} = plastic moment capacity at each column base, θ_p = target plastic drift based on an assumed yield drift (θ_y) as shown in Figure 16.b (an elastic-perfectly plastic behaviour is assumed here), and the design yield base shear (V_y) is obtained as

$$\frac{V_y}{W} = \frac{-\alpha + \sqrt{\alpha^2 + 4\gamma C_e^2}}{2}, \text{ where } \alpha = \left(\sum_{i=1}^{n} \lambda_i h_i\right)\frac{8\theta_p \pi^2}{T^2 g} \tag{14}$$

where, h_i = i^{th} floor height, θ_p = target plastic drift based on an assumed yield drift (θ_y). The factor λ_i (= F_i/V_y) represents the shear force distribution in the SPSW system as discussed by Ghosh et al. (2009). The required plate thickness at each story is obtained using the following equation

$$t_i = \frac{2P_i}{0.95F_y L} = \frac{2V_i}{0.95F_y L} \tag{15}$$

where, V_i = i^{th} story shear demand, F_y = material yield strength and L = bay width. The base column moment capacity (M_{pc}) is obtained as per recommendations by Roberts (1995)

$$M_{pc} = \frac{F_y t_1 h_1^{\,2}}{16} \tag{16}$$

The design axial force (P_c) on the columns is calculated based on the moment equilibrium about the base of the SPSW system. The column sections are selected based on a standard P-M interaction.

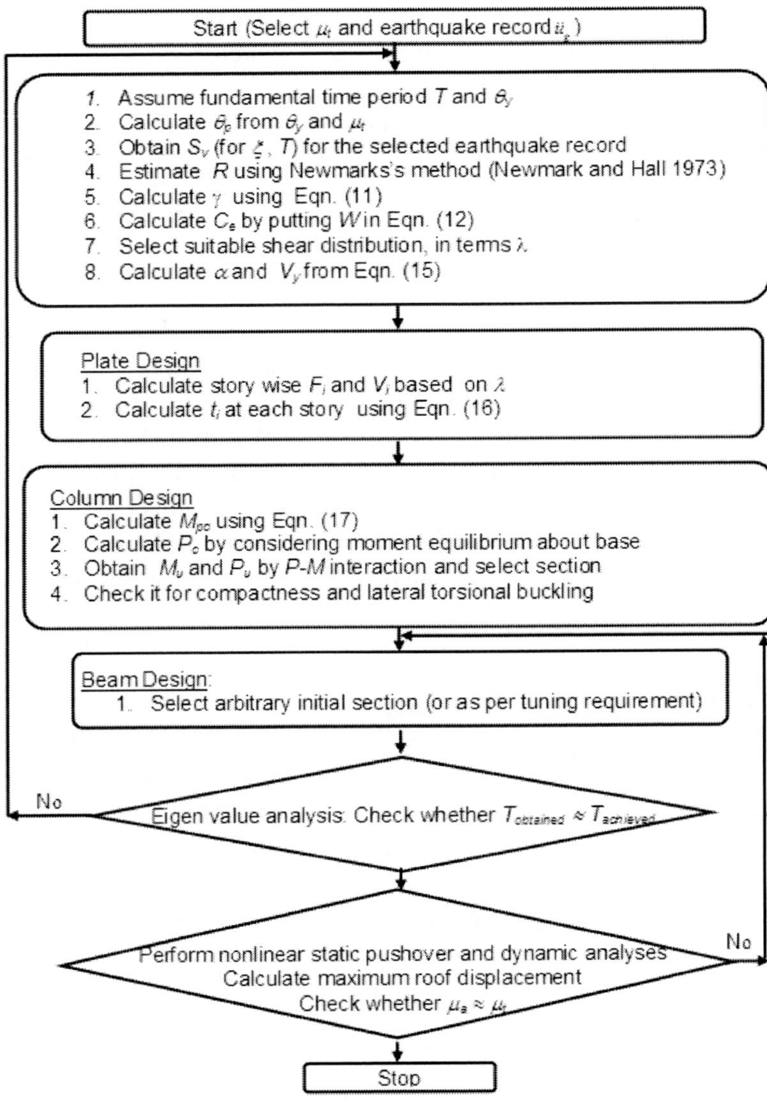

Figure 18. Flowchart for the performance-based design of steel plate shear wall with pinned beam-to-column connection as per Ghosh et al. (2009)

Ghosh et al., (2009) analytically have validated this method by designing a 4-story steel structure with pin-connected beams with one SPSW bay (Figure 17), subjected to various ground motion scenarios and for different target ductility ratios. This design is further modified by tuning the pin-connected beam members so as to achieve actual or achieved ductility ratio (μ_a) closer to the target ductility ratio (μ_t). A design flowchart is provided in Figure 18 giving the individual design steps. The analytical test results show that this design procedure is very effective in achieving the target ductility ratios as well as following the pre-selected yield mechanism. It dose not require any complicated analysis from the designer/practising engineer's part. The design procedure remains simple while satisfying an advanced performance based design.

Gupta et al. (2009) have successfully applied the PBSD method proposed by Ghosh et al. (2009) using standard hot rolled sections available in USA (AISC 2005b) and in India (BIS 1964). It is observed that due to the lack of available standard rolled Indian sections with large capacities, the application of SPSW gets limited to weak earthquake-large ductility designs. It was recommended that to utilize this advanced earthquake design methods, the range of available sections in India needs to be enhanced.

6. Design Code Provisions for SPSW

Steel plate shear walls figure prominently only in the design standards of two countries, Canada and USA. The first one to incorporate any specific provision for the design of SPSW was the Canadian standard CAN/CSA-S16.1 in 1994 (CSA 1994), although it had only mentioned about SPSW in an appendix. The 2001 edition of the same standard had detailed design specifications for steel plate shear walls, which are discussed in this section. In the USA, the AISC Seismic Provisions (AISC 2005a) was the first one to include guidelines for the design of SPSW systems. Later, AISC published a separate design guide for SPSW systems only. These two publications are also reviewed in this section.

6.1. Provisions in the Canadian Standard

The 2001 Canadian standard, CAN/CSA-S16-01 (CSA 2001) incorporated mandatory clauses on the design of steel plate shear walls. The CAN/CSA-S16-01 seismic design process for SPSW prescribed how to calculate the appropriate design base shear and its distribution over the height of the system. The preliminary sizing of elements of SPSW system was recommended to be based on the equivalent storey brace model proposed by Thorburn et al. (1983). After the preliminary design, any analysis was prescribed to be performed using a more refined model multi-strip model (Thorburn et al. 1983).

The Canadian standard had provisions for both limited ductility and ductile steel plate shear walls. For the limited ductility walls, no special requirements were made for beam-to-column connections and were assigned to a force reduction factor $R = 2$, while the ductile SPSW were with moment resisting beam-to-column connections and with the largest force reduction factor $R = 5$). For a ductile SPSW, the boundary elements were designed to remain elastic in order to resist the full tension field developed in infill plates. This ensures that the infill plate can yield in tension prior to plastic hinging of the boundary elements (providing for substantial energy dissipation in seismic applications). The possible resistance of shear wall V_{re} was expressed as

$$V_{re} = 0.5R_y F_y tL\sin 2\alpha \tag{17}$$

where R_y = ratio of the expected yield stress to the design yield stress (which is 1.1 for A572 Grade 50 steel). In order to ensure the ductile failure mode of SPSW this code recommended the use of a factor B (ratio of the probable shear resistance at the base of the wall for given plate thickness, to the factored lateral force at the base of the wall obtained from the calculated seismic load) to magnify the moments and axial forces of columns obtained from an elastic analysis. It should be noted that this magnification is not required if column forces and moments are obtained from a nonlinear pushover analysis.

6.2. Provisions in US Standards

In 2005, the special plate shear wall was added to the AISC Seismic Provisions (AISC 2005a). Section 17 of the Seismic Provisions contained the requirements for the design of SPSW. This document used the following terminology for various elements of a special plate shear wall: *vertical boundary element* (VBE) for a column, *horizontal boundary element* (HBE) for a beam, and *web* for the steel panel. According to the AISC Seismic Provisions (AISC 2005a) the HBE and VBE were designed to remain elastic under maximum forces that could be generated by the fully yielded webs. Thus the concept of capacity design was incorporated in this standard. For boundary elements, plastic hinging was permitted at HBE ends only. The nominal shear strength of a web V_n was calculated as

$$V_n = 0.42F_y t_w L_{cf} sin(2\alpha) \tag{18}$$

where L_{cf} = clear distance between VBE flanges, t_w = thickness of the web. The specification generally followed the LRFD format (and also the ASD format) of design equations in tune with other AISC specifications.

These provisions suggested that the ultimate strength of a web would fully develop only when the corresponding frame members were sufficiently rigid and strong to "anchor" the tension field developed. Thus, for a column (VBE), it was recommended that the moment of inertia I_c should be such that

$$I_c \geq \frac{0.00307 t_w h^4}{L} \tag{19}$$

where h is the story height between HBE centrelines and L is the width between VBE centrelines. It should be noted that there is no such specification similar to Equation (19) for the HBE at roof and foundation level. The required strength of a HBE should be greater of the forces corresponding to the expected yield strength in tension, of the web or of the panel at an angle α or that determined from load combination in the applicable building code assuming that the panel provides no support for gravity loads. The boundary elements are required to be proportioned in order to meet the strong-column-weak-beam requirement. In addition to this, the boundary members should satisfy the compact section requirements and need to be checked for lateral torsional buckling (LTB) and should provided with lateral bracing, if needed, in order to avoid LTB.

The recently published AISC Design Guide 20: Steel Plate Shear Walls (Sabelli and Bruneau 2007) has developed the AISC 2005 Seismic Provisions into a complete design methodology. The Design Guide 20 has also discussed the history and background of the design of steel plate shear walls. This guide has included design procedures as well as design examples for steel plate shear walls in both high-seismic ($R = 5$) and low seismic regions ($R = 3$). This design guide has been developed in accordance with the existing relevant standards ASCE 7-05 for minimum design loads in buildings (ASCE 2005), ANSI/AISC 360-05 for structural steel (AISC 2005b), and AISC Seismic Provisions (AISC 2005a). Overall, it has followed the capacity design philosophy and the LRFD format.

7. Recent Developments

In the last five or so years, research works on various aspects of steel plate shear walls have been reported in research publications as well as in publications focusing more on engineering practice. Research on steel plate shear wall is now being conducted in various countries around the world, such as: USA, Canada, Iran, Taiwan, UK, Korea, India, China and Turkey. This section gives a brief account of the various developments in research and applications of SPSW around the world in recent years.

7.1. Capacity Design of Boundary Elements

Elements of capacity design concepts were incorporated in the CAN/CSA-16 (CSA 2001) and later in the AISC Seismic Provisions (AISC 2005a). However, these capacity design concepts were not fully developed. These indirect capacity design approaches recommended a magnification of the moments and axial forces in columns (obtained from an elastic analysis of SPSW) by a factor. As per Berman and Bruneau (2008), these design approaches significantly underestimate the VBE design loads at upper stories and thus a capacity design is not achieved. Recent works on the development of capacity design procedure for SPSW systems focus specifically on the yielding hierarchy and the provisions to attain that. Astaneh-Asl (2001) discussed the preferred failure hierarchy in SPSW systems in detail, considering almost all the possible types of local and global failures. Based on a little limited failure considerations than these, Berman and Bruneau (2008) have recently developed a reasonably accurate and relatively efficient method for estimating the VBE demands with fully yielded infill panels under applied lateral loads. Their proposed procedure combines a linear elastic model of SPSW and plastic analysis concepts. A simple VBE free body diagram is then used to determine the design VBE axial forces and moments. Assuming a collapse mechanism with uniform drift with fully yielded web panels and plastic hinges at HBE ends, demands on boundary elements are calculated from the free body diagram as shown in Figure 19.a to Figure19.d. The fully yielded infill panels exert uniformly distributed transverse loads (along the plane of the plate) to the HBE and VBE. These forces are calculated from the ultimate shear force capacity of the infill panel. The axial loads in the HBE are calculated by developing a preliminary elastic model of the VBE. After estimating the axial load in HBE, the beam sections are selected and a reduced plastic moment capacity of the HBE due to the effect of axial force (P-M interaction) is calculated. The axial force in VBE can be found by considering the moment equilibrium about the base.

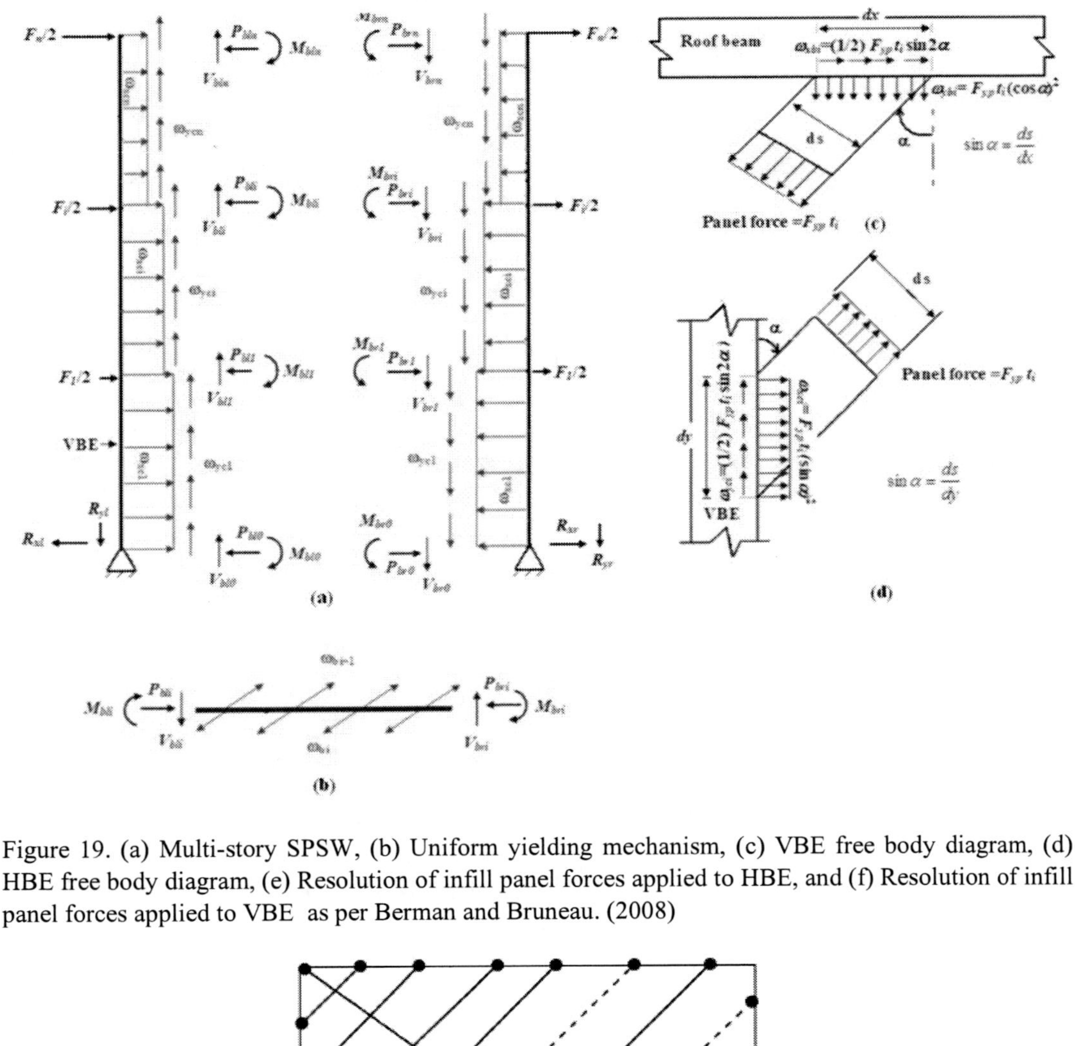

Figure 19. (a) Multi-story SPSW, (b) Uniform yielding mechanism, (c) VBE free body diagram, (d) HBE free body diagram, (e) Resolution of infill panel forces applied to HBE, and (f) Resolution of infill panel forces applied to VBE as per Berman and Bruneau. (2008)

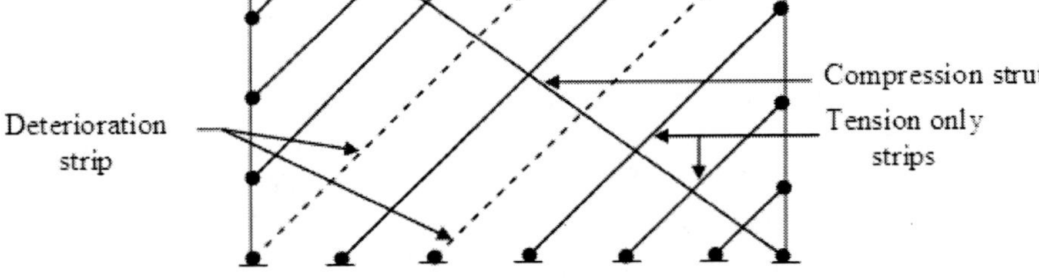

Figure 20. Modified strip model as per Shiskin et al. (2008)

Berman and Bruneau (2008) have designed a four-story SPSW with constant and variable infill panel thickness by both the code-prescribed and the proposed capacity design approaches. The adequacy of these two methods has been checked through nonlinear static analyses of the four-story SPSW systems designed. Axial force and moment diagrams for VBE obtained from the proposed design procedure are found to be in good agreement with those obtained from pushover analysis.

7.2. Modified Strip Model of SPSW

The multi-strip model developed for SPSW by Thorburn et al. (1983) neglected the pre-buckling shear resistance of infill panel. Driver et al. (1997) observed that the multi-strip model underestimates both the elastic stiffness and the ultimate capacity of the SPSW, because this model neglects the small contribution from the compression diagonal (before buckling) to the strength and stiffness of the infill panel. Moreover, this multi-strip model dose not represents the gradual deterioration in strength of SPSW at large inelastic displacement cycles.

In order to overcome this deficiency, Shishkin et al. (2008) have refined the original multi-strip model, termed as modified strip model (Figure 20), which incorporates a) tension strips with bilinear axial hinges, b) a diagonal compression strut with bilinear axial hinge in order to simulate the effect of the pre-buckling compression diagonal, and c) deterioration hinges in some tension strips to simulate tearing of the infill panel as observed in most of the cyclic load tests. The area of compression strut A_{cs} used in the modified strip model is calculated similar to that for the equivalent brace model of SPSW (Thorburn et al. 1983):

$$A_{CS} = \frac{tL sin^2\alpha}{2 sin\varphi sin2\varphi} \tag{20}$$

where φ is the acute angle of diagonal strut with the vertical. This modified strip model has been validated using the experimental results from tests on a four-story SPSW specimen (Driver et al. 1998a) and on a one-story SPSW specimen (Lubell et al. 2000). This model is found to yield results with varying levels of accuracy.

7.3. Tests on Full-Scale SPSW Specimens

In order to ensure experimentally the replaceability of infill panels after sustaining an earthquake as well as to explore the behaviour of the repaired SPSW in a subsequent earthquake, Qu et al. (2008) have tested two two-story full-scale SPSW specimens with reduced beam section details and a composite floor. This experimental program has been carried in two phases at the National Centre for Research on Earthquake Engineering (NCREE) in Taipei, Taiwan. The Phase I tests, have consisted of infill panels with horizontal tube restrainers on both sides to minimize the out-of-plane displacement and the buckling sound. In the Phase II tests, damaged infill steel plates have been replaced with new infill plates without the use of any restrainers. The Phase I test specimen has been tested under pseudo-dynamic loads using the Chi-Chi earthquake record scaled up to three levels of excitations representative of seismic hazards having 2%, 10%, and 50% probabilities of exceedances in 50 years. The ground accelerations have been scaled so that the spectral acceleration (with 5% damping) associated with the first mode period was equal to that in the design response spectra. In Phase II, the repaired SPSW specimen has been tested under pseudo-dynamic load corresponding to the Chi-Chi earthquake record scaled to a seismic hazard having a 2% probability of occurrence in 50 years.

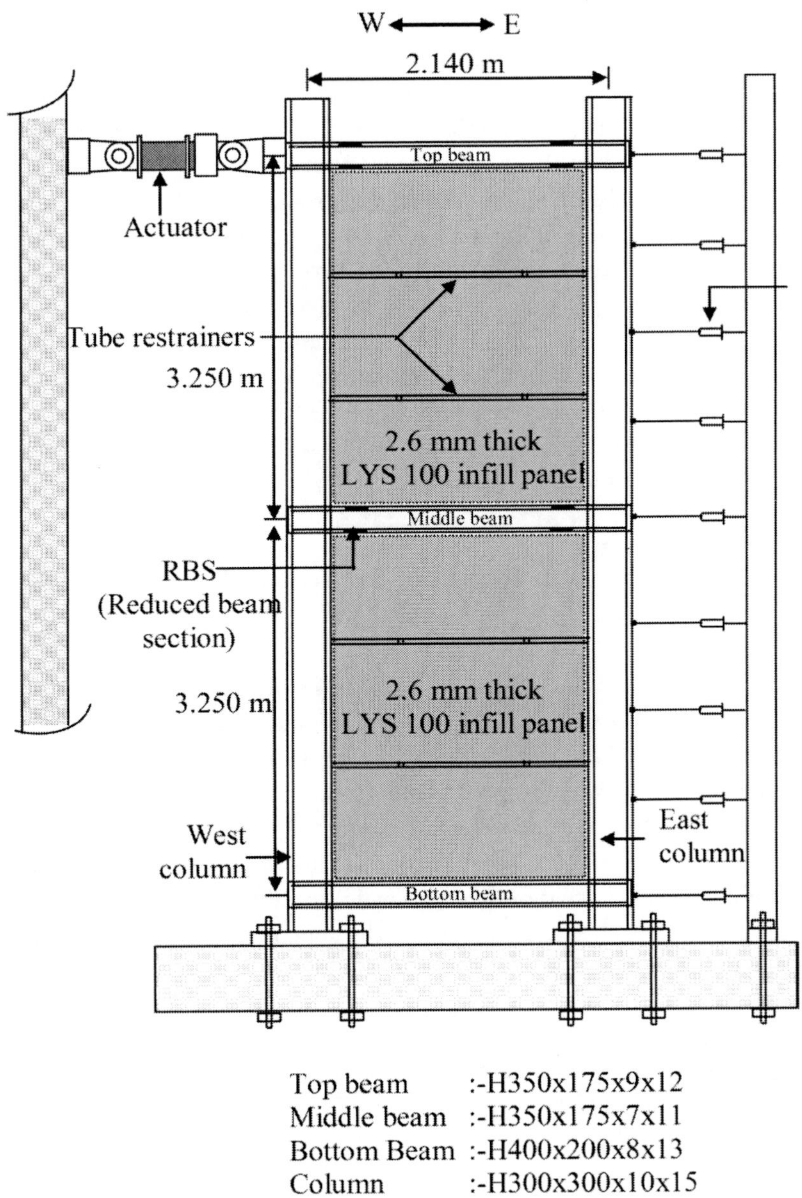

Figure 21. Cyclic load test on two-story narrow steel plate shear walls as per C.-H. Li et al. (2009)

Results of this experimental program show that; i) the horizontal restrainers are very effective in improving the serviceability of SPSW, ii) a SPSW repaired by replacing the infill panels buckled in a prior earthquake by new ones can sustain and dissipate significant amounts of hysteretic energy in a subsequent earthquake without severe damage to the boundary frame or overall strength degradation, iii) SPSW specimens as per AISC guidelines exhibit stable force-displacement behaviour and provide a significant energy dissipation capacity, exhibiting substantial redundancy. The experimental results have also been validated using the dual-strip or cross-strip model comprised of tension only strips as well as using three-dimensional finite element model in finite element package Abaqus/Standard.

Recently four two-story "narrow" (with aspect ratio of about 1:0.6) SPSW specimens have been cyclically tested to a roof drift of 0.05 radians in NCREE, Taiwan (Li et al. 2009). Figure 21 shows the test set-up of a sample specimen. Low yield strength steel plates of 2.6 mm thickness have been used in these specimens, along with reduced beam sections at the ends of floor beams. Two specimens have been constructed with horizontal tube restrainers that sandwich over the steel panel from the two sides using through bolts and have been pin-connected to the column flanges, while the other two have had no restrainers. The main purpose of this experimental program has been to investigate the seismic performance of narrow SPSW frames and the restrained SPSW frames. Test results have ascertained the effectiveness of the RBS and the welded beam-web-to-column connection. The horizontal steel tube restrainers have reduced the axial force demands in beams and columns. The restrained SPSW frames have experienced smaller out-of-plane deformations and dissipated more seismic energy than the unrestrained ones.

7.4. Comparison with Moment Resisting Frames and Concentrically Braced Frames

Park et al, (2007) have conducted an experimental study on three-story thin unstiffened SPSW specimens in order to explore the potential maximum ductility and energy dissipation capacity that can be attained. Test results have been compared with those obtained from tests conducted on steel concentrically braced frame (CBF) and steel moment resisting frame (MRF). The CBF and MRF have consisted of beams and columns with the same sizes as those used for the SPSW system. For comparison, the braces in the CBF have been designed to have the same steel weight as that of the infill panel in SPSW. Before testing the steel plate walls, a pushover analysis has been carried out on the specimens by using Abaqus/Standard in order to approximately estimate the yield displacement. All test specimens of SPSW, MRF and CBF have been subjected to a specified target displacement in the proportion of their predicted yield displacements. The shear-dominated steel plate wall has had displacement ductility 2.8 times that of the CBF and 3.3 times that of the MRF, whereas the energy dissipation have been 5.8 and 2.8 times, respectively.

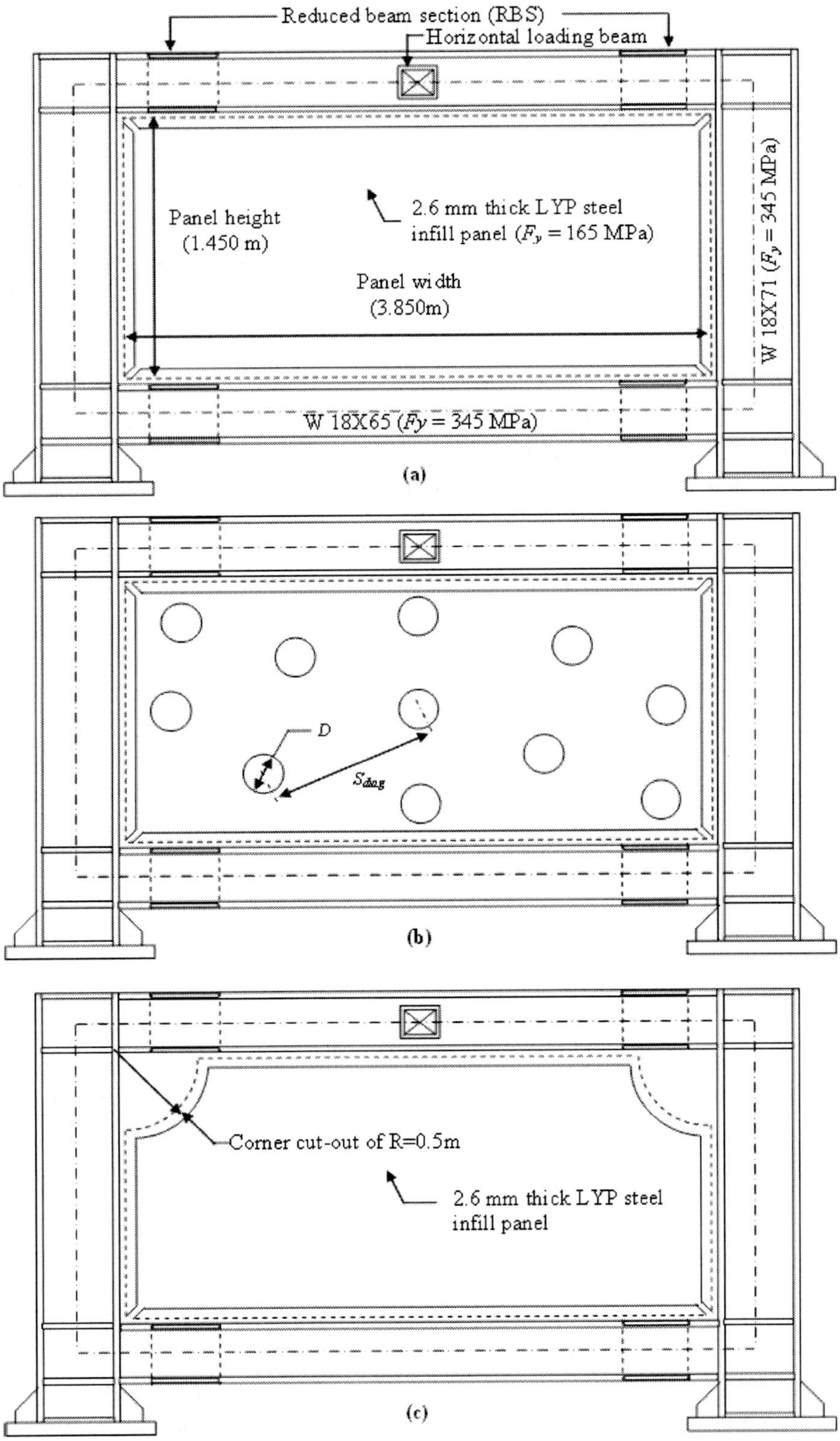

Figure 22. Single-bay, single-story SPSW specimens with (a) solid infill panel, (b) perforated infill panel and (c) corner cut-out infill panel tested as per Vian. (2005)

7.5. Special Perforated SPSW and SPSW with Cut-Out Corners

The infill panel material is an important factor in deciding the panel thickness which in turn governs the sizing of the boundary elements in an SPSW system. Sometimes, the available strength and stiffness of infill panel may be greater than that needed for a design. This results in heavy sections for boundary elements so as to develop a full tension field in the infill panel. Vian et al. (2009a) have made attempts to alter the solid infill panel system by using the diagonal patterned perforations in panel and using the reinforced cut-out corners. These configurations serve two main purposes: i) as an alternative to impractical thinner infill panel, ii) as an alternative to heavily stiffened openings in panel for service utility. Earlier, Roberts and Sabouri-Ghomi (1992) had investigated the cyclic performance of SPSW with centrally placed circular opening and proposed the strength and stiffness reduction factor for perforated panel. In both the configurations, Vian et al. (2009a) have introduced the reduced beam section (RBS) in SPSW at the end of "anchor beams" (at top and bottom levels). The objective of using RBS at the end of a boundary beam is to reduce the overall system demand on the vertical boundary elements. These two configurations of SPSW, using low yield stress (LYS) steel for the infill, have been studied experimentally and analytically. Three specimens (Figure 22) – with solid panel, perforated panel with perforations inclined at angle of 45°, and panel with reinforced cut-out corners – have been tested under quasi-static cyclic load. Each of these configurations has exhibited a ductile behaviour. In all cases the plastic hinges have developed in RBS. RBS for SPSW anchor beams has been recommended to effectively control boundary frame yielding during a significant earthquake.

Vian et al. (2009b) have also analytically validated the experimental results, using three-dimensional finite element analysis. Four-noded S4R shell elements have been used to model both boundary members and infill panels. The analytical results have shown good agreement in overall behaviour with experimental results. Based on this study, Vian et al., (2009b) have developed an equation for the shear strength $V_{yp.perf}$ of perforated panel:

$$V_{yp.perf} = \left(1 - \frac{D}{S_{diag}}\right) V_{yp} \tag{21}$$

where D is the diameter of perforations, S_{diag} is the diagonal spacing between perforations and V_{yp} is the shear strength of the solid panel.

7.6. Steel Plate Shear Walls with Various Configurations of Infill Plate

The structural capacity of steel plate walls with various infill plate configurations have been experimentally investigated by Choi and Park (2009) with cyclic load tests on five one-third scaled models of three-story steel plate shear wall systems. They have varied the following parameters for these tests: a) the connection type (bolted versus welded connection) between the boundary frame and the infill plate, b) length of the welded connection between the boundary frame and the infill plate (full connections versus partial connections), and c) opening in the infill plate (solid wall versus coupled wall with a opening in between). In all these specimens, an infill panel of aspect ratio 1:2.2 with 4 mm thick plates has been used.

The boundary elements have been designed as per the Korean standards. The SPSW specimen with partial weld connections between infill panel and boundary element has contained infill panel continuously welded to beams and partially welded to columns. The coupled wall specimen has consisted of two separated walls with a coupling beam. Results indicated that: i) the walls with bearing bolt-connected infill plates have exhibited large initial stiffness and load-carrying capacity as compare to those of the walls with weld-connected infill plates, ii) the steel plate wall with infill plates partially weld-connected exhibited an excellent deformation capacity equivalent to that of the solid wall with fully connected infill plates, although its load-carrying capacity and energy dissipation capacity have been relatively less, iii) the coupled SPSW also has exhibited a good deformation capacity, equivalent to the deformation capacity of the solid wall.

7.7. Use of Light-Gauge and Cold-Rolled Infill Panels

In case of low-rise buildings in low-seismic areas, the required plate thickness to resists the specified shear force works out to be very small than the standard available thicknesses of infill sheets in the market. In order to overcome this difficulty, the use of light-gauge, cold-formed steel panels has been proposed by Berman and Bruneau (2003b). A single-story SPSW specimen with light-gauge cold-formed infill panel of 1.0 mm thickness has been tested, under cyclic loading conforming to ATC standards (ATC 1992). They have reported that the specimen have reached a displacement ductility ratio of 12 and drift of 3.7%. It has also been found that the infill panel has provided approximately 90% of the initial stiffness of the system. Later, Berman and Bruneau (2005) have performed experiments on three light-gauge single-story steel plate shear wall systems. Two SPSW with flat infill panels (thickness = 0.9 mm) and one with corrugated infill panel (thickness = 0.75mm) have been tested under quasi-static conditions. SPSW specimens with both flat infill panel as well as corrugated infill panel have exhibited significant ductility and energy dissipation. Recently, Tipping and Stojadinović (2008) have conducted 44 cyclic load tests on corrugated sheet steel shear walls (CSSW). Their objective has been to establish relevant factors (R, C_d and Ω_0) that determine the seismic design strength. Based on these experiments, they have proposed an R value of 5.5, a C_d value of 3.25, and a Ω_o value of 2.5 for the corrugated metal shear walls.

7.8. Smart and Resilient Steel Walls

A NEESR (2009) sponsored project on "Smart and Resilient Steel Walls for Reducing Earthquake Impact" has started very recently under Prof. Jeffrey Berman's leadership. The goal of this project is to develop a smart and resilient steel plate shear wall (SR-SPSW) system with the potential to apply seismic design in areas of both moderate and high seismicity. The system strategically combines the benefits of self-centering and steel plate shear wall technologies to create a robust, ductile, and easily repairable system that will reduce life-cycle costs for buildings. This project will include large-scale testing using advanced experimental techniques and instrumentation in order to generate data to be used for developing numerical models to explore the physical behaviour of this new SR-SPSW system. This project is also proposed to fill critical knowledge gaps in SPSW system behaviour, including the understanding of coupled SPSW behaviour and the expected distribution of yielding in multi-storey SPSW so as to ensure that the new SR-SPSW system are implemented as successfully as conventional SPSW systems.

8. Application of SPSW

As mentioned earlier, heavily stiffened thick SPSW was used in construction of buildings in the early 1970s. At that time, the application of SPSW was for the seismic retrofit of existing structure as observed in the seven-story hospital building in California, USA, where steel plates were used in combination with steel bracings and RC shear walls in order to increase the seismic resistance of existing structure. The application of thin unstiffened SPSW in actual construction has increased significantly in recent years due to the extensive research and development in the last three decades. These thin unstiffened SPSW are used in high seismic areas of US, Canada, Japan, and Mexico. In this section, few examples of application of unstiffened thin SPSW in commercial and residential projects have been discussed.

The 16-storey Moffit Hospital building in San Francisco, USA was constructed with five steel plate shear walls having plate thicknesses between 10 to 32 mm as a lateral load resisting system (Robert 1995). A reinforced concrete shear wall was placed around the elevator core, where a steel wall may have resulted in vibration problems. The steel panels were covered on both sides with 250 mm of reinforced concrete through steel reinforcing ties in order to provide additional rigidity and fire resistance. The 23-story US Federal Courthouse in Seattle, USA building used thin and light SPSW panels instead of thick RC shear walls. The use of SPSW instead of RC shear wall resulted in 2% saving in footage area, 18% reduction in seismic weight and reduced construction time. In addition to this, the SPSW system had excellent post-buckling strength as observed from the experiments conducted by Astaneh-Asl and Zaho (2002). For the proposed US Federal Aviation control tower in Medford, USA, the application of SPSW is not only as the lateral load resisting system but also as the blast resisting system. This high-rise incorporates SPSW panels (20 ft long, 10 ft wide, and 1/8 in thick) which are designed using nonlinear time-history finite element analysis under blast impulse loading conditions. The most recent example of the use of thin SPSW is for the 56-story, LA Live Hotel and Residence in Los Angeles, USA. This project uses relatively lighter steel panels (1/4-3/8 in thick) where a relatively thicker RC shear wall (30 in thick) was required for the same design. The benefits of using this SPSW system are a) 35% reduction in seismic weight, b) increased floor area, c) reduced project completion time and d) better quality control.

SPSW have also been used in low-rise residential buildings with pre-engineering framing systems where the SPSW are shop-fabricated. Few examples are a) the 17,000 square feet residence in Atherton, USA with 14 gauge thick low yield stress panels, b) 9,000 square feet residence in San Mateo County, USA with 12 gauge thick panels, and c) the 23,000 square feet two-story structure in Los Altos, USA with significantly open floor plan.

9. Future Needs in SPSW Research

Steel plate shear wall systems have a huge potential of application in both moderately seismic and high seismic areas. The advantages offered by SPSW makes it more economical and superior in seismic performance than traditional lateral load resisting systems. Besides recent applications as discussed in the previous section, applications of SPSW have been minimum, which may have resulted due to i) conservative or over-designed SPSW with limited aspect ratios as per current code provisions, which hampers its economy, ii) cumbersome and time-consuming analysis techniques not suitable as a design tool for practicing engineers, iii) less flexural stiffness as compare to RC shear wall, which challenges its application in high rise buildings where wind load governs the design, iv) almost negligible out-of plane stiffness, which affect the application for structures susceptible to face impact and blast loading, and v) lack of knowledge of the behaviour of SPSW with non-traditional configurations. This section deals with future need in SPSW research in order to overcome the abovementioned limitations, and to get a wide acceptance similar to other lateral load resisting systems.

The current code provisions, Canadian Standards CAN/CSA-16 (CSA 2001) and AISC Seismic Provisions (AISC 2005a), incorporate the indirect capacity design provisions for SPSW. The provisions for VBE are as per strength and stiffness requirements, which lead to an overestimation of demands in the lower stories and underestimation in the upper stories. There is no such provision for HBE, and the sizing of HBE is based on flexural demands from infill panel forces. Balanced design provisions need to be developed for VBE and HBE both in accordance with capacity design concepts.

Considering the demands of performance-based design philosophy in current and future seismic design codes, it is required to develop the performance-based design procedure for SPSW in order to ensure that they can meet multiple performance objectives in an efficient and economic manner. Although Ghosh et al. (2009) had developed a performance-based seismic design method for SPSW, it needs to be generalized further considering all design aspects so that it could be incorporated in design standards/codes. For example, application of this or other PBSD methods to SPSW with rigid-connected beams, to high-rises with SPSW, suitability of different lateral force distributions in this PBSD, etc. need attention in future so that SPSW design can meet future seismic code requirements.

The demands in various elements of SPSW are evaluated using different analytical models, from the equivalent story brace to the detailed finite element model. It has been observed from various analytical and experimental studies that the simplified models have over predicted the ultimate strength and detailed finite element models proved to be more time-consuming and giving somewhat stiffer structures than the simplified models. For practicing engineers, it is important to have a simplified analytical model for design and analysis of an SPSW system, which will require less computational efforts without sacrificing the accuracy of results significantly. Thus, more refined simplified modelling as well as improvement in finite element using "super element" which is based on tension membrane formulation is required for future developments of SPSW.

For high-rise structures, flexural action dominates over the shear and SPSW that are more flexible cannot offer the required overturning stiffness. A practical solution for this problem is the use of coupled shear walls (with a coupling beam between two SPSW bays) (Sabelli and

Bruneau 2007). However, no experimental or detail analytical work has been carried regarding this, and future research on SPSW should focus on this area. Wind tunnel tests on higher story buildings with SPSW need to be conducted for checking the ability of high-rise SPSW to resist wind loads.

Steel is vulnerable to fire hazards, and like any other steel structure SPSW requires to be designed for fire loading. As opposed to other uses of structural steel, SPSW has a large exposed steel area which increases it vulnerability. The fire resistant design aspect for SPSW needs to be investigated through extensive experimental as well as analytical research programs, and proper fire design guidelines – not just fire proofing mechanism – need to be developed and standardized for these systems.

The almost negligible out-of-plane stiffness of SPSW limits its application to real structures that require to resist accidental loadings like blast and impact. Till now, very little work has been carried out on the blast resistance of SPSW (Warn and Bruneau 2009). The future generation SPSW needs to be designed accordingly to resist accidental loads like blast.

10. Conclusions

Thin unstiffened steel plate shear wall (SPSW) is a very effective lateral load resisting system and is rapidly gaining popularity as an appealing alternative to conventional lateral load systems in highly seismic areas. For the past three decades, significant amount of valuable research works have been performed on SPSW worldwide to evaluate the static and dynamic behavior of SPSW and in order to formulate efficient seismic design and analysis techniques. A detailed summary of these research and development activities on various aspects of SPSW systems is presented in the previous sections. In addition, this chapter also provides the current state-of-the-art and state-of-the-practice for SPSW along with future directions where research on SPSW for the next five to ten years should be headed to. In terms of design philosophy, the design methodology for SPSW systems should gradually move from elastic force-based design to capacity design to performance-based seismic design. Design methods need to be developed for non-seismic loading, such as, wind, blast, fire, etc. New modeling techniques are also necessary in order to make the analysis and design methods for SPSW convenient for practical purposes. It is expected that actual use of this relatively new lateral load resisting system will greatly increase in the coming decade, and this process can be accelerated and enhanced if the roadblocks, as mentioned earlier, are removed through an effective cooperation among researchers, code writers, practicing engineers and developers.

Acknowledgments

The authors acknowledge Prof. Subhash Goel, Prof. Michel Bruneau, Prof. Robert Driver, Prof. Gilbert Grondin, Prof. Abolhassan Astaneh-Asl, Dr. Rafael Sabelli, Prof. Adam Lubell, Prof. Jeffrey Berman, and Prof. Bing Qu for their valuable contributions regarding their recent research activities on SPSW. This review work is partially funded by the Department of Science and Technology (DST), Government of India. However, the opinions expressed here are of the authors and do not necessarily represent the views of DST.

References

Abaqus, Abaqus/Standard User's Manual. Version 6.4, Hibbitt, Karlsson, & Sorensen, Inc., (HKS), Pawtucket (RI), 1994.

AISC, ANSI/AISC 341-05, *Seismic Provisions for Structural Steel Buildings*, American Institute of Steel Construction Inc., Chicago, IL, USA, 2005a.

AISC, ANSI/AISC 360-05, *Specification for Structural Steel Buildings*, American Institute of Steel Construction Inc., Chicago, IL, USA, 2005b.

AISC, *Load and Resistance Factor Design Specification for Structural Steel Building* American Institute of Steel Construction Inc., Chicago, IL, USA, 1999.

ASCE, SEI/ASCE 7-05, *Minimum Design Loads for Buildings and Other Structures* (including Supplement No. 1), American Society of Civil Engineers, Reston, VA, USA, 2005.

Astaneh-Asl, A; Zhao, Q. Cyclic Behavior of Steel Shear Wall Systems, *Proceedings of the Annual Stability Conference*, Structural Stability Research Council, Seattle, WA, 2002.

Astaneh-Asl, A. Chapter 10: Steel Shear Walls; *Design of Steel Composite Structures, Including Seismic Effects*, Department of Civil and Environmental Engineering, University of California, Berkeley, USA, 2003.

Astaneh-Asl, A. Seismic Behavior and Design of Composite Steel Plate Shear Walls; *Steel Technical Information and Product Services Report,* Structural Steel Educational Council, Moraga, Canada, 2002.

Astaneh-Asl, A. Seismic Behavior and Design of Steel Shear Walls; Steel Technical Information and Product Services Report, *Structural Steel Educational Council*, Moraga, Canada, 2001.

Astaneh-Asl, A. Seismic Behavior and Design of Steel Shear Walls; *Steel Technical Information and Product Services Report*, Structural Steel Educational Council, Moraga, Canada, 2001.

Astaneh-Asl, A. Steel Plate Shear Wall, Proceedings of U.S.-*Japan Partnership for Advanced Steel Structure*, San Francisco, 2000.

ATC, *Guidelines for Seismic Testing of Components of Steel Structures*, Report 24, Applied Technology Council, Redwood City, CA, USA, 1992.

Basler, K. Strength of Plate Girders in Shear. *ASCE Journal of the Structural Division*, 1961, 87(7), 151-180.

Behbahanifard, M; Grondin, G; Elwi, A. *Experimental and Numerical Investigation of Steel Plate Shear Walls; Structural Engineering Report* No. 254, Department of Civil Engineering, University of Alberta, Edmonton, Alberta, Canada, 2003.

Behbahanifard, M; Grondin, G; Elwi, A. Experimental and Numerical Investigation of Steel Plate Shear Walls; *Structural Engineering Report* No. 254, Department of Civil Engineering, University of Alberta, Edmonton, Alberta, Canada, 2003.

Berman, JW; Bruneau, M. Capacity Design of Vertical Boundary Elements in Steel Plate Shear Walls. *AISC Engineering Journal*, 2008, First Quater, 57-71.

Berman, JW; Bruneau, M. *Experimental Investigation of Light-Gauge Steel Plate Shear Walls for the Seismic Retrofit of Buildings; Technical Report* MCEER-03-0001, Multidisciplinary Center for Earthquake Engineering Research, Buffalo, New York, USA, 2003b.

Berman, JW; Bruneau, M. Experimental Investigation of Light-Gauge Steel Plate Shear Walls. *ASCE Journal of Structural Engineering*, 2005, 131(2), 259-267.

Berman, JW; Bruneau, M. Plastic Analysis and Design of Steel Plate Shear Walls. *ASCE Journal of Structural Engineering*, 2003a, 129(11), 1448-1456.

Berman, JW; Bruneau, M. Steel Plate Shear Walls Are Not Plate Girders. *AISC Engineering Journal*. 2004, Third Quarter, 95-106.

Berman, JW; Bruneau, M; Lowes, l; Fahnestock, L; Okazaki, T. NEESR-SG: Smart and Resilient Steel Walls for Reducing Earthquake Impacts. George E. Brown, Jr. *Network for Earthquake Engineering Simulation (NEES) Research (NEESR)*, 2009, Award Abstract #0830294.

Berman, JW; Lowes, LN; Okazaki, T; Bruneau, M; Tsai, KC; Driver, RG; Sabelli, R. Research Needs and Future Directions for Steel Plate Shear Walls, *Proceedings of ASCE Structures Congress Structures* 2008: Crossing Borders, Vancouver, Canada, 2008.

Bhowmick, AK; Driver, RG; Grondin, GY. Seismic Analysis of Steel Plate Shear Walls Considering Strain Rate and P–delta Effects *Journal of Constructional Steel Research*, 2009, 65(5), 1149-1159.

BIS, SP. 6(1)-1964. *Handbook for Structural Engineers*, No. 1: Structural Steel Sections, Bureau of Indian Standards, New Delhi, India, 1964.

Bruneau, M; Bhagwager, T. Seismic Retrofit of Flexible Steel Frames Using Thin Infill Panels. *Engineering Structures*, 2002, 24(4), 443-453.

Bruneau, M; Berman, JW; Lopez-Gracia, D; Vian, DA. Review of Steel Plate Shear Wall Design Requirements and Research. *AISC Engineering Journal*, 2007, First Quater, 27-34.

Bruneau, M; Berman, J; Lopez-Garcia, D; Vian, D. Steel Plate Shear Wall Buildings: Design Requirements and Research, Steel Plate Shear Wall Buildings: *Design Requirements and Research*, Montreal, Canada, 2005.

Caccese, V; Elgaaly, M; Chen, R. Experimental Study of Thin Steel-Plate Shear Walls Under Cyclic Load. *ASCE Journal of Structural Engineering*, 1993, 119(2), 573-587.

Choi, IR; Park, HG. Steel Plate Walls with Various Infill Plate Designs. *ASCE Journal of Structural Engineering*, 2009, 135(7), 785-796.

CSA, CAN/CSA S16-01, *Limit States Design of Steel Structures*, Canadian Standards Association, Willowdale, Ontario, Canada, 2001.

CSA, CAN/CSA S16-94, *Limit States Design of Steel Structures*, Canadian Standards Association, Willowdale, Ontario, Canada, 1994.

Driver, RG; Grondin, GY; Behbahanifard, M; Hussain, MA. Recent Developments and Future Directions in Steel Plate Shear Wall Research, *Proceedings of the North American Steel Construction Conference*, Ft.Lauderdale, FL, 2001.

Driver, RG; Kulak, GL; Elwi, AE; Kennedy, DJL. Cyclic Tests of Four-Story Steel Plate Shear Wall. *ASCE Journal of Structural Engineering*, 1998a, 124(2), 112-120.

Driver, RG; Kulak, GL; Elwi, AE; Kennedy, DJL. FE and Simplified Models of Steel Plate Shear Wall. *ASCE Journal of Structural Engineering*, 1998b, 124(2), 121-130.

Driver, RG; Kulak, GL; Kennedy, DJL; Elwi, AE. Finite Element Modelling of Steel Plate Shear Walls, *Proceedings of the Structural Stability Research Council Annual Technical Session*, Toronto, Canada, 1997.

Driver, RG; Kulak, GL; Kennedy, DJL; Elwi, AE. Large-Scale Test on a Four Storey Steel Plate Shear Wall Subjected to Idealized Quasi-Static Earthquake Loading, *Proceedings of the 7th Canadian Conference on Earthquake Engineering*, Acapulco, Mexico, 1995.

Driver, RG; Kulak, GL; Kennedy, DJL; Elwi, AE. Seismic Behaviour of Steel Plate Shear Walls; *Structural Engineering Report No. 215*, Department of Civil Engineering, University of Alberta: Edmonton, Alberta, Canada, 1997.

Driver, RG; Kulak, GL; Kennedy, DJL; Elwi, AE. Seismic Performance of Steel Plate Shear Walls Based on a Large-Scale Multi-Storey Test, *Proceedings of the 11th World Conference on Earthquake Engineering*, Acapulco, Mexico, 1996.

Driver, RG; Kulak, GL; Kennedy, DJL; Elwi, AE. Seismic Behaviour of Steel Plate Shear Walls; Structural Engineering Report No. 215, *Department of Civil Engineering, University of Alberta*, Edmonton, Alberta, Canada, 1997.

Elgaaly, M; Liu, Y. Analysis of Thin-Steel-Plate Shear Walls. *ASCE Journal of Structural Engineering*, 1997, 123(11), 1487-1496.

Elgaaly, M. Thin Steel Plate Shear Walls Behaviour and Analysis. *Thin-Walled Structures*, 1998, 32, 151-180.

Elgaaly, M; Caccese, V; and Du, C. Post-Buckling Behavior of Steel-Plate Shear Walls under Cyclic Loads. *ASCE Journal of Structural Engineering*, 1993, 119(2), 588-605.

Fujitani, H; Yamanouchi, H; Okawa, I. Sawai, N; Uchida, N; Matsutani, T. Damage and Performance of Tall Buildings in the 1995 Hyogoken Nanbu Earthquake, *Proceedings of 67th Regional Conference* (in conjunction with ASCE Structures Congress XIV), Council on Tall Building and Urban Habitat, Chicago, 1996, 103-125.

Ghosh, S; Das, A; Adam, F. Design of Steel Plate Shear Walls Considering Inelastic Drift Demand. *Journal of Constructional Steel Research*, 2009, 65(7), 1431-1437.

Gupta, MK; Kharmale, SB; Ghosh, S. Ductility-Based Seismic Design of Steel Plate Shear Walls: Practical Application using Standard Sections. *International Journal of Advanced Structural Engineering*, 2009, 1(2), 93-110.

Habashi, HR; Alinia, MM. Characteristics of the Wall Frame Interaction in Steel Plate Shear Walls. *Journal of Constructional Steel Research*, 2010, 66(2), 150-158.

Kennedy, DJL; Kulak, GL; Driver, RG. Discussion of 'Post-Buckling Behavior of Steel-Plate Shear Walls Under Cyclic Loads' by M. Elgaaly, V. Caccese and C. Du. *ASCE Journal of Structural Engineering*, 1994, 120(7), 2250-2251.

Kharrazi, MHK; Prion, HGL; Ventura, CE. Implementation of M-PFI Method in Design of Steel Plate Walls. *Journal of Constructional Steel Research*, 2008, 64(4), 465-479.

Kuhn, P. *Stresses in Aircraft Structures*, McGraw-Hill Book Co; New York, USA, 1956.

Kulak, GL; Kennedy, DJL; Driver, RG. Discussion of 'Experimental Study of Thin Steel-Plate Shear Walls under Cyclic Load' by V. Caccese, M. Elgaaly and R. Chen. ASCE *Journal of Structural Engineering*, 1994, 120(10), 3072-3073.

Kulak, GL; Kennedy, DJL; Driver, RG; Medhekar, MS. Steel Plate Shear Walls: An Overview. *AISC Engineering Journal*, 2001, First Quarter, 38, 50-62.

Kulak, GL; Kennedy, DJL; Driver, RG; Medhekar, MS. Behavior and Design of Steel Plate Shear Walls, *Proceedings of the North American Steel Construction Conference*, Toronto, Canada, 1999.

Kurban, CO; Topkaya, C. A Numerical Study on Response Modification, Overstrength, and Displacement Amplification Factors for Steel Plate Shear Wall Systems. *Earthquake Engineering & Structural Dynamics*, 2009, 38(4), 497-516.

Li, CH; Tsai, KC. Experimental Responses of Four 2-story Narrow Steel Plate Shear Walls, *Proceedings of ASCE Structures Congress Structures* 2008: Crossing Borders, Vancouver, Canada, 2008.

LS-DYNA, LS-DYNA - Gen*eral Purpose Transient Dynamics Finite Element Program. Version 6.4*, Livermore Software Technology, Livermore, California, USA, 2003.

Lubell, AS. Performance of Unstiffened Steel Plate Shear Walls Under Cyclic Quasi-static Loading; M.S. Thesis, *Department of Civil Engineering*, University of British Columbia, Vancouver, Canada, 1999.

Lubell, AS. *Performance of Unstiffened Steel Plate Shear Walls Under Cyclic Quasi-static Loading*; M.S. Thesis, Department of Civil Engineering, University of British Columbia, Vancouver, Canada, 1999.

Lubell, AS; Prion, HGL; Ventura, CE; Rezai, M. Unstiffened Steel Plate Shear Wall Performance Under Cyclic Loading. *ASCE Journal of Structural Engineering*, 2000, 126(4), 453-460.

Mimura, H; Akiyama, H. Load-Deflection Relationship of Earthquake-Resistant Steel Shear Walls with a Developed Diagonal Tension Field. *Transactions of the Architectural Institute of Japan*, 1977, 260, 109-114.

Montgomery, CJ; Medhekar, M. Discussion on "Unstiffened Steel Plate Shear Wall Performance under Cyclic Load. *ASCE Journal of Structural Engineering*, 2001, 127(8), 973-973.

Park, HG; Kwack, JH; Jeon, SW; Kim, WK; Choi, IR. Framed Steel Plate Wall Behavior under Cyclic Lateral Loading. *ASCE Journal of Structural Engineering*, 2007, 133(3), 378-388.

Purba, R; Bruneau, M. Finite-Element Investigation and Design Recommendations for Perforated Steel Plate Shear Walls. *ASCE Journal of Structural Engineering*, 2009, 135(11), 1367-1376.

Qu, B; Bruneau, M; Lin, CH; Tsai, KC. Testing of Full-Scale Two-Story Steel Plate Shear Wall with Reduced Beam Section Connections and Composite Floors. *ASCE Journal of Structural Engineering*, 2009, 134(3), 364-373.

Rezai, M. *Seismic Behaviour of Steel Plate Shear Walls by Shake Table Testing*; PhD Dissertation, Department of Civil Engineering, University of British Columbia, Vancouver, Canada, 1999.

Rezai, M. *Seismic Behaviour of Steel Plate Shear Walls by Shake Table Testing;* PhD Dissertation, Department of Civil Engineering, University of British Columbia, Vancouver, Canada, 1999.

Rezai, M. *Seismic Behaviour of Steel Plate Shear Walls by Shake Table Testing*; PhD *Dissertation,* Department of Civil Engineering, University of British Columbia, Vancouver, Canada, 1999.

Rezai, M; Ventura, CE. Seismic Loading Behavior of Thin Steel Plate Walls, *Proceedings of ASCE Structures Congress Structures* 2008: Crossing Borders, Vancouver, Canada, 2008.

Rezai, M; Ventura, CE; Prion, HGL. Numerical Investigation of Thin Unstiffened Steel Plate Shear Walls, *Proceedings of the 12th World Conference on Earthquake Engineering*, Aukland, New Zealand, 2000.

Roberts, TM; Sabouri-Ghomi, S. Hysteretic Characteristics of Unstiffened Plate Shear Panels. *Thin-Walled Structures*. 1992, 12(2), 145-162.

Roberts, TM; Shahabian, F. Ultimate Resistance of Slender Web Panels to Combined Bending, Shear and Patch Loading. *Journal of Constructional Steel Research*, 2001, 57, 779-790.

Roberts, TM; Sabouri-Ghomi, S. Hysteretic Characteristics of Unstiffened Plate Shear Panels. *Thin-Walled Structures*, 1992, 12(2), 145-162.

Roberts, TM. Seismic Resistance of Steel Plate Shear Walls. *Engineering Structuresm* 1995, 17(5), 344-351.

Sabelli, R; Bruneau, M. Design Guide 20: *Steel Plate Shear Walls*", American Institute of Steel Construction, Chicago, IL, USA, 2007.

Sabouri-Ghomi, S; Roberts, TM. Nonlinear Dynamic Analysis of Steel Plate Shear Walls Including Shear and Bending Deformations. *Engineering Structures*, 1992, 14(5), 309-317.

Sabouri-Ghomi, S; Kharrazi, MHK; Mam-Azizi, SED; Sajadi RA. Buckling Behavior Improvement of Steel Plate Shear Wall Systems. *The Structural Design of Tall and Special Buildings*. 2007, 17(4), 823-837.

Sabouri-Ghomi, S; Ventura, CE; Kharrazi, HKM. Shear Analysis and Design of Ductile Steel Plate Walls. *ASCE Journal of Structural Engineering*, 2005, 131(6), 878-889.

Shishkin, JJ; Driver, RG; Grondin, GY. Analysis of Steel Plate Shear Walls Using the Modified Strip Model. *ASCE Journal of Structural Engineering*, 2009, 135(11), 1357-1366.

Shishkin, JJ; Driver, RG; Grondin, GY. Analysis of Steel Plate Shear Walls Using Conventional Engineering Software, *Proceedings of the 33rd Annual General Conference of the Canadian Society for Civil Engineering*, Toronto, Canada, 2005.

Takahashi, Y; Takemoto, Y; Takeda, T; Takagi, M. Experimental Study on Thin Steel Shear Walls and Particular Bracings Under Alternative Horizontal Load, Preliminary Report, *IABSE Symposium on Resistance and Ultimate Deformability of Structures Acted on by Well Defined Repeated Loads*, Lisbon, Portugal, 1973, 185-191.

Thorburn, LJ; Kulak, GL; Montgomery, CJ. *Analysis of Steel Plate Shear Walls; Structural Engineering Report No. 107*, Department of Civil Engineering, University of Alberta, Edmonton, Alberta, Canada, 1983.

Thorburn, LJ; Kulak, GL; Montgomery, CJ. *Analysis of Steel Plate Shear Walls; Structural Engineering Report No. 107*, Department of Civil Engineering, University of Alberta, Edmonton, Alberta, Canada, 1983.

Timler, PA; Kulak, GL. *Experimental Study of Steel Plate Shear Walls; Structural Engineering Report No. 114*, Department of Civil Engineering, University of Alberta, Edmonton, Alberta, Canada, 1983.

Timler, PA. Design Procedures Development, Analytical Verification and Cost Evaluation of Steel Plate Shear Wall Structures; Technical Report No. 98-01, *Earthquake Engineering Research, Facility*, Department of Civil Engineering, University of British Columbia, Canada, 1988.

Timler, PA; Kulak, GL. Experimental Study of Steel Plate Shear Walls; *Structural Engineering Report No. 114*, Department of Civil Engineering, University of Alberta, Edmonton, Alberta, Canada, 1983.

Timler, PA; Ventura, CE. Economical Design of Steel Plate Shear Walls from a Consulting Engineer's Perspective, *Proceedings of the North American Steel Construction Conference*, Toronto, Canada, 1999.

Timler, PA; Ventura, CE; Prion, H; Anjam, R. Experimental and Analytical Studies of Steel Plate Shear Walls as Applied to the Design of Tall Buildings. *The Structural Design of Tall Buildings*, 1998, 7(3), 233-249.

Tipping, S; Stojadinovic, B. Innovative Corrugated Steel Shear Walls for Multi-story Residential Buildings, *Proceedings of The 14th World Conference on Earthquake Engineering*, Beijing, China, 2008.

Topkaya, C; Atasoy, M. Lateral Stiffness of Steel Plate Shear Wall System. *Thin-Walled Structures*, 2009, 47(8-9), 827-835.

Topkaya, C; Kurban, CO. *Natural Periods of Steel Plate Shear Wall Systems Journal of Constructional Steel Research*, 2009, 65(3), 542-551.

Tromposch, EW; Kulak, GL. Cyclic and Static Behaviour of Thin Panel Steel Plate Shear Walls; *Structural Engineering Report No. 145*, Department of Civil Engineering, University of Alberta, Edmonton, Alberta, Canada, 1987.

Vian, D; Bruneau, M. Testing of Special LYS Steel Plate Shear Walls, *Proceedings of the 13th World Conference on Earthquake Engineering*, Vancouver, Canada, 2004.

Vian, D. Steel Plate Shear Walls for Seismic Design and Retrofit of Building Structures; *PhD Dissertation*, Department of Civil, Structural, and Environmental Engineering, University of Buffalo, New York, USA, 2005.

Vian. D; Bruneau, M; Purba, R. Special Perforated Steel Plate Shear Walls with Reduced Beam Section Anchor Beams II: Analysis and Design Recommendations. *ASCE Journal of Structural Engineering*, 2009b, 135(3), 588-605.

Vian. D; Bruneau, M; Tsai KC; Lin, YC. Special Perforated Steel Plate Shear Walls with Reduced Beam Section Anchor Beams I: Experimental Investigation. *ASCE Journal of Structural Engineering*, 2009a, 135(3), 211-220.

Wagner, H. Flat Sheet Metal Girders with Very Thin Metal Webs, Part I—General Theories and Assumptions; *Technical memorandum no. 604*, National Advisory Committee for Aeronautics, Washington, DC, USA, 1931.

Warn, GP; Bruneau, M. Blast Resistance of Steel Plate Shear Walls Designed for Seismic Loading. *ASCE Journal of Structural Engineering*, 2009, 135(10), 1222-1230.

Xue, M; Lu, LW. Interaction of Infilled Steel Shear Wall Panels with Surrounding Frame Members, *Proceedings of the Structural Stability Research Council Annual Technical Session*, Bethlehem, PA, 1994, 339-354.

Youssef, N; Wilkerson, R; Fischer, K; Tunick, D. Seismic Performance of a 55 Storey Steel Plate Shear Wall. *The Structural Design of Tall and Special Buildings*, 2009, DOI:10.1002/tal.545.

Zhao, Q; Astaneh-Asl, A. *Cyclic Tests of Steel Shear Walls*; Report Number UCB/CE-Steel-2002-01, Department of Civil and Environmental Engineering, University of California, Berkeley, USA, 2002.

In: Structural Steel and Castings
Editor: Lena M. Becker, pp. 107-145
ISBN: 978-1-61728-104-4

Chapter 3

Recent Advances on Design of Shear Connections

Yanglin Gong
Department of Civil Engineering, Lakehead University,
Thunder Bay, ON, Canada

Abstract

This chapter reports some recent studies on the analysis and design of shear connections. These new developments include: 1) the formulae for computing the moment of various shear connections (such as double-angle, single-angle, end-plate, and clipped end-plate shear connections) corresponding to the plastic limit state of its supported beam; 2) an experimental study on the double-angle shear connections with shortened outstanding-legs; 3) design method for bolted single-angle connections; and 4) resilience analysis of shear connections to resist progressive collapse under a double-span condition.

Introduction

Shear connections are used to transfer the gravitational loads on girders or beams to their supporting columns or girders. A shear connection is normally only designed to resist the shear force from its supported member. Since the moment in a shear connection can be ignored for the design of its connected members, the shear connection is also called simple connection in steel construction.

Depending on the shape of the connection steel, shear connections are commonly classified into six types, i.e., double-angle, single-angle, single-plate (or shear tab), end-plate, tee-type, and seated beam connections. The design of a shear connection typically involves checking the strength of the connection itself plus checking the strength of the connectors. Thus, one task in the design of a shear connection is to determine the structural demands on the connection steel and the associated connectors.

Moment of Shear Connections

Shear connections develop a moment in the plane of the beam web due to its rotational stiffness when the supported beam deflects under gravitational loads. Though this moment is usually negligible for the design of the connected beam and column, its influence on the components of a connection shall be taken into account generally.

The moment that a connection develops is directly a function of its rotation. Astaneh [1] pointed out that an end rotation of 0.03 radians is, in general, sufficient for the beam to reach its full plastic moment. Thus, for deriving moment equations in this section, a connection rotation at the level of 0.03 radians is taken as that corresponds to the ultimate limit state of the supported beam.

Thornton [2] gave Eq. (1) to calculate the maximum moment that a tee shear connection could develop. Figures 1a and 1b show the yieldline mechanism adopted by Thornton in which the second yieldline occurs along the toe of the stem fillet. Thornton assumed that the compression zone had a zero depth and the force in tension zone was uniformly distributed (T in Figure 1c. Note that the compression zone is not shown in the figure). Thornton's equation was

$$M_{th} = \frac{2m_p L^2}{b}\left[2+\left(\frac{b}{L}\right)^2\right] \tag{1}$$

where b is the distance between the two yieldlines at the upper end; L is the length of tee connection; and $m_p=F_y t^2/4$, the plastic moment of flange plate of tee section per unit length, in which F_y is the yield strength of steel and t is the thickness of tee flange. Note that Eq. (1) applied whether the tee flange was bolted (Figure 2a) or welded (Figure 2b) to the supporting member.

Thornton [3] later extended the application of Eq. (1) to double-angle and end-plate shear connections (Figure 2). For double-angle connections, Thornton assumed that both yieldlines were on the outstanding leg (Figure 2a). Thornton [2] concluded that Eq. (1) gave an upper bound solution since the predicted moment was almost always greater than the test moment. Though Eq. (1) is easy to use, the past studies [2, 3] also showed the equation often dramatically overpredicted the connection moment.

Recently, the author had conducted several experimental studies on both single and double-angle connections [4,5,6,7]. The yieldline mechanisms observed in these tests were not exactly the same as those described in Thornton's work. Figure 3 illustrates the yieldline pattern of a single-angle connection test by the author [4]. The inclined first yieldline is evident on the outstanding leg. However, unlike Thornton's model, the second yieldline occurred on the web-framing leg (called framing leg for simplicity hereafter) along the end line of the beam web.

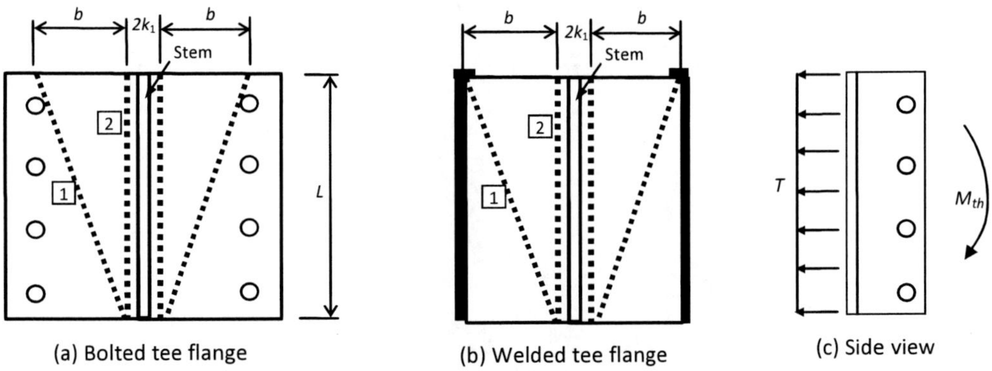

Figure 1. Thornton's yieldline mechanism for tee connections.

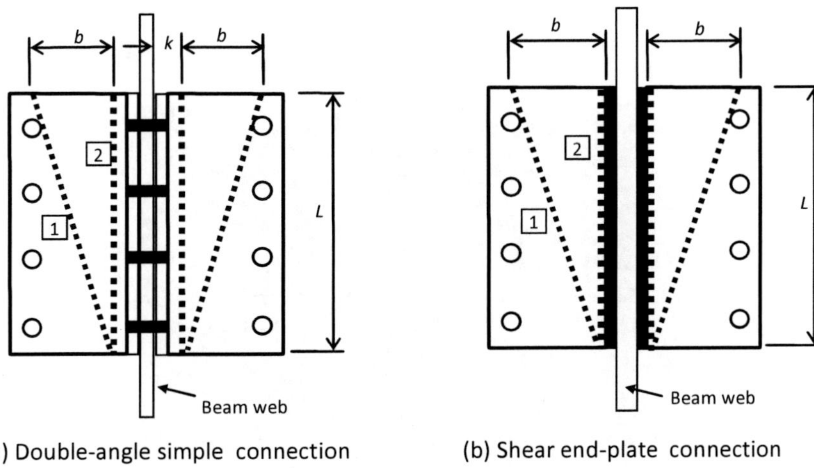

Figure 2. Thornton's yieldline mechanisms for double-angle and end-plate connections.

Single-Angle Shear Connections

In this section, a semi-analytical a is employed to derive the moment equations for a single-angle shear connection which has the yielding pattern as shown in Figure 3. Then the method will be extended to find moments for double-angle and end-plate shear connections. It is understood that the support for the connection is assumed to be rigid to allow the connection to develop moment.

A compression zone was observed on the back of the tested single-angle connections at the lower end where the angle was bearing against the supporting column. The compression zone was at the heel in the shape of a triangle, as shown in Figure 4. From measurements, the depth of the compression zone L_c was about 6% to 10% of the total angle length L.

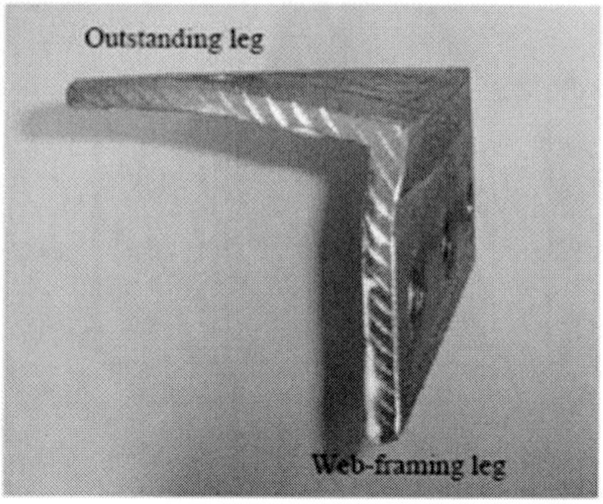

Figure 3. Top view of a single-angle connection after test.

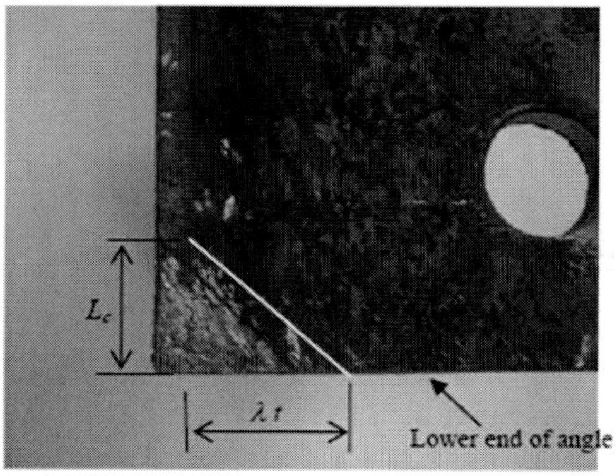

Figure 4. Compression zone.

The yieldline mechanism, as shown in Figure 5, is proposed for a single angle connection based on the plastic deformations observed in Figure 3. It is assumed that the neutral axis, or x-axis, is located at a distance of L_c above the lower end of the connection. The external couple, M_p, is the moment imposed to the connection through the web of the supported beam. The first yieldline is on the outstanding leg and is inclined from the upper end to the x-axis in order to be compatible with the connection rotation. The second yieldline is on the framing leg and is parallel to the heel. Corresponding to moment M_p, T_y is the horizontal tension force per unit length applied to the angle from the beam web, and C_y is the compression force per unit length in the compression zone between the connection and the supporting column. Then,

$$M_p = \int_0^{L_t} T_y y dy + \int_0^{L_c} C_y y dy = \sum T_y \bullet (\xi L_t) + \int_0^{L_c} C_y y dy \tag{2}$$

where y is the coordinate of angle (Figure 5a); L_t is the depth of the tension zone; $\sum T_y$ is the tensile force resultant and $\sum T_y = \int_0^{L_t} T_y dy$; ξL_t is the distance from the force resultant ΣT_y to the neutral axis; L_c is the depth of the compression zone; and $L=L_c + L_t$.

The determination of ξ value requires the knowledge of T_y distribution. Many factors may affect the T_y distribution, such as the amount of connection rotation, the fitting of the bolts between the angle and beam web, angle thickness, steel strength, etc. Thus, attempting a single exact value for ξ may not be able to reflect the nature of the diversity of shear connections. However, a plausible range of ξ value can be adopted as $0.4 \leq \xi \leq 0.65$, as 0.4 reflects a trapezoid T_y distribution and 0.65 a triangular distribution [8]. The equilibrium of forces in horizontal direction requires

$$\Sigma T_y = \int_0^{L_c} C_y dy \tag{3}$$

By assuming that the compression zone is a triangular area having a height of L_c and base of λt (product of coefficient λ and angle thickness t) (Figure 4) and that the compression bearing stress σ_c is uniformly distributed over the compression zone, we obtain

$$\int_0^{L_c} C_y dy = L_c \lambda t \sigma_c / 2 \tag{4}$$

and

$$\int_0^{L_c} C_y y dy = \lambda t \sigma_c L_c^2 / 3 \tag{5}$$

The average value of λ measured from the tested connections was 2.7. The L_c value was from 6% to 10% of length L. The bearing stress σ_c must have exceeded F_y as the permanent plastic deformation in compression zone was evident. An upper bound value for σ_c should be determined by considering the following three limit states at the lower end of the framing leg: 1) local yielding, 2) local crippling, and 3) local bearing of the compression zone. For local yielding of the framing leg under compressive stresses, Canadian standard CSA-S16 [9] stipulates that the nominal strength is $t(L_c + 4t)F_y$ (Clause 14.3.2). Let the compressive force resultant $\sigma_c(0.5 L_c \lambda t)$ be less than $t(L_c + 4t)F_y$, we get

$$\sigma_c \leq \frac{2(L_c + 4t)}{\lambda L_c} F_y \tag{6}$$

where F_y is the yield strength of the connection angle. For the crippling limit state, CSA-S16 gives the nominal strength as $0.6t^2 \sqrt{EF_y}$, which results in

$$\sigma_c \leq \frac{0.6t\sqrt{EF_y}}{\lambda L_c} \tag{7}$$

where E is the Young's modulus of the steel material. For bearing limit state, CSA-S16 requires $\sigma_c \leq 1.5F_y$ (Clause 13.10). If λ is taken as 2.7, L_c is assumed to be equal to λt, and E=200×10^3 MPa, then, for a connection having F_y from 250 MPa to 500 MPa, we can easily find that the upper bound value for σ_c is governed by the local bearing strength, i.e.,

$$\sigma_c \leq 1.5F_y \tag{8}$$

Denoting b_0 as the distance from the first yieldline at the upper end to the angle heel (Figure 5a) and Δ_t the horizontal deflection of the angle heel at the upper end (Figure 5c), the work of T_y is then written as

$$W_e = \int_0^{L_t} \left(T_y dy\right)\frac{\Delta_t}{L_t} y = \frac{\Delta_t}{L_t}\left(\sum T_y\right)\xi L_t \tag{9}$$

Note that actual dimension b_0 is influenced by the factors such as angle length, end distance of bolt holes, bolt diameters, and whether bolt washer is present or not. For simplicity, b_0 is approximately taken as the distance from the centre line of the bolt hole (or from the weld line if outstanding leg is welded to the supporting member, Figure 5d) to the heel of the angle. The internal work of the first yieldline is

$$W_{i,1} = m_{p1}\sqrt{L_t^2 + b_0^2}\left[\Delta_t\sqrt{1+\left(b_0/L_t\right)^2}\,/b_0\right] \tag{10}$$

where: m_{p1} is the plastic moment per unit length of outstand leg plate considering combined effect of shear force and moment. According to Drucker [10]

$$m_{p1}/m_p + \left(T_{ya}/V_0\right)^4 = 1 \tag{11}$$

in which m_p is the plastic moment per unit angle length under moment only; $V_0 = F_y t/\sqrt{3}$, the ultimate shear resistance of unit angle length under shear force only; and $T_{ya} = \sum T_y/L_t$, the average of T_y along length L_t. Note that the shear force herein of the outstanding leg is equal to the tension force T_y on the framing leg. The internal work of the second yield line is

$$W_{i,2} = m_{p2}L\Delta_t/b_0 \tag{12}$$

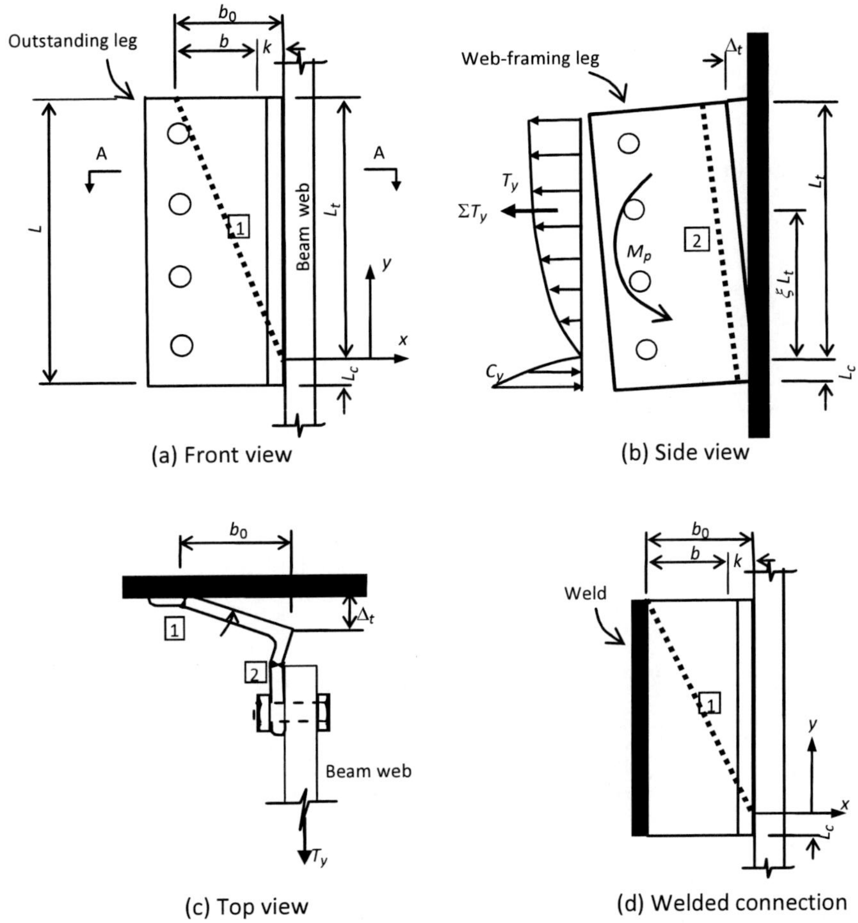

Figure 5. Yieldline mechanism for single angle connections.

where the rotation $\theta = \Delta_t / b_0$ is constant along the second yieldline; m_{p2} is the plastic moment per unit length of framing leg considering the combined effect of axial force and moment; and according to Chen and Sohal [11]

$$m_{p2} / m_p + \left(T_{ya} / T_0\right)^2 = 1 \tag{13}$$

in which T_0 is the tensile resistance per unit length of angle plate and $T_0 = F_y t$. It was observed from tests that compression zone depth is small and L_t is approximately 90% of L. Thus, without introducing much error, we replace L with L_t in Eq. (12) to simply the formulation, i.e.,

$$W_{i,2} = m_{p2} \Delta_t L_t / b_0 \tag{14}$$

Then, the virtual work principal, $W_e = W_{i,1} + W_{i,2}$, gives

$$\sum T_y = \frac{m_{p1}L_t}{\xi b_0}\left[1+\left(\frac{b_0}{L_t}\right)^2+\frac{m_{p2}}{m_{p1}}\right] \tag{15}$$

Substituting Eq. (15) into Eqs. (3) and (4) gives

$$L_c = \frac{2m_{p1}L_t}{\xi b_o}\left[1+\left(\frac{b_0}{L_t}\right)^2+\frac{m_{p2}}{m_{p1}}\right]\frac{1}{\lambda t\sigma_c} \tag{16}$$

Further, from $L=L_c+L_t$ we have

$$j = \frac{L_t}{L} = \frac{1}{1+\dfrac{2m_{p1}}{\xi b_0\lambda t\sigma_c}\left[1+\left(\dfrac{b_0}{L_t}\right)^2+\dfrac{m_{p2}}{m_{p1}}\right]} \tag{17}$$

From Eqs. (5) and (15), the connection moment Eq. (2) becomes

$$M_p = \frac{m_{p1}L_t^2}{b_o}\left[1+\left(\frac{b_o}{L_t}\right)^2+\frac{m_{p2}}{m_{p1}}\right]+\frac{\lambda t\sigma_c L_c^2}{3} \tag{18}$$

For regular single-angle connections such as the tested, the value of j in Eq. (17) can be estimated using $(b_0/L_t)^2 \cong 0$ due to $b_0 << L_t$ for the connections having more than 3 bolts. Furthermore, as illustrated later, m_{p1} and m_{p2} are normally very close to m_p; thus $m_{p2}/m_{p1} \cong 1$ by ignoring the combined effect of forces in Eq. (17). Let $\sigma_c=\zeta F_y$, Eq. (17) is reduced to

$$j = \frac{1}{1+(t/\xi\zeta\lambda b_0)} \tag{19}$$

Eq. (19) is used to estimate the depth of tension zone. Eq. (19) tells that the compression zone is deeper for thicker angles, which is in agreement with Lipson's test [15]. For the tested connections, b_0=65 mm, t=9.5 mm, and

$$j = \frac{1}{1+(0.146/\xi\zeta\lambda)} \tag{20}$$

In Eqs. (19) and (20), the value of λ can be determined appropriately by measuring the compression zone of the tested connections. To determine the values of ξ and ζ, the relationships between j and ξ and ζ are illustrated in Figure 6 for $0.4 \le \xi \le 0.65$ and $\zeta=1.0$ and 1.5. Fortunately, Figure 6 tells that the j value is not sensitive to the variation of ξ and ζ values. In the analysis hereafter, the value of λ is taken as 2.7 based on the test results. The values of ξ and ζ are arbitrarily taken as 0.55 and 1.1, respectively. From Eq. (20), j is equal to 0.92 for the tested connections, which agrees well with the test results with respect to the compression zone depth (i.e., L_c was 6 to 10 percent of L). The contribution of the compression zone to the connection moment can be calculated from the ratio:

$$\eta = \frac{\frac{\lambda t \sigma_c L_c^2}{3}}{\frac{m_{p1} L_t^2}{b_0}\left[1+\left(\frac{b_0}{L_t}\right)^2+\frac{m_{p2}}{m_{p1}}\right]} = \frac{\lambda t \sigma_c L_c^2}{3}\frac{b_0}{2m_{p1}L_t^2} = \frac{2\lambda}{3}\frac{\sigma_c}{F_y}\frac{b_0}{t}\left(\frac{L_c}{L_t}\right)^2 \tag{21}$$

where $(b_0/L_t)^2 \cong 0$, $m_{p2}/m_{p1} \cong 1$ and $m_{p1} = F_y t^2/4$ are employed. Applying Eq. (21) to the tested connections, i.e., $L_c/L_t = 0.087$, $b_0/t = 65/9.5 = 6.8$, and $\lambda=2.7$, we have

$$\eta = \frac{2\lambda}{3}\frac{\sigma_c}{F_y}\frac{b_o}{t}\left(\frac{L_c}{L_t}\right)^2 = 0.093\frac{\sigma_c}{F_y} \tag{22}$$

Therefore, the fraction of the compression zone moment to the total connection moment for the tested connections is thus found as

$$\eta_c = \eta/(1+\eta) \tag{23}$$

From Eq. (23), when $\sigma_c=F_y$, $\eta_c=9\%$; when $\sigma_c=1.5F_y$, $\eta_c=12\%$; i.e., the compression zone contributes about 9% to 12% of the total connection moment for the tested connections. It is also shown that some error in the determination of σ_c has negligible influence on the total moment. Hence, Eq. (18) is rewritten for single-angle connections as

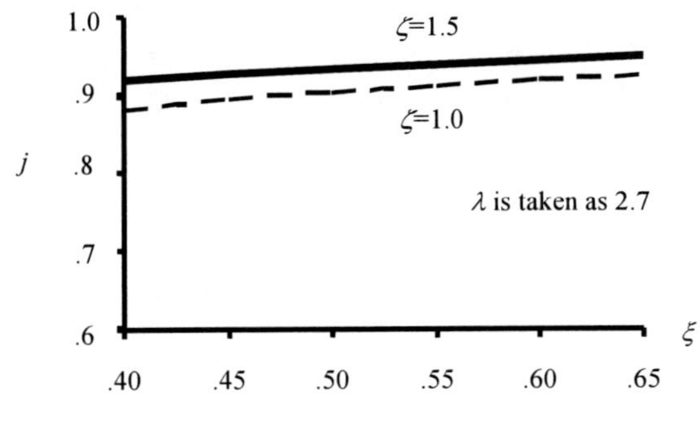

Figure 6. Neutral axis location versus tensile force resultant location.

$$M_p = \frac{m_{p1}L_t^2}{b_0}\left[1+\left(\frac{b_0}{L_t}\right)^2+\frac{m_{p2}}{m_{p1}}\right]+tL_c^2F_y \tag{24}$$

where λ and σ_c are taken approximately as 2.7 and $1.1F_y$, respectively.

Next, let us examine the effect of combined actions on the connection moment represented by Eqs. (11) and (13). From Eqs. (11) and (15), we have

$$\frac{T_{ya}}{V_0} = \frac{m_{p1}}{\xi b_0}\left[1+\left(\frac{b_0}{L_t}\right)^2+\frac{m_{p2}}{m_{p1}}\right]\frac{\sqrt{3}}{tF_y} \tag{25}$$

Let $(b_0/L_t)^2 \cong 0$, $m_{p2}/m_{p1} \cong 1$ and $m_{p1} \cong F_y t^2/4$, Eq. (25) is simplified as

$$\frac{T_{ya}}{V_0} = \frac{\sqrt{3}}{2}\frac{t}{\xi b_0} \le 2.2\frac{t}{b_0} \tag{26}$$

where $\xi \ge 0.4$ is adopted. In general, the ratio T_{ya}/V_0 is much smaller than 0.5. For example, for the tested connections, t/b_0=0.146 and $T_{ya}/V_0 \le 0.32$. Therefore, $(T_{ya}/V_0)^4$ is very small or in the other words, the combined effect of shear force and moment, Eq. (11), can be neglected for regular single-angle connections; i.e., $m_{p1}=m_p$ for regular connections. For framing leg, similarly,

$$\frac{T_{ya}}{T_0} = \frac{m_{p1}}{\xi b_0}\left[1+\left(\frac{b_0}{L_t}\right)^2+\frac{m_{p2}}{m_{p1}}\right]\frac{1}{tF_y} = \frac{t}{2\xi b_0} \le 1.25\frac{t}{b_0} \tag{27}$$

In general, the ratio T_{ya}/T_0 is less than 0.25. For example, for the tested connections, t/b_o=0.146 and $T_{ya}/T_0 \le 0.18$, then $(T_{ya}/V_0)^2 \le 0.033$. Therefore, the combined effect of axial force and moment, Eq. (13), can be neglected for regular single-angle connections as well (i.e., $m_{p2}=m_p$). Therefore, if the combined effects of actions are ignored, Eq. (24) can be reduced to

$$M_p = \frac{m_p L_t^2}{b_0}\left[2+\left(\frac{b_0}{L_t}\right)^2\right]+tL_c^2F_y \tag{28}$$

If the compression zone is assumed to be at the very lower end of the angle, i.e., L_c=0 and L_t=L, the connection moment M_p is then further simplified as

$$M_c = \frac{m_p L^2}{b_0}\left[2 + \left(\frac{b_0}{L}\right)^2\right] \tag{29}$$

For single-angle connections, Thornton equation should be rewritten as

$$M_{th1} = \frac{m_p L^2}{b}\left[2 + \left(\frac{b}{L}\right)^2\right] \tag{30}$$

Note that Eq. (29) is virtually the same as Thornton equation, except Eq. (29) uses dimension b_0 instead of b.

In summary, the procedure for computing the ultimate moment of a connection in which the combined effects of loads are neglected is

(1) assume λ=2.7, ζ=1.1 and ξ=0.55, calculate j by Eq. (19), then $L_t = jL$ and $L_c = L - L_t$;
(2) M_p is calculated by Eq. (28).

The more general procedure in which the combined effects of loads are not ignored is

(1) assume λ=2.7, ζ=1.1 and ξ=0.55, calculate j by Eq. (19), then $L_t = jL$ and $L_c = L - L_t$;
(2) an iterative method, which is adopted to calculate m_{p1}, m_{p2} and ΣT_y through Eqs. (11), (13) and (15), is as follows:
   (i) set iteration index i=1, and set T_{ya}=0;
   (ii) calculate m_{p1} by Eq. (11) and m_{p2} by Eq. (13);
   (iii) calculate ΣT_y by Eq. (15), and update $T_{ya} = \Sigma T_y / L_t$;
   (iv) If T_{ya} converge, then go to (3); otherwise, set $i=i+1$, go to (ii).
(3) M_p is calculated from Eq. (24).

For the general procedure considering combined effects of loads, it usually takes only 2 to 3 iterations to achieve the convergence of T_{ya} with a less than 0.5% error.

Double-Angle Shear Connections

Double-angle shear connections have one angle on each side of the beam web. Due to two angles, the moment Eqs. (24), (28) and (29) are modified accordingly into Eqs. (31), (32) and (33) respectively in the following; i.e., when considering combined effects of forces, we have

$$M_p = \frac{2m_{p1}L_t^2}{b_0}\left[1 + \left(\frac{b_0}{L_t}\right)^2 + \frac{m_{p2}}{m_{p1}}\right] + 2tL_c^2 F_y \tag{31}$$

When ignoring the combined effects of forces, we have

$$M_p = \frac{2m_p L_t^2}{b_0}\left[2 + \left(\frac{b_0}{L_t}\right)^2\right] + 2tL_c^2 F_y \tag{32}$$

If the compression zone is assumed to be zero depth, we have

$$M_c = \frac{2m_p L^2}{b_0}\left[2 + \left(\frac{b_0}{L}\right)^2\right] \tag{33}$$

The total tension zone force, from Eq. (15), is

$$\sum T_y = \frac{2m_{p1} L_t}{\xi b_0}\left[1 + \left(\frac{b_0}{L_t}\right)^2 + \frac{m_{p2}}{m_{p1}}\right] \tag{34}$$

End-Plate Shear Connections

An end-plate shear connection consists of a vertical plate shop-welded to the end of the web transverse to the longitudinal axis of the supported beam. The end of the beam is then connected to the supporting member by bolts. Comparing to the angle connections in previous sections, this type of connections has no framing leg and thus yieldlines can only occur on the end-plate. The moment equation for end-plate shear connections including compression zone is written as

$$M_p = \frac{2m_p L_t^2}{b}\left[2 + \left(\frac{b}{L_t}\right)^2\right] + 2tL_c^2 F_y \tag{35}$$

where dimension b is determined by assuming the second yieldline is at the toe of fillet weld; i.e., the weld size and the thickness of the beam web should be taken into account in computing b. The combined effect of shear and moment is neglected in Eq. (35). If the compression zone is assumed to be zero depth, then Thornton Eq. (1) should be used.

The clipped end-plate connection (Figure 7) is commonly used to facilitate erection for double connections. Since the plate lengths L_1 and L_2 at each side of the beam web are different, Eq. (35) should be modified into Eq. (36) for computing the ultimate moment of a clipped connection; i.e.,

$$M_p = \frac{m_p}{b}\left\{L_{1t}^2\left[2 + \left(\frac{b}{L_{1t}}\right)^2\right] + L_{2t}^2\left[2 + \left(\frac{b}{L_{2t}}\right)^2\right]\right\} + t\left(L_{1c}^2 + L_{2c}^2\right)F_y \tag{36}$$

where L_{1t}, L_{2t}, L_{1c} and L_{2c} are the tension and compression zone depths corresponding to L_1 and L_2, respectively. Similarly, Thornton Eq. (1) should be revised as

$$M_{th} = \frac{m_p}{b}\left\{L_1^2\left[2+\left(\frac{b}{L_1}\right)^2\right]+L_2^2\left[2+\left(\frac{b}{L_2}\right)^2\right]\right\} \tag{37}$$

Summary of Connection Moments

The basis for the moment equations derived in this section is the yieldline mechanism depicted in Figure 5, in which the second yieldline must be on the framing leg of the connection angle. For the connections of this case, Thornton equation does not apply or it will overpredict connection moment considerably. The comparisons between the theoretical analysis results and test data demonstrated that the yieldline mechanism adopted by this study is correct for angle connections having a thickness less than and equal to 10 mm.

The reasons for the second yieldline occurring on the framing leg include: 1) the outstanding leg of the angle deflect vertically under the shear load from the supported beam, which leads the framing leg to flex about the end of the beam in order to accommodate the bending curvature of the outstanding leg; 2) the setback of the beam web (dimension s_t in Figure 8) is usually much greater than the thickness of angle plate, which allows the formation of the second yieldline on the framing leg. In USA [12], the standard beam setback s_t is 13 mm. But in general, the actual setback is greater than 13 mm. Bolt gauge g_w is from 57 mm to 65 mm, and the edge distance of bolts at the beam end s_w is 38 mm or greater. Thus the maximum nominal beam setback s_t ranges from 19 mm to 27 mm, which is greater than the nominal heel radius of an angle whose thickness is $t \leq 9.5$ mm; 3) the slippage of bolts and bearing deformation of bolt holes at the framing leg increase beam setback; 4) the actual heel radius of connection angles are often smaller than the nominal value. For example, the measured heel radius of the tested connections by Gong [5] was 13 mm instead of the nominal value 19 mm; and 5) the moment M_c predicted by Eq. (33) is less than the moment M_{th} predicted by Thornton equation. The ratio of M_{th} to M_c is approximately equal to b_0/b. Therefore, as long as actual conditions permit, the connection will form the second yieldline on the framing leg instead of on the outstanding leg. The second yieldline may even form within the heel radius, as long as the increased yielding moment at the hinge does not result in M_c being greater than M_{th}.

The derivation of moment equations took into account the following parameters: plate thickness, connection length, steel strength, connection rotation, combined effects of forces, beam web thickness, and fillet weld size. The proposed method enable a designer to determine various demands related to connection design. For examples, compression zone depth can be estimated using Eq. (19); tension zone force can be determined through Eq. (15). This study also demonstrated that the evaluation of connection moment is not sensitive to the distribution of forces in the tension zone.

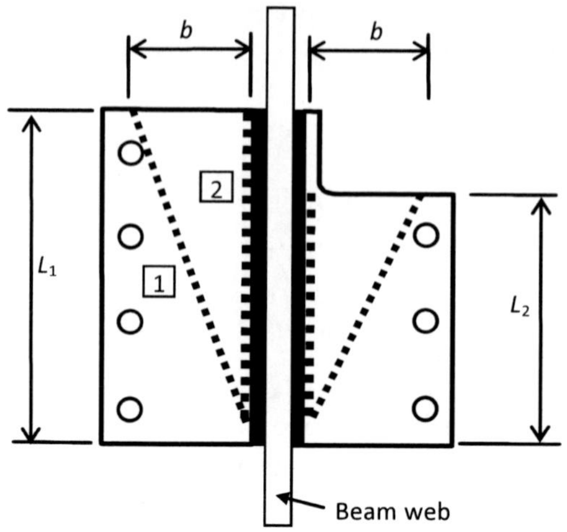

Figure 7. Clipped end-plate connections.

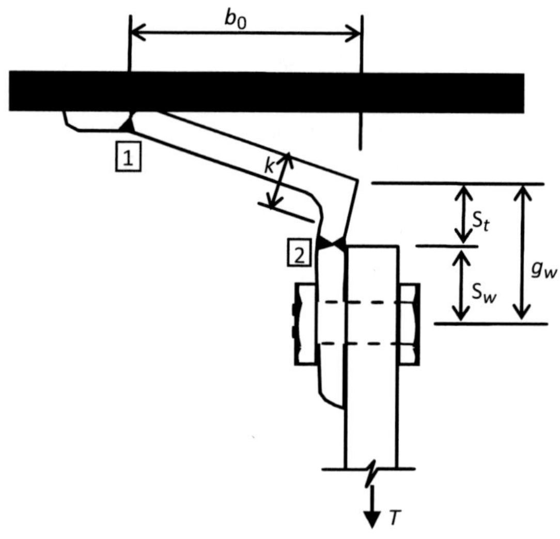

Figure 8. Beam setback and yieldlines on connection angle.

Equations (24) and (31) are most general and applicable to any design situations of angle connections. For the angles having thickness $t \leq 9.5$ mm, numerical calculations showed that the combined load effects could be ignored safely, and thus Eq. (28) or (32) can be employed to simplify the calculation. Finally, by assuming a zero depth compression zone, the connection moment can be estimated easily and conservatively by Eq. (29) or (33).

The study on both regular and clipped end-plate shear connections showed that the moment equations developed herein could provide a more accurate prediction than Thornton's equation. Thornton moment is about 10 to 20 percent greater than the moment calculated from Eqs. (35) or (36). For the design of end-plate connections, Thornton equation is still applicable (though being more conservative) since the actual yieldline mechanism was consistent with Thornton model. In fact, Thornton equation can be viewed as a special case of Eq. (35) by assuming zero depth of compression zone.

The moment equations provide a means for directly computing the connection moment corresponding to the strength limit state of the supported beam. When applying these equations to connection design, the actual steel yield strength should be used instead of the specified yield strength. This actual material strength can be obtained from mill test or using the expected strength of the material. Note that the actual yield strength is used to compute moment only and thus will not compromise the safety of the designed structure.

Since the conclusion on the yieldline mechanism is based on the experimental tests having an angle thickness within the range of 6 to 10 mm, caution is required to apply the conclusion to the connections with an angle thickness of greater than 10 mm. The primary concern is whether the beam setback is large enough for the second yieldline to form on the framing leg.

Double-Angle Shear Connections with Short Outstanding Legs

Simple connections not only require sufficient strength to resist the end shear of the supported beam but also must have enough rotation capacity (or ductility) to accommodate the end rotation demand of the beam. However, in actual practice the commonly used simple connections are proportioned for shear only, and their rotational capacities are generally perceived as non-problematic. For a double-angle connection, when the angles are welded to the support and bolted to the beam, it is also called a "knife angle connection". To ensure connection flexibility, the angle thickness should not be greater than necessary, with a minimum thickness of 6 mm [13] and a maximum thickness of 16 mm [12].

The standard knife connections in the Handbook of Steel Construction tables [13] use angles with outstanding legs having a width of either 76 mm or 89 mm. This leg width creates a problem when a beam is to be connected to a column with a face width less than, say 152 mm. Typically, the common practice is either to weld the angles to the column by flare bevel groove welds or to shorten the outstanding legs in order to accommodate fillet welds (Figure 9). This practice reduces the ductility of the connections since the distortional deformations of the outstanding legs are reduced.

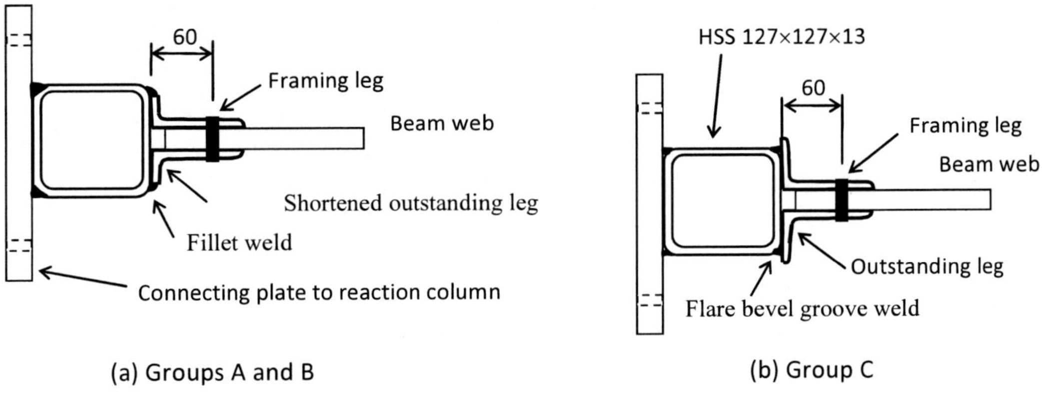

Figure 9. Test specimens.

This section presents an experimental study of 12 full-scale connection specimens, which confirmed the acceptability of the aforementioned practices for the connections having 6 or less bolts in one vertical row.

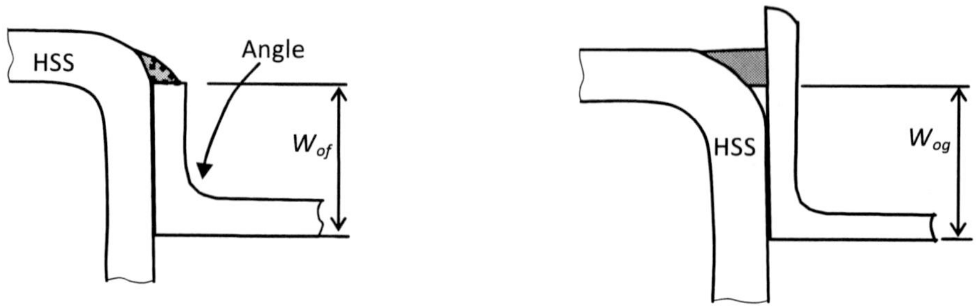

(a) Groups A and B (b) Group C

Figure 10. Definition of width for the outstanding leg.

Test Program

The specimen consisted of a pair of angles welded to a short length of HSS column as shown in Figure 9. Three groups of four specimens were tested, labeled A, B and C, with the specimens in each group containing 3, 4, 5 or 6 bolts in one vertical row as shown in Table 1. The angles back-to-back spacing was sufficient to accommodate a test beam with a 32 mm thick web that could be reused in test. To obtain outstanding leg widths that would be expected on an HSS 102×102 column for a beam with a typical web thickness, the column shape was increased to an HSS 127×127. Based on measurements of the fabricated specimens, the outstanding leg width (W_{of} and W_{og} in Figure 10) varied between 31 to 32 mm for the groups A and B specimens and 34 to 37 mm for the group C specimens.

Groups A and B used L102×76×9.5 and L102×76×6.4 angles, respectively. The outstanding legs of these two groups were sheared to fit within the face of the HSS column. The nominal fillet weld size was 6 mm and 5 mm for groups A and B, respectively, with a top return of twice the weld size. Group C used L102×76×9.5 angles, and the outstanding legs were welded to the column using flare bevel groove welds with no return at top. The flare bevel groove welds were made to be flush with the side face of HSS column. The welding procedure used was gas metal arc welding, and matching electrodes were ER49S (the specified ultimate strength X_u=490 MPa). A nominal angle thickness of 9.5 mm was selected for groups A and C since it is commonly used in practice. Group B used thinner angles to investigate whether reduction in angle thickness could increase connection ductility. All the connection angles were of CSA-G40.21 [14] 300W steel, while all the HSS columns were of G40.21 350W steel. The thickness of the tube wall was 12.4 mm in average, and the outside corner radius was 22 mm.

The framing legs supported the test beam using ASTM A325 ϕ19 mm high strength bolts (F_u=825 MPa). The diameter of the bolt holes was 20.6 mm. The vertical center-to-center distance of the bolt holes was 76 mm. The distance between the boltline and the column face was 60 mm (Figure 9). The length of connection angles is given in Table 1. In order to reuse the reaction column, the HSS column of each specimen was welded to a 19 mm thick plate (Figure 9), which in turn was bolted to the reaction column during testing (Figure 11). The average measured yield and tensile strengths were F_y=396 MPa and F_u=537 MPa, respectively, for the angles with nominal thickness of 9.5 mm, while F_y=364 MPa and F_u=514 MPa for the angles with nominal thickness of 6.4 mm. The weld size recorded in Table 1 was the average value for a specimen.

The test setup is shown diagrammatically in Figure 11. The main components of the setup are a permanent test beam, a reaction column, a beam brace, a support seat, three jacks, and jack reaction frames (not shown in the figure). The test beam and column were fabricated of such proportions that their deformations during testing were insignificant. The specimen was attached to the flange of the reaction column using pretensioned high strength bolts. Jack 1 applied the bulk of the shear load to the specimens. Jack 2 was used to apply a rotation demand to specimens by pushing down the far end of the test beam. Jack 3 supported the test beam and limited the magnitude of beam rotation. The applied load at each jack was monitored by a load cell, and the shear load applied to the specimen was calculated from the equilibrium of the test beam. A support seat was placed under the HSS column such that the majority of the shear load applied to a specimen during testing was directly taken down to the rigid floor through the seat. The brace prevented the test beam from twisting out-of-plane during loading.

Six linear displacement sensors, labeled as LP1 through LP6), were installed to measure the displacements of the specimen and test beam. LPs 1 to 4 were mounted horizontally on both sides of the specimen, while LPs 5 and 6 were installed vertically under the test beam. Measurements from LPs 1 to 4 were used to compute connection rotations. LP5 measured the vertical displacement under the boltline, while measurements from LPs 5 and 6 permitted the computation of beam rotations.

Table 1. Double-angle connection specimens

Specimen ID	Number of bolts	Nominal angle length (mm)	Connection angle thickness (mm)	Measured weld size* (mm)	Width of angle outstanding legs (mm)
A1	3	230	9.5	7.0	31
A2	4	310	9.5	7.5	32
A3	5	390	9.5	7.4	31
A4	6	470	9.5	6.7	32
B1	3	230	6.4	5.3	31
B2	4	310	6.4	5.6	32
B3	5	390	6.4	6.3	31
B4	6	470	6.4	5.1	31
C1	3	230	9.5	13.8	36
C2	4	310	9.5	13.8	37
C3	5	390	9.5	13.8	35
C4	6	470	9.5	13.8	34

* For groups A and B, weld size refer to fillet weld leg. For group C, weld size refers to the effective throat computed as (5/8)R, where R=radius of HSS corner=22 mm [Table J2.2, ANSI/AISC 360-05]

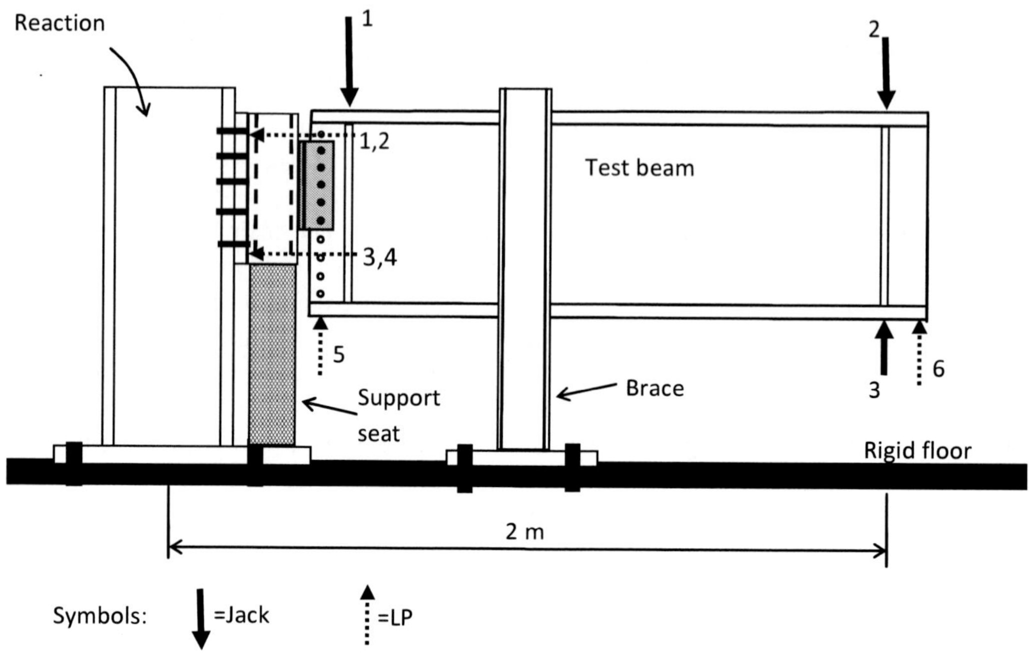

Figure 11. Elevation of test setup.

To simulate the actual conditions, the test procedure required that the shear load be imposed in combination with a realistic rotation on the connection. The maximum rotation demand for shear connections was taken as 0.03 radians. A complete test process was divided into two stages. Stage one, a rotation not less than 0.03 radians was applied to the specimen by pushing down the far end of the beam using jack 2 while jack 1 at the near end was unloaded. During stage two, jack 1 load was increased monotonically until the connection failed or the capacity of jack 1 was reached while jacks 2 and 3 were holding the far end of the beam in position.

Test Results

At the early phase of rotational loading in stage one, the connection specimens experienced bolt slip before the bolts came to bear against holes in the angles and beam web. The pretension force of the bolts was deliberately controlled to be small during snug-tightening such that bolt slip generally took place at the beginning of rotational loading to facilitate monitoring the bolt slip. After bolt slip, further rotational loading caused the top of the connection angles to pull away from the face of the HSS column. No strength failure was observed during stage one loading for any of the specimens.

The maximum applied rotation during stage one is recorded in column 6 of Table 2. The relationship between the bending moment at the boltline and the connection rotation for some specimen during stage one is given in Figure 12. These moment-rotation curves are used to quantify the connection rotation due to the bolt slip in the analysis of rotational capacity.

During the shear loading in stage two, four failure modes, i.e., rupture of fillet welds, rupture of bolt group, rupture of angle net section at boltline, and yielding of angle gross section, were observed for 8 specimens, while the remaining 4 specimens did not fail due to jack 1 loading capacity being reached. The maximum value of the applied shear load and the corresponding failure mode are presented in Table 2.

Table 2. Testing results of double-angle shear connections

Specimen ID	Failure mode	Shear load at failure[ᴲ] (kN)	Predicted Strength (kN)	Ratio of test strength to predicted	Maximum rotation (rad.)	Rotation at failure (rad.)
A1	Fillet welds shear	819	754	1.09	0.037	0.035
A2	Fillet welds shear	1163	1086	1.07	0.034	0.023
A3	Not observed	>1467	1343	>1.09	0.036	0.031[#]
A4	Not observed	>1466	1462	>1.00	0.034	0.031[#]
B1	Fillet welds shear	516	568	0.91	0.038	0.034
B2	Fillet welds shear	790	816	0.97	0.036	0.027
B3[*]	Gross section yield	1138	1112	1.02	0.034	0.028
	Fillet welds shear	1161	1144	1.01		
	Net section rupture	1161	1156	1.00		
B4	Fillet welds shear	1053	1113	0.95	0.035	0.033
C1	Bolt group shear	905	846	1.07	0.037	0.031
C2	Bolt group shear	1297	1128	1.15	0.037	0.023
C3	Not observed	>1543	1410	>1.09	0.027	0.023[#]
C4	Not observed	>1552	1692	>0.92	0.034	0.030[#]

[*] Connection B3 experienced three failure modes.

[#] Connection did not fail during test. The rotation is the value corresponding to the maximum shear load.

[ᴲ] The shear load for connections A3, A4, C3 and C4 is the maximum load reached during test.

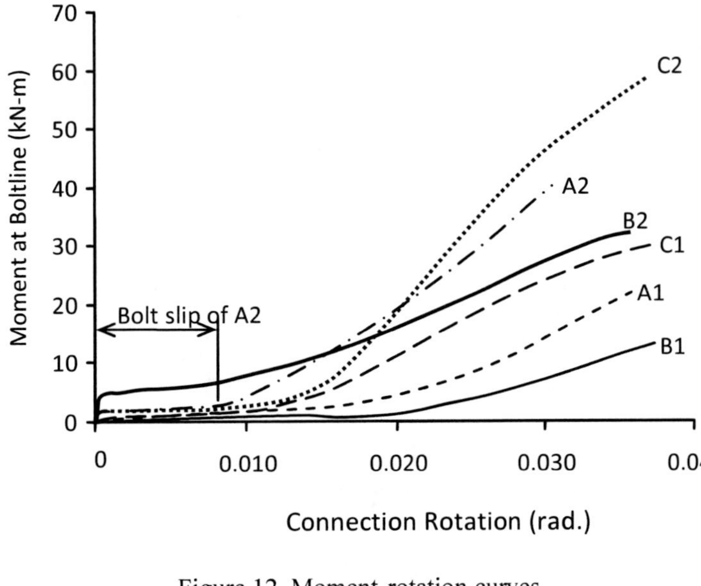

Figure 12. Moment-rotation curves.

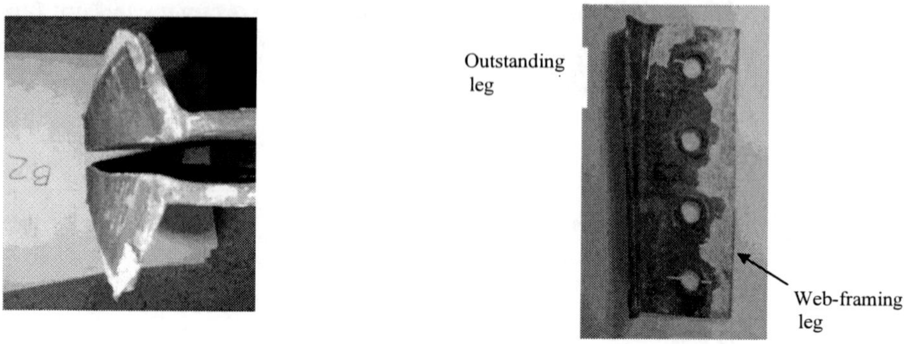

(a) Top view (b) side view

Figure 13. Deformations of angle legs after test for specimen B2.

The most common mode of failure was rupture of connectors under shear force. Specimens A1, A2, B1, B2 and B4 failed by the rupture of fillet welds with the failure surfaces in the weld metal. Specimens C1 and C2 failed by the rupture of bolt group. Figure 13 shows the connection angles of specimen B2 after test. Permanent shear deformation and local bearing deformation around bolt holes can be clearly seen in Figure 13b. The large distortional deformation of the outstanding legs is evident in Figure 13a. Specimen B3 was the only connection to experience multiple failure modes. When B3 approached the end of stage two loading, a large yield flow was detected by LP5, i.e., the vertical shear deflection kept growing while the shear load was increasing slowly. However, to reach a failure mode of rupture, the loading was continued until the top portion of the fillet welds was seen to crack and pulled away from the HSS column (failure surface was in the weld metal), then immediately, the angles were seen to fracture from the second bolt from the top to the angle bottom along the inner edge of bolt holes. Upon inspecting recorded data, the connection angles were deemed to become fully plastic at a shear load of 1138 kN, and to fracture at the peak load of 1161 kN (Table 2).

The vertical deflection of connections during stage two loading caused the applied connection rotation to decrease. This explains why the connection rotation at the instance of failure, recorded in the last column of Table 2, is smaller than the corresponding maximum rotation experienced in stage one loading.

Analysis of Strength

The formulae of the nominal shear strength V_r of the connections corresponding to various failure modes are given as, respectively [9]:

$$V_r = 0.67 A_w X_u \tag{38}$$

$$V_r = 0.6 n m A_b F_{ub} \tag{39}$$

$$V_r = 0.6A_{nv}F_u \tag{40}$$

$$V_r = 0.6A_{gv}F_y \tag{41}$$

where Eq. (38) computes the shear rupture strength of weld group with rupture surface in weld metal; Eq. (39) the shear rupture strength of bolt group; Eq. (40) the shear rupture strength of connection angles along the net section of boltline; and Eq. (41) the shear yielding strength of angle gross section. The symbols in these equations are A_w=effective throat area of weld; X_u=ultimate strength of electrode, taken as 490 MPa for the test; n=number of bolts; m=number of shear planes, which was 2 for the connections; A_b=cross-sectional area of one bolt, taken as 285 mm^2; F_{ub}=tensile strength of A325 high-strength bolt, which was 825 MPa; A_{nv}=net area in shear; F_u=tensile strength of angle steel, which was 537 MPa and 514 MPa for 9.5 mm and 6.4 mm thick angles, respectively; A_{gv}=gross area in shear; and F_y=yielding strength of angle steel, which was 396 MPa and 364 MPa for 9.5 mm and 6.4 mm thick angles, respectively.

The minimum value of the four nominal strengths was taken as the predicted strength. The ratio of the test strength to the predicted strength is given in the 4th column of Table 2. The observed failure of each broken connection was consistent with the predicted failure mode. The average strength ratio of fillet welds was 0.99 for specimens A1, A2, B1, B2, B3 and B4, with a range of 0.91 to 1.09 and a standard deviation of 0.08, which are consistent with Lesik and Kennedy [15]. For specimens C1 and C2, the test strengths of the bolt groups are greater than predicted. For specimens A3, A4, C3 and C4, no failure was observed during the test due to the capacity of jack 1 was reached. For specimen C4, the maximum applied load was only 8% less than the predicted strength; Since no obvious sign of pending failure was observed, it is reasonable to assume that C4 specimen has shear strength at least as high as predicted. For specimen B3, its nominal weld shear strength and net section rupture strength are essentially equal, while its nominal yielding strength of gross sections is slightly smaller. During test, it was observed that the angles underwent large inelastic deformation before the weld shear and net section rupture occurred simultaneously.

Analysis of Rotational Capacity

Efforts were made hereafter to quantify the major sources of rotational capacity for the test specimens. Three major sources, i.e., bolt slip, hole bearing deformation, and distortion of outstanding legs, which are all available for the test connections regardless of the type of failure mode, are analyzed and discussed. It is believed that such analyses provide insights into the nature of connection rotation. Note that the quantifications of connection rotation are restricted to stage one rotational loading only.

Connection Rotation due to Bolt Slip

Since a bolt hole is larger than the bolt diameter, the slip of bolts is unavoidable. In practice, bolt slip is included in the rotational capacity of connections. The amount of bolt slip is dependent on the hole clearance and the relative geometrical location of holes. The maximum possible slip, s_b, of a bolt in its hole is written as

$$s_b = (d_{hc} - d) + (d_{hw} - d) \tag{42}$$

where d_{hc} is the diameter of holes on connection angles, and d_{hw} is the diameter of holes on the web of beam. For this study, d_{hc}=d_{hw}=20.6 mm, d=19 mm, and s_b=3.1 mm. Therefore, the maximum possible connection rotation, θ_s, due to the slip of bolts can be computed as

$$\theta_s = 2s_b / L_b \tag{43}$$

where L_b is the distance between centers of the top and bottom bolts. For the specimens, the computed θ_s value is given in Table 3. Equation (43) predicts that the connection rotation from bolt slip decreases as the depth of the connection increases. Since the bolts and their holes cannot all be expected to be in their most favorable locations for maximizing rotation due to slip, the amount of connection rotation that actually takes place due to bolt slip is usually much less than that computed from Eq. (43).

The moment-rotation curves for stage one (Figure 12) were used to quantify the connection rotation due to bolt slip. The flat segment on the curve in the early loading was due to the bolt slip. The small moment during the bolt slip was caused by the friction force between the angles and the beam web. The slip rotation is determined (see an example for specimen A2 in Figure 12) and recorded in Table 3, where one can see that the average test rotation due to bolt slip is approximately 40% of the maximum possible bolt-slip rotation. In laboratory specimens, this percentage of slip is consistent with the data reported by Kulak [16].

Table 3. Connection rotation due to bolt slip

Specimen ID	Maximum possible rotation due to bolt slip θ_s (rad)	Actual rotation due to bolt slip (rad.)	Ratio of the actual rotation to the maximum possible rotation
A1	0.040	0.015	0.38
A2	0.027	0.008	0.30
A3	0.020	0.009	0.45
A4	0.016	0.009	0.56
B1	0.040	0.017	0.43
B2	0.027	0.007	0.26
B3	0.020	0.009	0.45
B4	0.016	0.006	0.38
C1	0.040	0.010	0.25
C2	0.027	0.012	0.44
C3	0.020	0.009	0.45
C4	0.016	0.008	0.50
Average			0.40

Connection Rotation due to Bearing Deformation of Bolt Holes

Bolt holes are subjected to bearing stresses, which cause local bearing deformation in the connected materials. For the test specimens, the history of bearing deformation of the bolt holes on web-framing leg has the following characteristics: In stage one the connection was subjected to an increasing negative moment. The upper bolts pulled away from the angles while the lower bolts were pushed into the angles. Bearing deformation of holes in this stage was dominated by horizontal local plastification, while vertical bearing deformation was negligible. In stage two the test beam became a simply-supported beam. While the shear load increased at the connection, a positive bending moment was superimposed at the boltline section, resulting in the upper bolts being pushed into the angles and lower bolts being pulled away from the angles. Therefore, the actual bearing deformation in stage two was the resultant of vertical bearing due to shear load and horizontal bearing due to superimposed moment.

The above characteristic of bearing deformation is demonstrated using a connection angle from specimen B2. Figure 14 shows a close-up view of the top hole of the B2 angle. The bolt in the figure was placed to demonstrate the movement of the bolt relative to the hole during the loading process. Figure 14(a) illustrates the position of the bolt at the end of stage one loading (the bolt pulled away from the left and was bearing against the right-hand side of the hole), while Figure 14(b) shows the position of the bolt at the end of stage two loading. The combined effect of vertical bearing and horizontal pushing (the superimposed positive moment pushed the bolt into the left) in stage two is evident for the bolt.

(a)

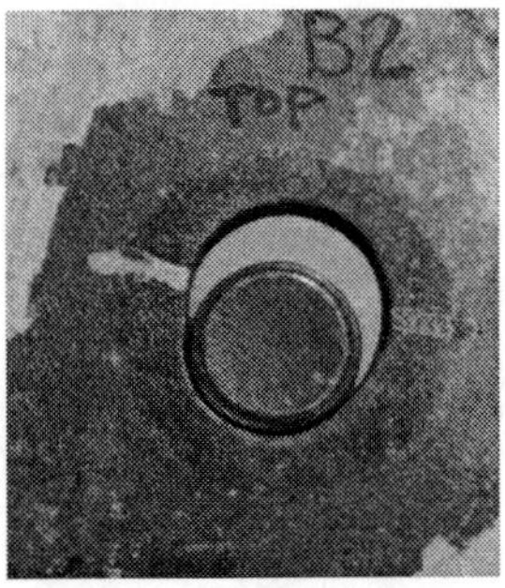

(b)

Figure 14. Top bolt position during testing (a) at the end of rotational loading (b) at the end of shear loading.

The diameters of the bolts and holes were 19.05 mm and 20.6 mm, respectively. After bearing deformations took place, the clearance between the bolt face and hole edge became greater than the fabrication hole clearance of 1.6 mm, which directly contributed to the rotation of the connection. Through measuring the enlarged hole diameter in the horizontal direction, the connection rotation resulting from the bearing deformation can be quantified.

Table 4 records the enlarged horizontal dimension of the top and bottom bolt holes on the connection angles after testing. The connection rotation resulting from bearing deformation is then calculated as

$$\theta_{hb} = \frac{(d_{tb} - d_{hc}) + (d_{bb} - d_{hc})}{L_b} \tag{44}$$

where the hole diameter before testing is d_{hc}=20.6 mm.

Table 4. Connection rotation due to bearing deformation

Specimen ID	Horizontal dimension of top hole (mm)	Horizontal dimension of bottom hole (mm)	Rotation due to bearing deformation (rad.)
A1	21.01	20.92	0.005
A2	21.07	21.02	0.004
A3	21.30	21.55	0.005
A4	21.73	21.77	0.006
B1	21.16	20.80	0.005
B2	21.23	21.48	0.007
B3	Not available	Not available	Not available
B4	21.73	21.31	0.005
C1	21.16	21.20	0.008
C2	21.06	20.98	0.004
C3	21.20	21.12	0.004
C4	21.80	21.57	0.006

Note: Measurements for B3 were not available because the net section fractured through bolt holes.

Equation (44) does not account for the contribution from bearing deformation of the holes on the test beam web. Also, elastic bearing deformation is ignored since the measured hole size is based on the permanent deformation only. Since d_{hc} is taken as the nominal diameter of bolt holes, Eq. (44) is accurate only when both the top and bottom bolts are in full contact with hole edges before any hole yielding occurs. In other words, any clearance between the bolts and the holes, which result from the slipping and bearing of the bolt group within the connection, is ignored in Eq. (44). Hence, Eq. (44) gives the lower bound of the rotational contribution from hole bearing deformation.

The connection rotation resulting from bearing deformation is given in the fourth column in Table 4. A larger rotational contribution from bearing deformation would be expected if the enlargement of bolt holes on the web of the supported beam is also taken into account. This would be the case for an actual beam having a thin web.

Connection Rotation due to Distortion of Outstanding Legs

Distortion of the outstanding legs at the completion of test can be seen clearly from the photo of the connection angles (Figure 13). This deformation is caused by the bending moment applied to the connection in the plane of the beam web. The distortion of outstanding legs results in the top of angles moving away from the supporting column. The longer the angle length or the greater the number of bolts in the connection, the more severe the distortional demand imposed on the outstanding legs of connection angles, since the rotational contributions from the bolt slip decreases.

Discussions on Rotational Capacity

In addition to the three previously discussed sources of connection rotational capacity, other sources such as deformation of bolts, web-framing legs and welds (Lesik and Kennedy [15]) also contribute to the rotational capacity. In particular, the plastic shear deformation of bolts may contribute a rotational capacity comparable to that of hole bearing deformation if bolt group shear governs the connection shear strength. However, bolt shear deformation is not singled out in the above analysis since stage one rotational loading did not involve strength failure plus bolt shear deformation was not available for the connections which did not fail by bolt shear.

The following observations are made regarding the rotational capacity of the test specimens.

1) Bolt slip made a significant contribution to the rotational capacity. For 3-bolt connections the bolt slip accounted for about 40% of the maximum connection rotation. For 4- and 5-bolt connections, the bolt slip contribution reduced to about 26% of the maximum rotation in average, while for the deeper 6-bolt connections bolt slip contribution was approximately 22% of the maximum rotation.
2) Bearing deformation of bolt holes also made a significant contribution to connection rotation. For the test specimens, the rotation from this source accounted for about 10% to 21% of the maximum rotation. If bearing deformation within the holes on a thin web of actual beams is also included, this source can make an even greater contribution. For a deeper connection, such as specimen A4 or C4, the rotation contributed by the bearing deformation is similar in magnitude to that attributed to bolt slip.
3) Despite that bolt slip and bearing deformation made considerable contributions to the maximum connection rotation, the distortion deformation of the outstanding legs still made the largest contribution to the rotational capacity. This was especially true for deeper connections.
4) In stage one loading, all the specimens except C3 underwent a connection rotation at least equal to 0.034 radians. While under large shear load in stage two loading, most of specimens had an imposed rotation greater than 0.03 radians, with the minimum value equal to 0.023 radians (Table 2). Such a level of connection rotation is generally large enough for a beam to reach full yielding at its maximum moment section.

Summary of the Test of Double-Angle Shear Connections

All the tested connections were deemed satisfactory as they were able to undergo the required rotation and were brought to their theoretical shear failure strengths without any premature failure. Therefore, the practice of either shortening the outstanding legs or using flare bevel groove welds are both acceptable for double-angle shear connections with (1) outstanding legs not shorter than 30 mm; (2) angle thickness from 6.4 mm to 9.5 mm; (3) the number of bolts not greater than 6, and (4) twice weld size top return for fillet welds and no top return for flare bevel groove weld.

Single-Angle All-Bolted Shear Connections

An all-bolted single-angle shear connection has both legs bolted to the connected members. The general practice is to use an equal leg angle having a leg width of 102 mm. The gauge on both legs is 65 mm, which provides adequate clearances for inserting and tightening the bolts. The practice in USA [12] also permits the use of unequal leg angles if both legs have a different number of bolt lines. The required angle thickness is 9.5 mm for 19 and 22 mm diameter bolts. For current practice, when computing the load carrying capacity of an all-bolted single-angle connection, the effect of eccentricity should be considered for the bolt group at the outstanding leg; however, this effect is not considered for the bolt group at the web-framing leg as long as only one vertical row of bolts is used [12].

The difference in considering eccentricity effects results in different computed resistances for the bolt groups at the outstanding and web-framing legs; for example, if the number of bolts at both legs is identical, the bolt group at the outstanding leg has a smaller computed resistance due to its nonzero eccentricity. Note that compared with Lipson's welded-bolted single-angle connections [18], all-bolted single-angle connections possess a much smaller rotational stiffness due to the flexure of the outstanding leg. The practice of ignoring the eccentricity effect for the bolts at the web-framing leg might overestimate the load carrying capacity for all-bolted connections. Thus, an experimental verification of the practice is required.

On the other hand, when using the instantaneous centre of rotation method [19] to compute the shear strength of the bolts at an outstanding leg, it is assumed that the outstanding leg is able to undergo rotation in its own plane without restraint in order to reach the ultimate limit state of the bolt group. However, in reality, the floor structure system often provides certain torsional restraints to the supported beam, preventing the beam and its end connections from free-twisting. For example, a composite steel beam is prevented from undergoing lateral torsion due to the restraints from the floor slab through shear connectors. It is believed that this type of lateral or torsional restraint to the supported beam would impact the ultimate resistance of its single-angle shear connections, and that an experimental investigation is desirable.

This section describes the testing of 19 full-scale single-angle shear connections for the purpose of addressing the aforementioned concerns. Through analysis of the test results, some suggestions are proposed to improve the current practice for the design of single-angle all-bolted simple connections.

Test Program

Each connection specimen consisted of an equal leg angle $L102\times102\times9.5$ (leg width 102 mm and thickness 9.5 mm) and several 19 mm diameter ASTM A325 high-strength bolts. All the specimens had only one vertical row of bolts at each leg with the number of bolts at one leg from 2 to 8 inclusive. The vertical distance between bolt holes was 76 mm and the vertical end distance of boltline was 32 mm. The gauge, or distance from the angle heel to both boltlines, was 65 mm (Figure 15). The bolt holes on the angles were 20.6 mm diameter standard holes. The length of each specimen is given in Table 5. The angle material was grade 300W steel of Canadian standard CSA-G40.21. The actual average yield and tensile strengths were F_y=445 MPa and F_u=565 MPa, respectively.

The specimens were divided into group N and group X (Table 5). For each specimen in group N, all the bolts were intercepted by the shear plane at thread, while for each specimen in group X, the bolts at outstanding leg were intercepted by the shear plane at shank but the bolts at web-framing leg were intercepted at thread. In Table 5, the first letter of specimen identification (ID) represents group name. The digit following the group letter is the number of bolts at one leg. The last letter of an ID identifies the individual specimen.

The 19 mm diameter ASTM A325 bolt has a specified design shear resistance of 99 kN when the shear plane intercepts its threads according to Canadian standard [13]; however, it was desirable to know the actual shear resistance of a single bolt for the purpose of analysis. To this end, three single bolt coupons were tested in double shears using a compression jig, which gave the shear resistance for one shear plane as R_1=138 kN.

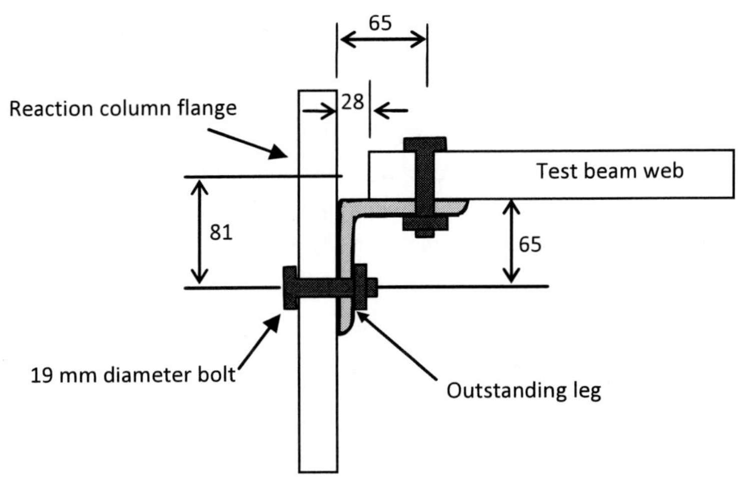

Figure 15. Geometry of connection angles.

The same test set-up, as shown in Figure 11, was adopted for the test program. The I-shaped test beam had a 32 mm thick web plate with 20.6 mm diameter drilled holes. The overall depth of the beam was 824 mm and the width of the flanges was 200 mm. The brace for beam was made of two wide flange sections which allowed the test beam to be placed between them. The clear distance between the two brace sections was 206 mm, which was 6 mm wider than the flanges of the test beam to facilitate assembling. Figure 16 shows a picture of the setup taken from the far end of the test beam. In addition to showing jacks 2 and 3, this picture illustrates how the brace provided lateral supports to the beam during test; i.e., after some beam rotation occurred, the beam flanges would bear against the two brace sections which then prevented the beam from further twisting.

Figure 16. Picture of test setup.

The geometrical relationship between the connection angle and the test beam is shown in Figure 15. The nominal load eccentricity for the bolts at the outstanding leg was 81 mm, which was the distance from the centerline of the bolt holes to the centerline of the beam web. The setback of the test beam was 28 mm, which was larger than common practice of 13 mm, in order to avoid the bottom flange of the beam bearing onto the reaction column during rotational loading. A complete test process was divided into two stages. For stage one, a rotation over 0.03 radians was applied to the specimen. During stage two, the load of jack 1 was increased monotonically until the connection failed while jacks 2 and 3 held the far end of the beam.

Test Results of Shear Strength

All the connections were loaded to the shear rupture of bolts in stage two. The failure mode and the corresponding failure shear load V_a for each connection were recorded in the third and fourth columns of Table 5, respectively. For group N, all the connections failed by shear rupture of the bolt group at the outstanding leg except specimen N2C, which failed by the rupture of the bolt group at web-framing leg. For group X, all the connections failed by shear rupture of the bolts at the web-framing leg. Table 5 also gives the connection rotation and the gross shear stress of the connection angle at the instance of failure.

Table 5. Test specimens and results of all-bolted single-angle connections

Specimen ID#	Angle length L (mm)	The leg at which shear rupture of bolts occurred	Shear load at failure V_a (kN)	Gross shear stress of angle at failure* (MPa)	Maximum rotation (rad.)	Rotation at failure (rad.)
N2C	140	web-framing leg	211	159	0.046	0.035
N2D	140	outstanding leg	192	144	0.046	0.034
N3C	216	outstanding leg	350	171	0.042	0.033
N3D	216	outstanding leg	349	170	0.044	0.035

Table 5. Continued

Specimen ID[#]	Angle length L (mm)	The leg at which shear rupture of bolts occurred	Shear load at failure V_a (kN)	Gross shear stress of angle at failure* (MPa)	Maximum rotation (rad.)	Rotation at failure (rad.)
N4B	292	outstanding leg	494	178	0.043	0.034
N4C	292	outstanding leg	510	184	0.044	0.036
N5A	368	outstanding leg	624	178	0.044	0.038
N6A	444	outstanding leg	738	175	0.036	0.030
N7A	520	outstanding leg	814	165	0.032	0.026
N8A	596	outstanding leg	940	166	0.035	0.030
X2A	140	web-framing leg	212	159	0.045	0.032
X2B	140	web-framing leg	221	166	0.054	0.041
X3A	216	web-framing leg	373	182	0.044	0.036
X3B	216	web-framing leg	371	181	0.044	0.034
X4A	292	web-framing leg	528	190	0.044	0.036
X5B	368	web-framing leg	669	191	0.041	0.033
X6B	444	web-framing leg	838	198	0.036	0.028
X7B	520	web-framing leg	950	192	0.035	0.028
X8B	596	web-framing leg	1090	193	0.037	0.030

[#] The digit in ID gives the number of bolts at one leg.
*The gross section shear stress is equal to V_a divided by the gross sectional area (length L times thickness t).

Table 6. Analysis of group N connections

Specimen ID	AISC C value	Predicted shear resistance R_{out} (kN)	Ratio of test strength to predicted strength V_a/R_{out}
N2C	0.84	116	>1.82
N2D	0.84	116	1.66
N3C	1.68	232	1.51
N3D	1.68	232	1.50
N4B	2.73	376	1.31
N4C	2.73	376	1.36
N5A	3.81	526	1.19
N6A	4.88	673	1.10
N7A	5.97	824	0.99
N8A	7.03	970	0.97

Analysis of Bolt Group at Outstanding Leg

As illustrated in Figure 15, the load eccentricity for the bolt group at the outstanding leg is taken as 81 mm. The ultimate shear strength of the bolt group shall be calculated using the instantaneous centre of rotation method, which is

$$R=CR_1 \tag{45}$$

where C is the strength coefficient, a function of eccentricity, bolt pitch and the number of bolts; R_1 is the shear strength of a single bolt, and R_1=138 kN from single bolt test. The value of the C coefficient is given in Table 6 for each connection in group N. Then, for each connection, the predicted ultimate resistance R_{out} using Eq. (45) and the ratio of test strength V_a to the predicted strength R_{out} is given in the third and fourth columns in Table 6, respectively. The strength ratios reveal that the shallower a connection was the greater the discrepancy of the predicted strength for the connection. This discrepancy was attributed to the fact that the brace provided torsional support to the test beam. Since the spacing between the two brace sections was 6 mm wider than the beam flange, the beam should have undergone an unrestrained rotation of 0.0073 radians (=6mm/823mm) before its flanges came to bear against the brace. During testing the eccentrically loaded bolts at the outstanding leg moved within their bolt holes due to slip and hole bearing deformations. The maximum twisting of the test beam caused by the slip of bolts at the outstanding leg is estimated to be

$$\theta_s = 2(d_h - d)/L_b \tag{46}$$

where hole diameter d_h was 20.6 mm; bolt diameter d was 19 mm; the distance between the two extreme bolts was L_b=76(n – 1) mm for a n-bolt connection. For examples, θ_s was 0.042 and 0.0084 radians for 2-bolt and 6-bolt connections, respectively. Therefore, for the connections having less than six bolts, the beam flanges would bear against the brace before bolt slip will take place completely. These lateral support forces applied to the beam flanges from the brace effectively reduced the eccentricity of the bolt group at the outstanding leg. However, for 7-bolt and 8-bolt connections, θ_s was 0.007 and 0.006 radians, respectively, which is smaller than the unrestrained rotation 0.0073 radians. Hence, the lateral support from the brace became unnecessary and the instantaneous centre of rotation method gave an excellent estimation of the ultimate strength.

The test confirmed that lateral or torsional restraints to the supported beam had a significant influence on the load carrying capacity of its connections. If torsion of a connection could be completely prevented, then load eccentricity would be totally eliminated. In this case, a connection with more bolts at the outstanding leg becomes unnecessary.

For a floor structure system that is able to provide partial restraints to the supported beam, the strength of the bolts at the outstanding leg can be computed by modifying Eq. (45) as

$$R=\beta C R_1 \text{ but } R \leq nR_1 \tag{47}$$

where β is the coefficient considering lateral support effect for the beam and n is the number of bolts at the outstanding leg. For the tested connections, the value of β, through regression analysis, is formulated as

$$\beta = V_a / R_{out} = 2.4n^{-0.43} \quad (n \leq 6) \tag{48}$$

The application of Eq. (47) requires some caution. The primary concern is whether explicit lateral or torsional restraints are available. Generally speaking, lateral supports for beams often exist. It appears the β value from Eq. (48) can be safely used for the cases where the supported beam is well restrained such as the beams in a composite construction. For connection N2C, the lateral support effect was so great that the strength of the bolts at the outstanding leg became greater than that of the bolts at the web-framing leg. By contrast, the rest of the specimens in group N failed by the rupture of the bolt group at the outstanding leg due to its larger load eccentricity.

Analysis of Bolt Group at Web-Framing Leg

Group X connections were specially designed to investigate the behaviors of the bolt group at the framing leg, as all the connections in this group failed by the shear rupture of the bolts at this leg. Without considering load eccentricity, as suggested by the current design practice, the predicted ultimate shear resistance is R_{web} ($R_{web}=nR_1$), and its ratio with test strength is given in Table 7. Obviously, the connections having more than 4 bolts had a predicted strength near the test strength. The connections having 2 and 3 bolts had a test strength significantly smaller than the predicted. This observation indicates that load eccentricity of the bolts at the web-framing leg should not be ignored when designing shallow connections (herein 2 and 3 bolts).

Table 7. Analysis of group X connections

Specimen ID	Predicted strength without eccentricity R_{web} (kN)	Ratio of V_a/R_{web}	Predicted strength for eccentricity of 32 mm R_{wr} (kN)	Ratio of V_a/R_{wr}
X2A	276	0.77	207	1.02
X2B	276	0.80	207	1.07
X3A	414	0.90	358	1.04
X3B	414	0.90	358	1.04
X4A	552	0.96	504	1.05
X5B	690	0.97	646	1.04
X6B	828	1.01	786	1.07
X7B	966	0.98	925	1.03
X8B	1104	0.99	1063	1.03

For simplicity, an eccentricity that is equal to half the distance from the boltline to column face is adopted to predict the strength of the bolts at the web-framing leg. For the tested connections, the eccentricity is 32 mm. The corresponding strength R_{wr} of each connection is given in Table 7. A comparison with test results indicates that such an eccentricity yields an excellent predicted strength for all the connections with consistent margin of safety.

The connection moment computed using Eq. (29) can be employed to estimate eccentricity for the bolts at the web-framing leg. For example, for the 3-bolt connection X3B subjected to the ultimate shear load V_a=371 kN with an angle length L=216 mm and thickness t=9.5 mm. Herein b_0=69 mm and F_y=445 MPa, the connection moment M_p=14.3 kN-m at the limit state. The moment inflection point of the beam is (14.3 kN-m/371 kN)=0.0385 m away from the column face (i.e., the point of zero moment is between the boltline and column face). Therefore, the eccentricity of the bolts at the web-framing leg is equal to (65 – 39)=26 mm, which gives C=2.68 and a predicted shear strength of 369 kN, which is virtually equal to the actual strength 371 kN.

For those connections beyond the scope of the tested specimens, the above example demonstrates a theoretical method for calculating the strength of the bolts at the framing leg.

Summary of All-Bolted Single-Angle Connections

The ultimate strength of the bolts at the outstanding leg is significantly affected by the torsional or lateral restraints of the supported beam. The test confirmed that the lateral support reduced load eccentricity and thus increased the strength of the bolt group at the outstanding leg. The empirical Eq. 48 was proposed for computing the extent of this strength increase. It was also proven that the tension force in the bolts at the outstanding leg, caused by the connection moment, can be ignored when calculating the ultimate shear strength of the bolt group.

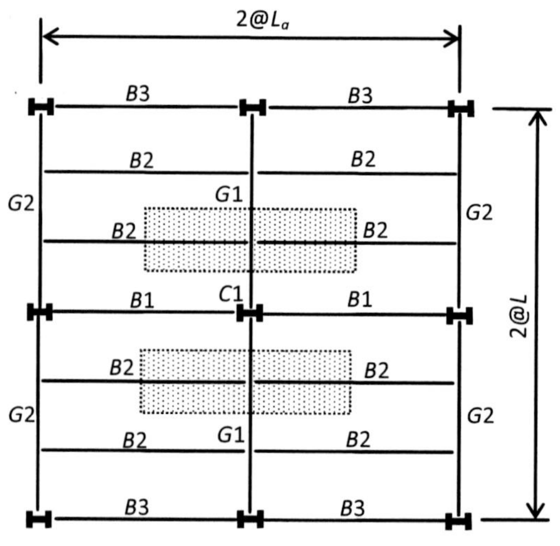

Figure 17. Typical floor framing plan.

The strength of the bolts at the framing leg can be determined properly without considering load eccentricity for the connections having more than 4 bolts; however, for shallow connections with only 2 or 3 bolts, consideration of load eccentricity is required. The recommended load eccentricity for the tested shallow connections is taken as a half of the bolt gauge. Alternatively, this recommended eccentricity can be applied for all of the connections since it yields an accurate yet conservative shear strength for all of the connections with consistent margin of safety. Also, the moment analysis of the connection provides a direct means for computing the eccentricity of those single-angle connections beyond the scope of this test program.

Resilience Analysis of Shear Connections

Traditionally, simple connections are designed for shear force only while it was long been recognized that these shear connections can resist certain amount of tensile force in the longitudinal direction of the supported beam. The tensile force resistance allows the development of a horizontal tying force (called catenary action) which helps to preserve the integrity of the structures. In Canada, the steel structure standard CSA-S16-01 [9] explicitly stated that connections shall be designed to provide resistance to progressive collapse as a consequence of a local failure. However, little specific design requirement was given in the standard. Instead, the standard stated that "the requirements of this standard generally provide a satisfactory level of structural integrity for steel structures".

The British steel design code BSI 2000 [20] explicitly requires that connections, regardless of height or span length, must be capable of transferring a minimum typing force of 75 kN at floor level. The experimental test of various shear connections by Guravich and Dawe [21] demonstrated that this tension force requirement could be easily met by a shear connection designed based on Canadian standard.

However, when assessing the integrity of a steel structure against progressive collapse, the scenario of a sudden column loss may impose a much greater demand on the strength and ductility of its shear connections than that was traditionally regarded acceptable (such as 75 kN typing force by British code), since the girders or beams need to structurally span two bays under sustained gravity loads (i.e., the so-called double span condition). According to *the guideline* for a p*rogressive collapse analysis and design by GSA [22],* the resilience of a connection refers to its ability to withstand destructive loading conditions that accompany a column removal (caused by accidental events such as explosion, impact and fire) without premature rupture. Providing connection resilience is considered essential to mitigating the progressive collapse in steel frameworks.

This section proposes an algorithm for evaluating the demands on steel connections under a double span condition. The effectiveness of the method is verified by using an existing example. The assumptions for the method include: 1) The supporting member of the supported beam/girder is able to provide support for catenary action of the beam; 2) the removal of a vertical element consists of the removal of member itself, and this removal should not impede into the connection or horizontal elements that are attached to the vertical element at the floor level; and 3) this study focus on girder or beam connections under a double span condition. The membrane and catenary action of floor slab is not directly considered, though the floor slab is assumed to be able to carry part of the sustaining gravity loads.

The double span condition created by the loss of an interior column is illustrated in Figures 17 and 18. The typical floor framing plan (Figure 17), which is employed as the basis for computing the gravity loads for the considered structural members, is adopted from Foley et al. [23]. In Figure 17, the interior column C1 supports girders G1 and beams B1. The connections of concern are the simple connections at the ends of girder G1 or beam B1. During the loss of interior column C1 (Figure 18), girders G1 and beams B1 need to sustain point-in-time gravity loading of the floor without collapse.

In the event of the ineffective interior column, the free body diagram of the girders G1 is drawn in Figure 19, where the connections are modeled as a dimensionless spring possessing rotational and longitudinal stiffness. From the equilibrium of the forces in the vertical direction for the two span condition (Figure 19a), we have

$$P = 2T\sin\theta + 2Y \tag{49}$$

where P is the point-in-time concentrated gravity load; T is the catenary tension force; and Y is the vertical shear force at the end of the girder. From the equilibrium of the moment for the single span girder in Figure 19b, we have

$$Y = (M_1 + M_2)/L \tag{50}$$

where M_1 and M_2 are the moments at connection 1 (on the lost column) and connection 2, respectively (M_1 is positive while M_2 is negative); and L is the nominal length of the girder. Note that in Eqs. (49) and (50), it is assumed that the end connections have zero length in the longitudinal direction of the girder. The girder is also assumed to remain straight during catenary action (i.e, the curvature of the girder under gravity loading is ignored). Thus, both connections 1 and 2 have identical rotation θ, which is equal to δ/L (δ is the vertical deflection of the girders at the location of the lost column, as depicted in Figure 18).

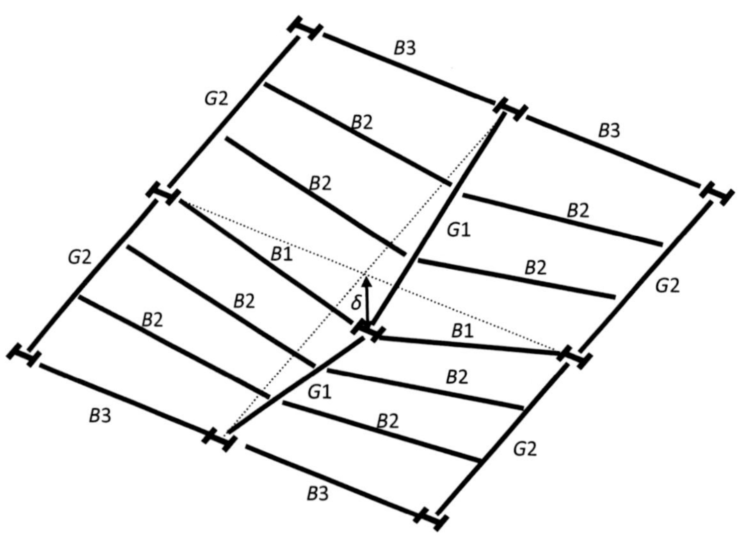

Figure 18. Double-span condition due to the loss of an interior column.

Since the connection moment is a function of connection rotation θ, we have

$$M_i = M_i(\theta) \quad (i = 1, 2) \tag{51}$$

Substituting Eq. (50) into Eq. (49) gives

$$P = 2T\sin\theta + 2(M_1 + M_2)/L \tag{52}$$

Because $\sin\theta \approx \theta$ when $0 \le \theta \le 0.4$ radians, Eq. (52) can be rewritten as

$$T = [P - 2(M_1 + M_2)/L]/2\theta \tag{53}$$

From Figure 19b, we have

$$\theta = \delta/L = \sqrt{L_1^2 - L^2}/L = \sqrt{(L_1/L)^2 - 1} \tag{54}$$

where L_1 is the length of the girder under catenary tension force T, including the elongation of the girder and tensile deformation of the end connections, i.e.,

$$L_1 = L + \frac{TL}{EA} + \Delta_1 + \Delta_2 \tag{55}$$

where TL/EA is the elongation of girder and Δ_i is the axial tensile deformation of connection i. The axial deformation of a connection is a function of tensile force T and is found as

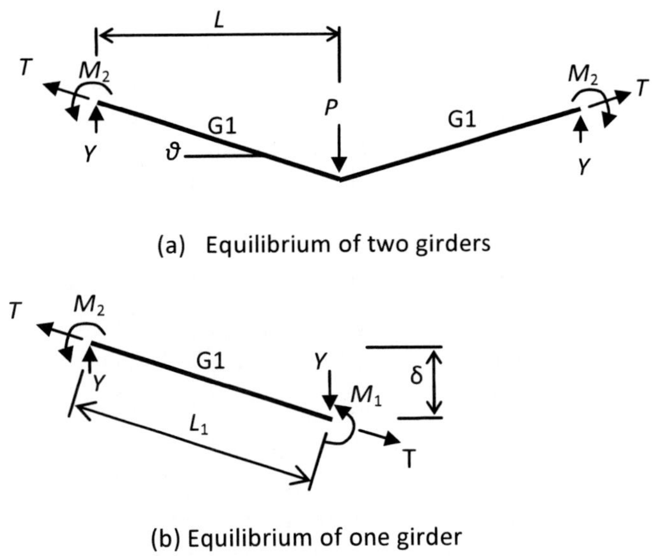

Figure 19. Equilibrium of double-span condition.

$$\Delta_i = \Delta_i(T) \quad (i = 1, 2) \tag{56}$$

Then, from Eq. (55) we have

$$\left(\frac{L_1}{L}\right)^2 = \left[1 + \left(\frac{T}{EA} + \frac{\Delta_1 + \Delta_2}{L}\right)\right]^2 \tag{57}$$

Substituting Eq. (57) into Eq. (54) gives

$$\theta = \sqrt{2\left(\frac{T}{EA}+\frac{\Delta_1+\Delta_2}{L}\right)+\left(\frac{T}{EA}+\frac{\Delta_1+\Delta_2}{L}\right)^2} \tag{58}$$

Equations (51) and (56), which should be determined based on experimental tests, represent the rotational and axial behaviors of the shear connections. The analysis procedure should account for the ultimate capacities of the connection rotation and axial deformation.

An iterative procedure for evaluating the demands on shear connections is described as follows:

1) Prepare data such as *P, L, E, A* and connection properties Eqs. (51) and (56); set iteration index j=1.
2) Assign an initial value to θ; the θ value at this moment is denoted as θ_{j1}.
3) Find M_1 and M_2 for connections 1 and 2 corresponding to the θ_{j1} value through Eq. (51). If θ_{j1} value exceeds the ultimate rotational capacity of a connection, the corresponding M is then taken as zero.
4) Compute catenary tension force T from Eq. (53).
5) Determine the axial deformations Δ_1 and Δ_2 through Eq. (56). If T from step 4 exceeds T_{ui}, the maximum tensile resistance of connection i, the corresponding Δ_i would be taken as Δ_{ui}, the ultimate axial deformation of connection i.
6) Compute the connection rotation θ from Eq. (58), where T was obtained from step 4 whether T exceed T_u or not; denote the new θ value as θ_{j2} .
7) Compute ratio $r_j = \theta_{j2}/\theta_{j1}$. If $0.99 \le r_j \le 1.00$, then θ value has converged and go to step 9; Otherwise, go to step 8.
8) Determine the θ value for the next iteration using the following strategy:
   i) For j = 1, set $\theta_{(j+1)1} = (\theta_{j1}+\theta_{j2})/2$, and go to step 3.
   ii) For $j \ge 2$, if $r_j > 1$ and $r_{j-1} < 1$ or $r_j < 1$ and $r_{j-1} > 1$, then $\theta_{(j+1)1} = (\theta_{(j-1)2}+\theta_{j2})/2$; else, $\theta_{(j+1)1} = (\theta_{j1}+\theta_{j2})/2$. Then, go to step 3.
9) Output analysis results θ, T, M_1, M_2, Y, Δ_1 and Δ_2.

In the above algorithm, the point-in-time load P is calculated using tributary area of girders G1 corresponding to the connections at the interior column C1 (the shaded areas in Figure 17). Note that Figure 19 and Eqs. (49) through (58) are also equally applicable to two beams B1. In this case, load P should be computed based on the tributary area of beam B1. Provided that support for catenary action is available, the proposed algorithm can be easily extended to the analysis of girder G2 under the removal of an edge column.

The algorithm is illustrated and verified using Foley's example [23] as follows. The floor dimensions are L=L_a=9.15 m. The specified dead load is 3.83 kPa, and the specified occupancy live load is 2.4 kPa. The girder G1 section is W530×101 with a cross-sectional area A=12900 mm^2, moment of inertia I=617×10^6 mm^4, plastic modulus Z=2620×10^3 mm^3, and Young's modulus E=200,000 MPa. The yield strength of girder steel is F_y=345MPa. All the four connections at the ends of girders G1 are assumed to have the identical connection behaviors. The moment-rotation curve is assumed to be elastic-perfectly plastic, as shown in Figure 20a, where the ultimate moment of the connections is assumed to be $\eta_M(ZF_y)$ (i.e., the product of coefficient η_M and the plastic moment of girder). The elastic rotational stiffness of the connections is assumed to be $\alpha_M(EI/L)$ (i.e., the product of coefficient α_M and girder flexural stiffness EI/L). The axial tensile force and deformation of the connections is assumed to be bilinear (Figure 20b). The T_y is assumed to be $\eta_N(AF_y)$ (i.e., the product of coefficient η_N and the yield resistance of girder) and the ultimate tensile resistance is taken as T_u=1.02T_y. The elastic tensile stiffness of the connections is taken to be $\alpha_N(EA/L)$ (i.e., the product of coefficient α_N and tensile stiffness of beam EA/L). It was unknown what was the ultimate tensile deformation employed by Foley et al. [23]. However, in this example, Δ_u corresponding to T_u is arbitrarily taken as Δ_u=30 mm.

For the sake of comparison with Foley's results, the same point-in-time floor load, which was 2.38 kPa, for the girder and beam system must be used. Load P is thus computed based on the tributary area of girder G1 (Figure 17) as P=(2.38)(9.15/3)(9.15)(2)=133 kN. The gravitational loads supported by beams B1 are not included in P since beams B1 are also under a double span condition. Note that 2.38 kPa is even less than the specified dead load of the floor 3.83 kPa since Foley et al. [23] found the composite floor slab alone was able to carry 2.4 kPa by membrane action.

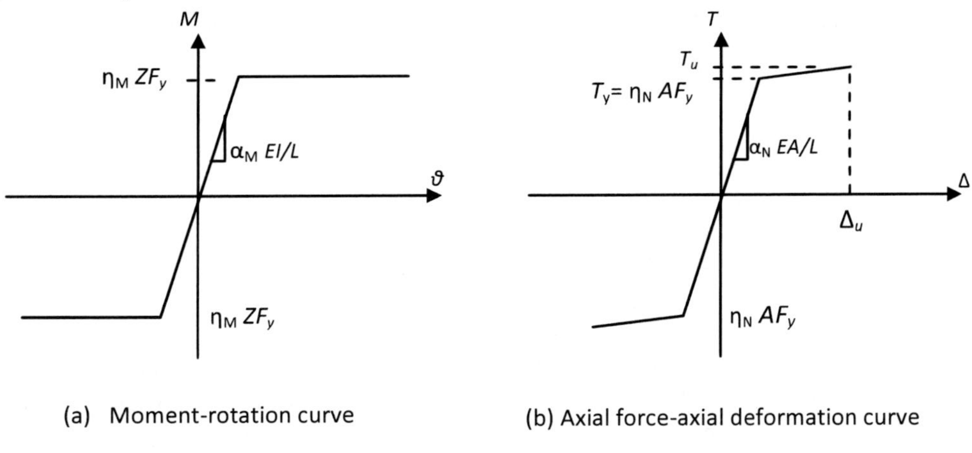

Figure 20. Assumed connection behaviors for the numerical example.

The analysis results are θ=0.074 radians (compared with θ=0.073 radians from Foley's 3-dimensional analysis), M_1=M_2=90.4 kN-m, and T=634 kN (which is less than the tensile capacity T_u=908 kN). Tension deformation of connections are Δ_1=Δ_2=11.2 mm and girder tension elongation is 2.2 mm. The deflection δ is 0.677 m, and the support shear force Y=19.8 kN (Note that force T has a vertical component which is also a shear force for the connection).

The algorithm was numerically tested using various initial θ values (θ must be nonzero as per Eq. (53)). The algorithm is found to be very effective and stable, with one analysis being always completed within a fraction of a second using an Intel Core Duo CPU computer.

The primary difference between the aforementioned algorithm and Foley's analysis method is that this algorithm is a 2-dimensional equilibrium analysis while Foley's method was a 3-dimensional model. Though Foley's method might be more sophisticated, the proposed algorithm herein is certainly much easier to use and more cost-effective while still able to provide an accurate result.

The resilience analysis and design of shear connections is a relatively new topic in steel construction, and much more works are needed for further development. This section is to develop a method for the determination of structural demands for simple connections. Herein, the portion of the total gravity load carried by the girder and beam system was assumed to be pre-determined, as literature review revealed that floor slab itself could carry a large portion of gravity load (guideline is required for how to determine the portion of floor load which is carried by girder-and-beam system). The 2-dimensional analysis model was proven to be as accurate as Foley's 3-dimensional model in spite of its simplicity.

For traditional shear connection design, shear force is usually the only demand needs to be calculated and the accompanying rotational demand is generally assumed to be nonproblematic. However, for connection resilience design, shear force becomes secondary important. Instead, axial force, axial deformation, and connection rotation become the primary demands. Especially, the ductility demands for catenary action need to be addressed explicitly.

Since both rotational demand and axial force demand could be several times as large as that for conventional shear connection design, the assumption that shear connections can generally maintain the integrity of a steel structure cannot be applied to double-span conditions without detailed calculation.

References

[1] Astaneh, A. Demand and supply of ductility in steel shear connections. *Journal of Constructional Steel Research*, 1989, 14, 1-19.

[2] Thornton WA. A rational approach to design of tee shear connections. *Engineering Journal*, 1st quarter, 1996, 34-37.

[3] Thronton, WA. Strength and ductility requirements for single shear connections with shear and axial Load. In: *Proceedings of National Steel Construction Conference*, American Institute of Steel Construction, Chicago, IL, May 7-9, 1997, 38-1 to 38-17.

[4] Gong, Y. Single-angle all-bolted shear connections. *Journal of Constructional Steel Research*, 2009, 65(6), 1337-1345.

[5] Gong, Y; Gillies, A. Double-angle shear connections with short outstanding legs. *Canadian Journal of Civil Engineering*, 2008, 35, 786-795.

[6] Gong, Y. Double-angle shear connections with small hollow structural section columns. *Journal of Construction Steel Research*, 2008, 64(5), 539-549.

[7] Gong, Y. Design moment of shear connections at the ultimate limit state. *Journal of Constructional Steel Research*, 2009, 65(Nos. 10-11), 1921-1930.

[8] Kennedy, DJL; Hafez, MA. Study of end plate connections for steel beams. *Canadian Journal of Civil Engineering*, 1984, 11, 139-149.

[9] CSA. CAN/CSA-S16-01, *Limit states design of steel structures*. Toronto (Ontario): Canadian Standards Association, 2001.

[10] Drucker. The effect of shear on the plastic bending of beams. *Journal of Applied Mechanics*, 1956, December, 509-514.

[11] Chen, WF; Sohal, I. *Plastic design and second-order analysis of steel frames*. Springer-Verlag, 1995.

[12] AISC. *Steel construction manual*, 13th ed. Chicago (Ill): American Institute of Steel Construction, 2005.

[13] CISC. Handbook of steel construction. 9th ed. *Canadian Institute of Steel Construction*, Toronto, Ontario, Canada. 2006.

[14] CSA. CAN/CSA-G40.21-04, *Structural quality steel*. Canadian Standards Association, Toronto, Ontario. 2004.

[15] Lesik, DF; Kennedy, DJL. Ultimate strength of fillet welded connection loaded in plane. *Canadian Journal of Civil Engineering*, 1990, 17, 55-67.

[16] Kulak, GL. High strength bolting for Canadian engineers. *Canadian Institute of Steel Construction*, Toronto, Ontario, Canada. 2005.

[17] McMullin, KM; Astaneh, A. *Analytical and experimental studies of double-angle framing connections*. Report No.UCB/SEMM-88/14, Department of Civil Engineering, University of California (Berkeley), 1988.

[18] Lipson, SL. Single-angle welded-bolted connections. *Journal of the Structural Division*, 1977, ASCE, 103(3), 559-571.

[19] Crawford, SF; Kulak, GL. Eccentrically loaded bolted connections. *Journal of the Structural Division*, 1971, ASCE, 97(3), 765-783.

[20] BSI. *Structural use of steel work in building*. Part I: Code of practice for design in simple and continuous construction: hot rolled section. Standard BS5950-1, British Standards Institution, London, UK. 2000.

[21] Guravich, SJ; Dawe, JL. Simple beam connections in combined shear and tension. *Canadian Journal of Civil Engineering*, 2006, 33, 357-372.

[22] GSA, *Progressive collapse analysis and design guidelines for new federal office buildings and major modernization projects*, www.oca.gsa.gov. The U.S. General Services Administration, 2003.

[23] Foley, CM; Barnes, K; Schneeman, C. Quantifying and enhancing robustness in steel structures: part 2 – floor framing systems. *Engineering Journal*, Fourth quarter, 2008, 267-286.

In: Structural Steel and Castings
Editor: Lena M. Becker, pp. 147-183
ISBN: 978-1-61728-104-4

Chapter 4

Fatigue of Structural Connections: A Recent Research on Multiaxial Fatigue

Mitao Ohga[1,*], Siriwardane Sudath Chaminda[1], Ranjith Dissanayake[2] and Abheetha Peiris[3]

[1]Department of Civil and Environmental Engineering, Ehime University, Bunkyo-cho 3, Matsuyama 790-8577, Japan

[2]Department of Civil Engineering, University of Peradeniya, Peradeniya 20400, Sri Lanka

[3]Department of Civil Engineering, University of Kentucky, Kentucky, USA

Abstract

The effect of stress concentration in connections between primary members of structures was found to be one of the main reasons for fatigue damage. Most of these connections are subjected to multiaxial fatigue in both high cycle and low cycle ranges. In case of high cycle fatigue, the equivalent effective stress with Miner's rule is the most popular multiaxial fatigue criterion in service life estimation of structures. However under many variable amplitude-loading conditions, Miner's predictions have been found to be unreliable since it does not properly take into account the loading sequence effect. Recently, a number of damage models have been proposed to capture the loading sequence effect of variable amplitude loads more precisely.

The scale of metal grains of a metallic aggregate is generally defined to be within the mesoscopic scale (grain scale). Many of the more precise high cycle fatigue theories involve grain scale than macroscopic scale. In the high cycle fatigue region, some grains undergo local plasticity while the rest of the matrix behaves elastically. Recently, one mesoscopic scale fatigue model was developed to obtain more precise estimation to multiaxial fatigue and it has been found that this model is applicable to common structures since the associated material parameter determination procedures are dependent on commonly available fatigue test data. However, experimental comparisons of this theory (model) have exhibited a certain amount of deviation for multiaxial fatigue life due to variable amplitude loading conditions since the considered damage law is Miner's rule.

* E-mail address: sira@cee.ehime-u.ac.jp. (Corresponding author)

Therefore the main objective of this chapter is to propose a new mesoscopic scale (grain scale) fatigue model to obtain a more precise estimation to multiaxial high cycle fatigue life for variable amplitude proportional loading. The accumulated inelastic meso-strain is considered as the damage variable in this model, and a new damage indicator is defined as an alternative to the damage factor of the previous model. Initially chapter describes the existing multiaxial high cycle fatigue models and their shortcomings. Then, the chapter describes the new fatigue model. The verification of the proposed fatigue model is performed by comparing the experimental fatigue lives with theoretical predictions. The proposed model is applied to estimate the fatigue life of a riveted bridge connection and obtained results are compared with previous estimations based on other known models. Finally, it is shown that the proposed multiaxial fatigue model gives a much more realistic fatigue life to variable amplitude proportional loading situation where detailed stress histories are known.

1. Introduction

The stress concentration effect in connections between the primary members of structures was found to be one of main reasons for fatigue damage (Fisher et al., 1984; Imam et al., 2005; Alampalli et al., 2006). Most of such connections are subjected to multiaxial fatigue. The equivalent effective stress with Miner's rule is the most popular multiaxial fatigue criterion in high cycle fatigue life estimation (Imam et al., 2005 & 2007; Suresh, 1998). However under many variable amplitude-loading conditions, Miner's predictions have been found to be unreliable since it does not properly take into account the loading sequence effect (Mesmeque et al., 2005; Siriwardane et al., 2008). Recently, a number of damage models have been proposed to capture the loading sequence effect of variable amplitude loads more precisely. Most of these are based on continuum damage mechanics (Li et al., 2006), strain energy or critical plane (Carpinteri et al., 2003). Model parameters of continuum damage mechanics and energy based criterions do not depend on generally available fatigue data of materials. Also these depend on complex material tests. In case of recently developed critical plane approaches, the determination of critical plane orientation in 3D state is much more complicated work for practicing engineering. On the other hand, corresponding cycle counting technique for random loading is totally different from well known cycle counting techniques such as rain flow or reservoir counting, etc. Therefore, most of these models are unable to be used for bridge life estimation.

However, it was realized that recently developed-sequential law (Mesmeque et al., 2005) based approach is more applicable to the connections since the method depend on available fatigue test data of materials and cycle counting approaches. As a result, in recent paper by authors (Siriwardane et al., 2008), sequential law associated method has been modified to apply for multiaxial fatigue life estimation of bridge connections. However, these studies have realized that even though the modified sequential law produces better results in multiaxial fatigue than Miner's rule, still these results exhibits much deviation to real fatigue life. Therefore, it is required to have a more precise fatigue model of which the model parameters and cycle counting techniques do not depend on complex procedures.

It has been found that more precise high cycle fatigue theories exist in mesoscopic scale than previously discussed macroscopic scale (Morel, 1998; Papadopoulos, 1999). The mesoscopic scale is defined at the grain level. In the high-cycle fatigue regime, some grains undergo local plasticity while the rest of the matrix behaves elastically. Recently, one mesoscopic scale fatigue model (Jabbado et al., 2008) was developed to obtain more precise estimation to multiaxial fatigue and it has been found that this model is applicable to common structures since associated material parameter determination procedures are dependent on commonly available fatigue test data. However, experimental comparisons of this model have exhibited a certain amount of deviation for multiaxial fatigue life due to variable amplitude loading conditions since the considered damage law is the Miner's rule.

Therefore, the major objective of this chapter is to introduce a new mesoscopic scale (grain scale) fatigue model to obtain a more precise estimation to multiaxial high cycle fatigue life for variable amplitude proportional loading while providing a comprehensive understanding about existing high cycle fatigue models which are commonly used to estimate fatigue of structural connections. Initially chapter describes the existing multiaxial high cycle fatigue models and their shortcomings. Then the chapter describes the new fatigue model in detail. The verification of the proposed fatigue model is performed by comparing the experimental fatigue lives with theoretical predictions. The newly proposed model is applied to estimate the fatigue life of a riveted bridge connection and obtained results are compared with previous estimations based on other known models. Finally comparisons of the results are made and hence validity and applicability of the proposed model is discussed.

2. Existing Multiaxial High Cycle Fatigue Models

The available multiaxial high cycle fatigue models, which are widely used to estimate fatigue life of existing structures are described by illustrating the current state of research. Initially the section describes commonly used fatigue model to estimate service life of connections. Then the deficiencies are comprehensively discussed. Final subsection describes briefly the recent multiaxial damage models, which are originated to overcome these shortcomings of previous model, and their deficiencies in comprehensive manner.

2.1. Commonly Used Multiaxial High Cycle Fatigue Model

The Miner's rule is extended to evaluate the fatigue damage due to multiaxial stresses. For this extension, the simple as well as the commonly used approach has been considered. It basically consists of transforming cyclic multiaxial stresses into equivalent uniaxial stress (generally called as effective stress which characterizes the deformation of the material) amplitude or range, which should generate same fatigue life as that due to the multiaxial stresses. The equivalent stress amplitude or range is then used to enter a uniaxial *S-N* curve to determine the damage indicator D_i of the Miner's rule. The usual methods for making such transformations are extensions of von Mises (octahedral shear stress) or Tresca (maximum shear stress) yield criteria for proportional loading conditions (Suresh, 1998; Mesmacque et al., 2005). Here, von Mises yield criteria is used for the transformation. Hence multiaxial stress fields at critical locations are considered in terms of equivalent von Mises stress amplitudes or ranges. The following steps are performed to predict the multiaxial fatigue life.

Step1. Stress evaluation: Initially the von Mises stress (effective stress) histories due to passage of loadings for critical location are obtained. Let σ_x, σ_y and σ_z denote the three normal components of stress applied to fatigue specimen and let τ_{xy}, τ_{yz} and τ_{zx} denote the three shear stress components. Hence, effective stress can be determined from the following general relation.

$$\sigma_e = \frac{1}{\sqrt{2}}\sqrt{(\sigma_x - \sigma_y)^2 + (\sigma_y - \sigma_z)^2 + (\sigma_z - \sigma_x)^2 + 6(\tau_{xy}^2 + \tau_{yz}^2 + \tau_{zx}^2)} \tag{1}$$

Then the von Mises stress histories are converted into effective stress ranges and effective mean stress by using the rainflow or reservoir cycle counting method. The effective stress range and mean stress are determined respectively by following equations which are derived only for the proportional loading conditions.

$$\sigma_{r,e} = \sigma_{\max,e} - \sigma_{\min,e} \quad ; \quad \sigma_{m,e} = (\sigma_{\max,e} + \sigma_{\min,e})/2 \tag{2}$$

where $\sigma_{\max,e}$ and $\sigma_{\min,e}$ are the maximum and minimum effective stress in considered effective stress range. The the modified Goodman equation (Suresh, 1998) was used to do the mean stress correction of each effective stress range using following relation.

$$\sigma_{r,e} = \sigma_{r,e}\Big|_{\substack{\sigma_{1m}=0 \\ \sigma_{2m}=0 \\ \sigma_{3m}=0}} \left\{1 - \frac{\sigma_{m,e}}{\sigma_u}\right\} \tag{3}$$

The σ_u is the ultimate tensile strength of the material. Hence, the effective stress ranges are transferred to equivalent fully reversed uniaxial cyclic stress ranges.

Step 2. Determination of fatigue curve: Then select the code provided curve that describes the classification of no details (class for plain section, e.g. Class B curve in the BS 5400 Part 10) or fully reversed tension-compression fatigue test based partially known *S-N* curves for life calculation.

Step 3. Application of damage law: Finally these equivalent fully reversed cyclic stress ranges are utilized instead of uniaxial stress of the cumulative damage law of Miner's rule (Suresh, 1998). In case of most of structures, the loadings vary during certain time interval (named as duration of loading block) and the similar pattern is repeating during particular period (number of years). The number of yearly repetition of various stress ranges (n) for each time period can be found using both stress cycle distributions and the frequency of loadings during the loading block. For the particular i^{th} period of age, Miner's summation of fatigue damage,

$$D_i = \left(\sum n/N\right)_i * \text{(No. years for period where similar loading block is repeating)} \tag{4}$$

Then the Miner's summation of the total cumulative fatigue damage for all periods is,

$$D = \sum D_i \text{ : where } i = 1 \text{ to } k \text{ which is equal to number of divided periods of age} \quad (5)$$

Therefore remaining Miner's summation of fatigue damage at present is,

$$D_r = (1 - D) \quad (6)$$

Hence, the remaining life (T_r) of each member can be obtained as,

$$T_r = D_r / D_{present} \text{ years; where } D_{present} = (\sum n/N)_k \quad (7)$$

In this calculation it is assumed that the future sequence and the magnitude of loadings are the same as that for the present period.

Here it is important to note that the described procedure is only applicable when the critical points are subjected to proportional loading conditions.

2.2. Deficiencies of Commonly Used Multiaxial High Cycle Fatigue Model

The above model (section 2.1) has been used for several decades. However, application of this model to the structures is found to be very few since secondary stress based fatigue estimations are somewhat complicated in the sense of evaluation of the correct stress histories. Indeed consideration of the current method, recently researchers have highlighted that some of deficiencies of the current method cause to produce unrealistic output and also cause to reduce common applications too. Following content describes these deficiencies comprehensively.

- An accurate secondary stress evaluations are also based on the far field stress field (measured primary stresses). Generally, past strain measurements are available for major structures. Most of the old structures do not have past strain measurements. Therefore, application of above method is limited to particular structures where the information of previous loading histories is available. This reason hinders the application of this model for common structures.
- However, under many variable amplitude-loading conditions, Miner's rule based life predictions have been found to be unreliable (Suresh, 1998; Mesmacque et al., 2005). Therefore, it is uncertain to use the Miner's rule for these estimations.
- Here it is important to note that the described procedure is limited to critical points, which are subjected to proportional loading conditions (Suresh, 1998). It is rare, but it is feasible that there is a possibility of non-proportional loadings around the structural connections in some situations. This reason is also made a limitation to apply this model in the field.
- One of the major drawbacks of this effective stress approach is that the differing effects of axial tension and compression mean stresses in a multiaxial fatigue test

may not be accurately captured. In addition, the orientation of fatigue cracks with respect to the loading axes is not quantitatively determined from such criteria (Suresh, 1998). Furthermore, a wide variety of experimental observations reveal that the normal stress also plays a critical role in influencing fatigue lives in multiaxial loading (Carpintri et al., 2003).

These drawbacks create an uncertainty about the commonly available multiaxial high cycle fatigue models to estimate life of existing structural connections under service loadings. Recently considerable amount of researches have been conducted to overcome these problems totally or partly. Therefore, the following section describes the recent multiaxial fatigue models which are originated to overcome these shortcomings.

2.3. Recent Multiaxial High Cycle Fatigue Models and Their Deficiencies

Recently, a number of damage models have been proposed to capture the loading sequence effect of variable amplitude loads more precisely. Most of these are based on continuum damage mechanics (Li et al., 2006), strain energy or critical plane (Carpinteri et al., 2003). Model parameters of continuum damage mechanics and energy based criterions do not depend on generally available fatigue data of materials. Also these depend on complex material tests.

Recently a fatigue model is proposed to estimate fatigue life under variable amplitude load based on critical plan approach (Carpintri et al., 2003). In this model, a specific cycle counting method is proposed. The scalar value of the normal stress vector perpendicular to the critical plane is taken as the cycle counting variable since the direction of such a vector is fixed with respect to time (conversely to the time-varying direction of the shear stress vector). This model consists of an equivalent stress, of which amplitude is defined through a nonlinear combination of normal ($N^*_{\max,z}$) and shear ($C^*_{a,z}$) stress components acting on the critical plane as follows.

$$\sigma_{eq,a,z} = \sqrt{(N^*_{\max,z})^2 + \left(\frac{\sigma_{af}}{\tau_{af}}\right)^2 (C^*_{a,z})^2} \tag{8}$$

where the σ_{af} is the normal stress fatigue limit of material either fully reversed axial loading or bending. The τ_{af} is the shear stress fatigue limit for fully reversed torsion. Then, the cumulative damage model is used to process the above equivalent stress amplitude and to determine fatigue damage as,

$$D = \sum_{z=1}^{Z} (D_Z)^q; \quad D_Z = \frac{1}{2N_0\left(\frac{\sigma_{af}}{\sigma_{eq,a,z}}\right)^{m_\sigma}} \quad for \quad \sigma_{eq,a,z} \geq c\sigma_{af}; \quad q = 1 + \frac{0.25}{(1-c)}(\sigma_{eq,a,z} - \sigma_{af}) \tag{9}$$

The symbol Z refers to the total number of reversals at particular time (i.e. may be the time taken for one block). The N_0, σ_{af} and m_c are parameters obtained from the considered stress-life curve considered (N_0 is the number of cycles at fatigue limit). The c is a safety coefficient ($0 \leq c < 1$). Note that, by posing q equal to unity, Eq. (9) would correspond to the classical Miner's rule.

However in this critical plane model, the determination of critical plane orientation in 3D state is much more complicated work for practicing engineers. On the other hand, corresponding cycle counting technique for random loading is totally different from well known cycle counting techniques such as rainflow or reservoir counting, etc. Therefore, most of these models are unable to be used for bridge life estimation. As a result, the applications are found to be very few in civil engineering structures.

Previous researchers have highlighted that more precise high cycle fatigue theories exist in mesoscopic scale (grain scale) than previously discussed macroscopic scale (Morel, 1998; Papadopoulos, 1999). The mesoscopic scale is defined at the grain level. In the high cycle fatigue regime, some grains undergo local plasticity while the rest of the matrix behaves elastically. Recently, one mesoscopic scale fatigue model (Jabbado et al., 2008) was developed to obtain more precise estimation to multiaxial fatigue and it has been found that this model is applicable to most of the civil engineering structures since associated material parameter determination procedures are dependent on commonly available fatigue test data. However, experimental comparisons of this model have exhibited a certain amount of deviation for multiaxial fatigue life due to variable amplitude loading conditions since the considered damage law is Miner's rule.

3. Proposed Multiaxial High Cycle Fatigue Model

This section proposes a new grain scale (mesoscopic scale) fatigue model to obtain a more precise estimation to multiaxial high cycle fatigue life of structures under variable amplitude random loadings. The mesoscopic scale is defined at the grain level. In infinite fatigue life, the material behaves elastically (or is in an elastic shakedown regime) at the macroscopic and mesoscopic scales (Dang Van., 1999). In finite fatigue life, the material behaves elastically (or is in an elastic shakedown regime) at the macroscopic level and experiences plastic shakedown at the mesoscopic scale. This mesoscopic stabilized behaviour is similar to the macroscopic one in low cycle fatigue. In simply, it is expressed that in the high-cycle fatigue regime, some grains undergo local plasticity while the rest of the matrix behaves elastically. For this regime, plastic strain and dissipated energy are relevant damage variables for estimating the fatigue life. In this model, the accumulated grain-scale plastic strain (inelastic meso-strain) is considered as the damage variable. A new damage indicator is defined to be an alternative to damage factor of previous model.

Initially, the relation to obtain the accumulated plastic strain per stabilized cycle is indicated. Then the section describes mesoscopic law of fatigue life for constant amplitude loading and newly proposed damage indicator respectively. A model parameter determination procedure is also clearly stated at end of the section.

3.1. Accumulated Plastic Meso-Strain Per Stabilized Cycle

The fatigue mechanisms at the grain scale is based on local (micro) plasticity and other damage processes such as cavitations, micro-cracking and void growths, depending on the material (ductile or brittle). These phenomena are described here by an inelastic strain tensor called plastic strain and hydrostatic stress since micro-crack and void growths depend on it. But the classical metal plasticity is insensitive to hydrostatic stress. In order to take these complex mechanisms into account, an elasto-plastic model is considered by accounting hydrostatic stress which is dependent on yield stress. More precisely, instead of considering crystal plasticity models, a mesoscopic model is considered, which governs a von Mises yield criterion with linear kinematic hardening and a hydrostatic stress dependent yield stress. Considering this mesoscopic scale elasto-plastic behavior, the accumulated plastic meso-strain (ε_s^{pc}) per stabilized cycle for constant amplitude proportional loading is found to be (Jabbado et al., 2008),

$$\varepsilon_s^{pc} = \frac{4}{\sqrt{3}} \frac{2k^* - k_{\max} - k_{\min}}{c} \tag{10}$$

where $c = b + 2\eta$. The b and η are the mesoscopic linear hardening modulus and the shear modulus respectively. The k^* is the radius of the smallest hypersphere which contains the entire history of the macroscopic deviatoric stress amplitude of the stabilized cycle ($\sigma'_{ij,a}$) as follows,

$$k^* = \sqrt{\tfrac{1}{2} \sigma'_{ij,a} \sigma'_{ij,a}} \tag{11}$$

The $k_{\max}$ and $k_{\min}$ are the maximum and the minimum values of mesoscopic yield stresses that can be reached during the loading cycle. The instantaneous hydrostatic stress does not have the same effect in tension as in compression. Generally tension progressively opens the microcracks, while compression closes the microcracks and therefore has less effect on damage. Therefore mesoscopic yield stress depends on the instantaneous hydrostatic stress which has similar value of macroscopic hydrostatic stress when simplified macro-meso relationship is adopted (Jabbado et al., 2008). Hence the maximum and the minimum values of mesoscopic yield stresses are derived as,

$$k_{\min} = \beta - \alpha p_a \text{ and } k_{\max} = \beta + \gamma p_a \tag{12}$$

where p_a is the macroscopic hydrostatic stress amplitude of a stabilized cycle. The α, β and γ are material parameters and these have to be determined using two fatigue limits (e.g. torsion and bending fatigue limits) and one *S-N* curve of particular material. The determination procedure has been described in detail at the section 3.4.

3.2. Mesocopic Law of Fatigue Life

The framework of this study concerns the finite life regime in high cycle fatigue. In this field, the mesoscopic law of fatigue life for constant amplitude proportional loading situation is described as a power relationship combining the accumulated plastic meso-strain per stabilized cycle (ε_s^{pc}) with the number of cycles to crack nucleation N (Jabbado et al., 2008).

$$N = A(\varepsilon_s^{pc})^{-\xi} \tag{13}$$

where ξ and A are material parameters to be determined from fatigue tests.

3.3. Damage Indicator for Multiaxial Fatigue

The hypothesis behind this fatigue law is that if the physical state of damage is the same, then fatigue life depends only on the loading condition. Suppose a component is subjected to a certain accumulated plastic meso-strain $(\varepsilon_s^{pc})_i$ per stabilized cycle of n_i number of cycles at load level i, N_i is the fatigue life (number of cycles to crack nucleation) corresponding to $(\varepsilon_s^{pc})_i$ (Figure 1). Hence, the residual life at load level i can be obtained as (N_i-n_i). The accumulated plastic meso-strain $(\varepsilon_s^{pc})_{(i)eq}$, which corresponds to the failure life (N_i-n_i) is named as i^{th} level damage accumulated plastic meso-strain. Hence, the new damage indicator, D_i is stated as,

$$D_i = \frac{(\varepsilon_s^{pc})_{(i)eq} - (\varepsilon_s^{pc})_i}{(\varepsilon_s^{pc})_u - (\varepsilon_s^{pc})_i} \tag{14}$$

where $(\varepsilon_s^{pc})_u$ is the intercept in Figure 1 with the ordinate at one-quarter of first fatigue cycle i.e. accumulated plastic meso-strain corresponding to tensile fracture (ductile failure) (Suresh, 1998). Here it is assumed that tensile fracture is initiated at the peak positive value of first cycle of the applied stress. Further it is considered that the simple tensile test represents one-quarter of a single completely reversed fatigue cycle.

Assuming the end of ith loading level damage D_i has been accumulated (occurred) due to the effect of $(\varepsilon_s^{pc})_{i+1}$ loading cycles, the damage is transformed to load level $i+1$ as below.

$$D_i = \frac{(\varepsilon_s^{pc})'_{(i+1)eq} - (\varepsilon_s^{pc})_{i+1}}{(\varepsilon_s^{pc})_u - (\varepsilon_s^{pc})_{i+1}} \tag{15}$$

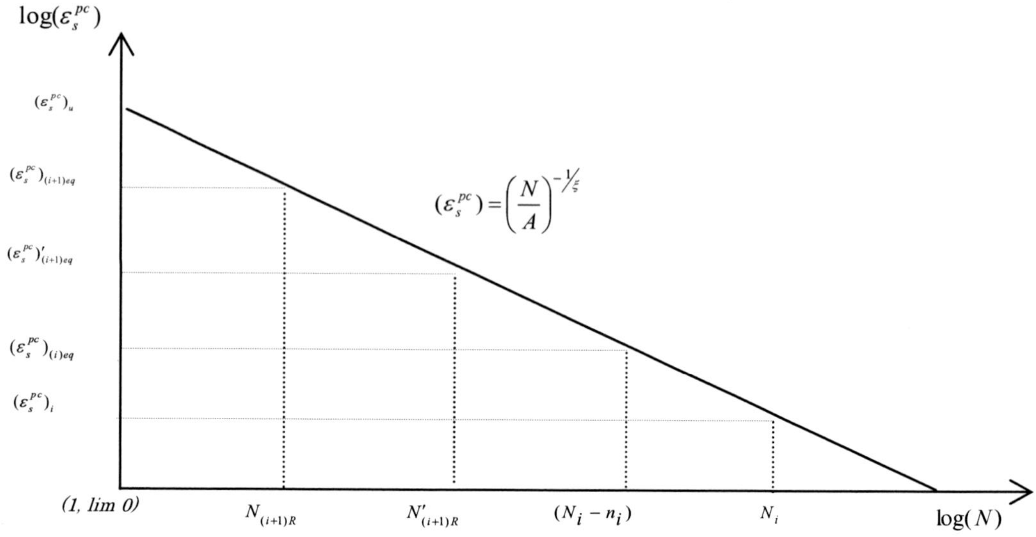

Figure 1. Schematic representation of mesoscopic law of fatigue life for constant amplitude loading.

The $(\varepsilon_s^{pc})'_{(i+1)eq}$ is damage equivalent accumulated plastic meso-strain at loading level *i+1* and it can be calculated from Eq. (15) as,

$$(\varepsilon_s^{pc})'_{(i+1)eq} = D_i[(\varepsilon_s^{pc})_u - (\varepsilon_s^{pc})_{i+1}] + (\varepsilon_s^{pc})_{i+1} \tag{16}$$

Thus the corresponding equivalent number of cycles to failure $N'_{(i+1)R}$ can be obtained from the Eq. (13) as shown in Figure 1. The $(\varepsilon_s^{pc})_{i+1}$ is the accumulated plastic meso-strain at the level *i+1* and supposing that it is subjected to $n_{(i+1)}$ number of cycles, then the corresponding residual life at load level *i+1*, $N_{(i+1)R}$ is calculated as,

$$N_{(i+1)R} = N'_{(i+1)R} - n_{(i+1)} \tag{17}$$

Hence damage accumulated plastic meso-strain $(\varepsilon_s^{pc})_{(i+1)eq}$, which corresponds to $N_{(i+1)R}$ at load level *i+1*, can be obtained from the Eq. (13) as shown in Figure 1. Then the cumulative damage at the end of load level *i+1* is defined as,

$$D_{(i+1)} = \frac{(\varepsilon_s^{pc})_{(i+1)eq} - (\varepsilon_s^{pc})_{i+1}}{(\varepsilon_s^{pc})_u - (\varepsilon_s^{pc})_{i+1}} \tag{18}$$

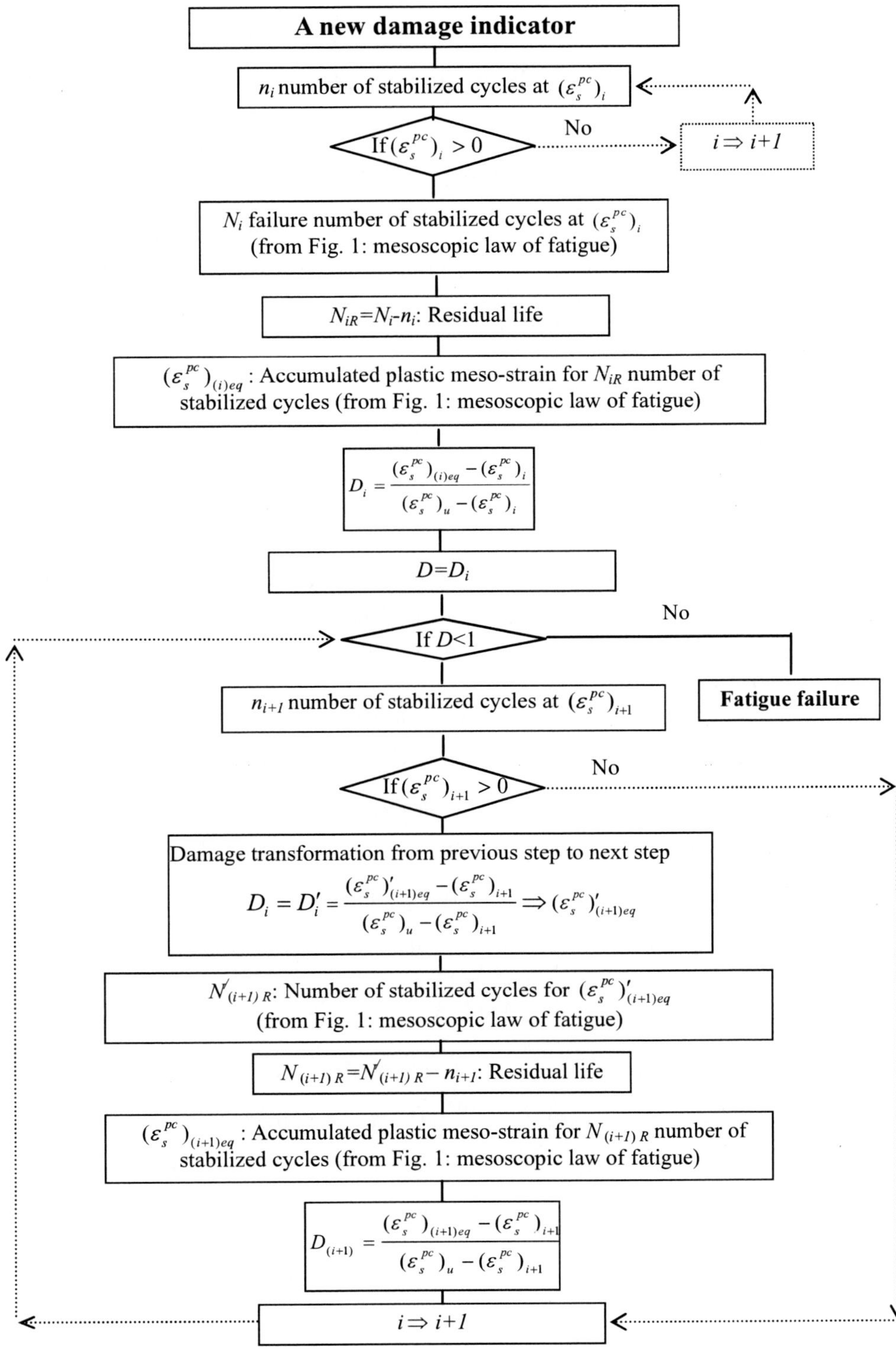

Figure 2. Flow chart for proposed damage indicator based algorithm.

At the first cycle the damage accumulated plastic meso-strain $(\varepsilon_s^{pc})_{(i)eq}$ is equal to $(\varepsilon_s^{pc})_i$ and the corresponding damage indicator becomes D_i=0. Similarly at the last cycle, the damage indicator becomes D_i=1 when $(\varepsilon_s^{pc})_{(i)eq}$ is equal to $(\varepsilon_s^{pc})_u$. Therefore, the damage indicator is normalized to one (D_i=1) at the fatigue failure of the material. Hence, the above procedure is followed until D_i=1. The proposed damage indicator based algorithm is more clearly summarized in the flow chart shown in Figure 2. Here, the defined fatigue failure is time taken for initiation of crack at the location of maximum stress of the structural component. In the case of structural components, it can probably be taken as the time taken for initiation of crack near a connection.

3.4. Determination of Model Parameters

The mesoscopic law for constant amplitude loading (Eq. (13)) depends on two material parameters ξ and A. The accumulated plastic meso-strain (ε_s^{pc}) per stabilized cycle in Eq. (10) depends on four parameters α, β, γ and c. However all six parameters are similar to previous mesoscopic model (Jabbado et al., 2008). Therefore, the determination method is also similar to the previous model and those are identified by using two fatigue limits (for example torsion and bending fatigue limits) and one *S-N* curve.

In the fully reversed torsion fatigue limit, the accumulated plastic meso-strain per stabilized cycle is zero ($\varepsilon_s^{pc} = 0$), as the grain asymptotically behaves elastically. Following the Eq. (11), the radius of the smallest hypersphere (k^*) can be derived as the value of torsion fatigue limit (t_{-1}). Since the material is in pure torsion, the macroscopic hydrostatic stress amplitude is also zero ($p_a = 0$). Substituting these values to Eq. (10), parameter β can be obtained as,

$$\beta = t_{-1} \tag{19}$$

Same as above, in the fully reversed bending fatigue limit (f_{-1}), the accumulated plastic meso-strain per stabilized cycle is zero ($\varepsilon_s^{pc} = 0$), as the grain asymptotically behaves elastically. Following the Eq. (11), the radius of the smallest hypersphere is derived as $k^* = f_{-1}/\sqrt{3}$. The macroscopic hydrostatic stress amplitude can be obtained as, $p_a = f_{-1}/3$. Substituting these values to Eq. (10), following relation of parameters α, β and γ can be derived,

$$\alpha - \gamma = 2\left(\frac{3\beta}{f_{-1}} - \sqrt{3}\right) \tag{20}$$

In the finite life domain of fully reversed *S-N* curve, the accumulated plastic meso-strain (ε_s^{pc}) per stabilized cycle can be obtained from Eq. (10) as,

$$\varepsilon_s^{pc} = \frac{2\dfrac{f}{\sqrt{3}} - 2\beta + \dfrac{f}{3}(\alpha - \gamma)}{c} \times \frac{4}{\sqrt{3}} \tag{21}$$

Here the f is a considered stress amplitude in finite life domain in the fully reversed *S-N* curve. Using the Eq. (13), the calculated fatigue life (N_{cal}) can be obtained as,

$$N_{cal} = A\left[\frac{2\dfrac{f}{\sqrt{3}} - 2\beta + \dfrac{f}{3}(\alpha - \gamma)}{c} \times \frac{4}{\sqrt{3}}\right]^{-\xi} \tag{22}$$

Table 1. Life times and nominal maximum stresses of variable amplitude multiaxial fatigue tests

Material	Designation for test	Maximum bending stress	Maximum Shear Stress	Experimental fatigue life (s)	Average fatigue life(s)
18G2A Steel	NWL 1	219	219	99600	130630
				78240	
				134700	
				148720	
				155520	
				167020	
	NWL 4	360	180	48060	42795
				41400	
				45240	
				37200	
10HNAP Steel	KWL 2	212	212	100560	98948
				95520	
				106240	
				92940	
				99300	
	KWL 4	342	171	84600	90705
				99900	
				93900	
				84420	
	KWL 8	320	160	123660	123275
				122890	

The Eq. (22) depends on the all six parameters. The best values of these parameters are determined by minimizing the error between the calculated (N_{cal}) and experimental ($N_{\exp}$) fatigue lives. To find best values, an optimization problem has to be performed by considering objective function as,

$$F_{Objective} = \sum abs\left(\frac{N_{cal}}{N_{\exp}} - 1 \right) \tag{23}$$

and constrains as the Eq. (19) & (20). To solve this problem efficiently, authors recommend to use the genetic algorithm employed toolbox in Mathlab.

4. Experimental Verifications of Proposed Model

In favor of verification, the models predicted fatigue lives were compared with previously proposed models and the experimentally observed fatigue lives of some of fatigue tests under both variable amplitude and random loading situations.

4.1. Comparison with Experimental Life under Variable Amplitude Load

In this section, the multiaxial fatigue test results of two materials are compared with the theoretically predicted fatigue lives under variable amplitude loading situations. The two materials are 18G2A and 10HNAP steels. The first material is a low alloy high-strength steel for building welded structures. The other material is a low-alloy structural steel of higher resistance to atmospheric corrosion. The tests were performed in the high cycle fatigue region under variable amplitude combined bending and torsion loading. Histories of bending and torsion moments have been independently generated. Then the nominal normal and shear stress histories were obtained using the section modulus of bending and torsion (Marciniak et al., 2008). The obtained normal and shear stresses were both in-phase loading (proportional loading). The dominant loading frequency was 20 Hz. The time interval of loading repetition (duration of loading block) was equal to 2000 seconds. Fatigue tests were performed under two proportional loading combinations for the 18G2A steel (Table 1). For the 10HNAP steel, three proportional loading combinations were tested as shown in Table 1. The fragments of normal and shear stress histories, obtained across the critical cross section during a selected loading combination are drawn in Figure 3 to illustrate the variable amplitude nature of loading histories. The stress histories of all these proportional loading combinations exhibit nearly zero mean stress effect. This provides evidence to conclude that each loading cycle of entire history generate a stabilized mesoscopic stress/ strain cycle for all the combinations.

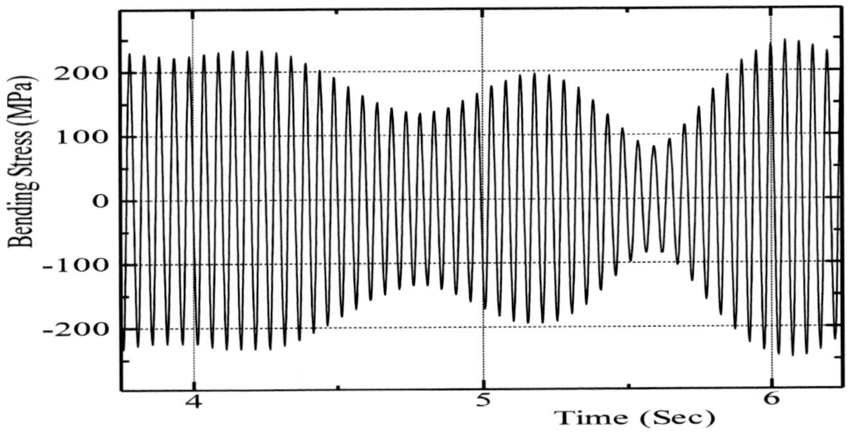

(a)

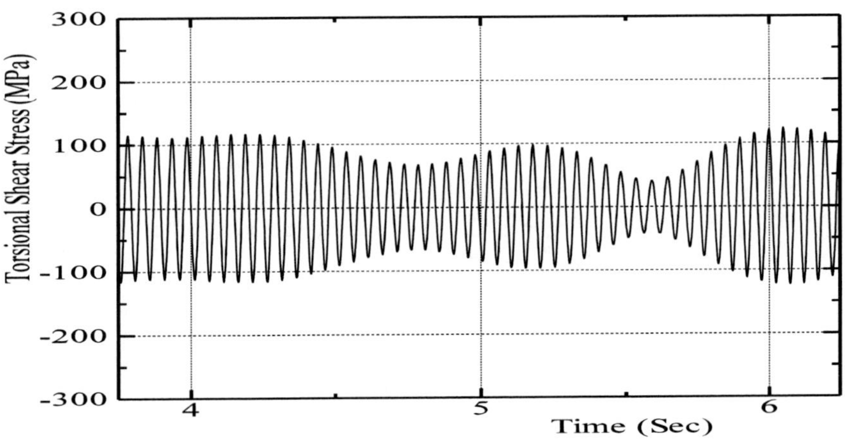

(b)

Figure 3. Fragments of nominal stress history of multiaxial variable amplitude loading test for: (a) bending and (b) torsion for NWL 4.

The partially known fully reversed tension-compression *S-N* curves and fatigue limits for tension-compression and torsion of above materials were obtained from different research papers (Korolczuk et al., 2005; Kluger et al., 2004). Using these data and following the guidelines given in section 3.4, a constrained minimization problem was performed to estimate the parameters $\alpha, \beta, \gamma, \xi$, c and A. The obtained values for $\alpha, \beta, \gamma, \xi$, c and A of 18G2A steel were 1.62, 170 MPa, 0.0841,0.9827, 49213 MPa and 987 respectively. The corresponding values for 10HNAP steel were 0.9016, 182 MPa, 0.0324, 0.9001, 48734 MPa and 1370 respectively (Jabbado et al., 2008). The proposed fatigue model is applied to obtain the theoretical fatigue life of each mentioned loading combinations. The comparisons between the calculated lives and the experimental results are shown in Table 2. In addition to this, experimental lives are compared with two previous theoretical approaches as shown in Table 2. The first approach is based on equivalent effective stress with Miner's rule (Suresh,

1998). The second approach is based on previous mesoscopic approach (Jabbado et al., 2008). Finally these results are plotted as shown in Figure 4 and it shows that there is better agreement between the values from the proposed model and the real fatigue life of these materials than the lives predicted by previous models. This verification reveals the validity of the proposed mesoscopic scale fatigue model in predicting the multiaxial fatigue life under variable amplitude proportional loading conditions.

Table 2. Comparison of the theoretical fatigue lives with experimental lives of variable amplitude loading tests

Test	Fatigue life (sec)			
	Experimetal	Effective stress based approaches	Mesoscopic plastic strain based approaches	
		Miner's rule	Jabbado et al.	Proposed model
NWL1	130630	159770	147265	123568
NWL4	42795	67502	58243	49628
KWL2	98948	44579	202229	116975
KWL4	90705	22265	93497	78239
KWL8	123275	45314	162519	109076

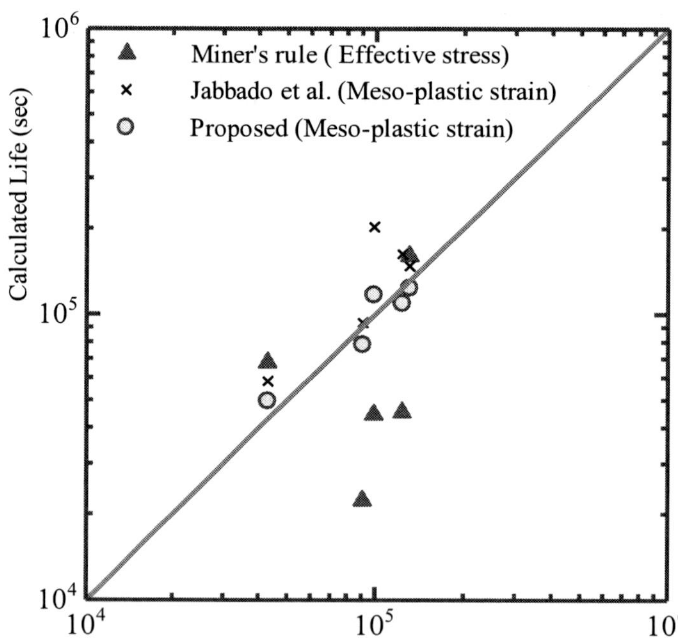

Figure 4. Comparison of calculated life versus experimental life of variable amplitude loading fatigue tests.

4.2. Comparison with Experimental Life under Random Load

The multiaxial fatigue test results under random loading of 10HNAP material are compared with the theoretically predicted fatigue lives in this section. The considered material is a low-alloy structural steel of higher resistance to atmospheric corrosion. The rounds specimens were tested in high cycle fatigue region under combination of proportional bending, $M_B(t)$, and torsional, $M_T(t)$, moments. The 14 specimens, which are subjected to $\alpha = \pi/4$ (i.e. $tg\alpha = M_T(t)/M_B(t)$), were selected for this comparison (Jabbado et al., 2008). The dominant sampling frequency of loading history was 266.67 Hz. The time interval of loading repetition (duration of loading block) was equal to 184.32 seconds. The proportional random histories of bending normal stress and torsional shear stress for these tests are obtained by multiplying the random loading sequence by loading factors corresponding to bending and torsion moments. The details of the considered test histories (i.e. maximum bending stress, shear stress and RMS value of the total moment) and experimental fatigue lives are shown in Table 3. The fragments of normal and shear stress histories, obtained across the critical cross section during a selected loading combination are drawn in Figure 5 to illustrate the random nature of loading histories. The obtained stress histories are complex and also of irregular shape. These stresses should be reduced into a series of equivalent stabilized cycles (zero mean stress cycles) which give the same damage to the material. In order to achieve this goal, initially the famous rainflow cycle counting technique (Suresh, 1998; Downing et al., 1982; Liu, 2005) is used to identify the stress ranges and sequences of closed stress cycles (Suresh, 1998; Christ, 1996). Then modified Goodman relation (Suresh, 1998) is used to transfer these counted cycle to mean stress zero equivalent stabilized cycle.

Table 3. Life times and nominal maximum stresses of random multiaxial fatigue tests

Test	Maximum bending stress	Maximum shear stress	Root mean square of total moment (Nmm)	Experimental fatigue life (sec)
T-1	480.46	240.24	14653.9	15379
T-2	458.40	229.20	13981.0	21465
T-3	440.49	220.25	13434.7	25350
T-4	427.39	213.70	13035.2	45949
T-5	406.38	203.19	12394.4	62434
T-6	389.81	194.92	11889.2	75225
T-7	332.48	166.24	10140.5	115009
T-8	325.59	162.80	9930.51	136794
T-9	321.97	160.99	8029.96	244757
T-10	320.90	160.45	9787.45	221370
T-11	273.63	136.82	9819.72	405444
T-12	267.25	133.63	7768.94	288080
T-13	263.29	131.64	8151.07	203365
T-14	254.72	127.36	8345.63	251723

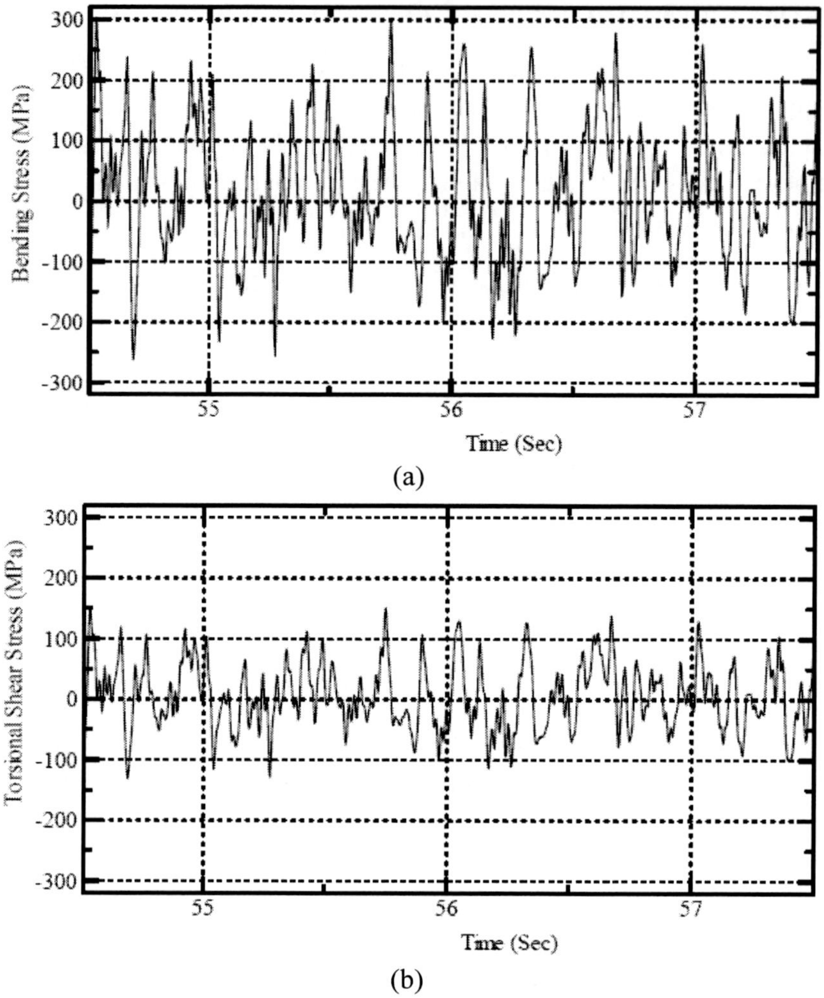

Figure 5. Fragments of nominal stress history of multiaxial random loading test for: (a) bending and (b) torsion for T-7.

The partially known fully reversed tension-compression *S-N* curves and fatigue limits for tension-compression and torsion of above materials were obtained from different research papers similar to section 4.1. Using these data and following the guidelines given in section 3.4, a constrained minimization problem was performed to estimate the parameters $\alpha, \beta, \gamma, \xi$, c and A. The obtained values for $\alpha, \beta, \gamma, \xi$, c and A of for 10HNAP steel were 0.9016, 182 MPa, 0.0324, 0.9001, 48734 MPa and 1370 respectively (Jabbado et al., 2008). The proposed fatigue model is applied to obtain the theoretical fatigue life of each mentioned loading combinations. The comparisons between the calculated lives and the experimental results are shown in Table 4. In addition to this, experimental lives are compared with three previous theoretical approaches as shown in Table 4. The first approach is based on equivalent effective stress with Miner's rule (Suresh, 1998). The second approach is based on critical plane which is used to define an equivalent stress amplitude (refer chapter 2 or Carpinteri et al., 2003). Here, the critical plane is determined from the weighted mean position of the principle stress directions.

The last approach among the previous approaches is based on previous mesoscopic approach (Jabbado et al., 2008). Finally these results are plotted as shown in Figure 6 and it shows that there is better agreement between the values from the proposed model and the real fatigue life of these materials than the lives predicted by previous models. This verification reveals the validity of the proposed mesoscopic scale fatigue model in predicting the multiaxial fatigue life under random proportional loading conditions.

Table 4. Comparison of the theoretical fatigue lives with experimental lives of random loading tests

Test	Fatigue life (sec)				
	Experi-metal	Effective stress based approaches	Critical plane based approach	Mesoscopic plastic strain based approaches	
		Miner's rule	Carpinteri et al.	Jabbado et al.	Proposed model
T-1	15379	2623	27267	24331	16230
T-2	21465	4367	34874	28737	26026
T-3	25350	15231	41535	35514	30196
T-4	45949	17329	48045	36113	44332
T-5	62434	21255	59063	40037	48651
T-6	75225	22526	69183	45147	54340
T-7	115009	126359	121058	83634	117312
T-8	136794	289927	127660	110339	139512
T-9	244757	589918	133544	132637	227712
T-10	221370	789250	134703	138332	243317
T-11	405444	1267589	223020	523037	282265
T-12	288080	1853850	243024	653026	300870
T-13	203365	2264051	255872	805770	333542
T-14	251723	2392560	282261	1283158	380023

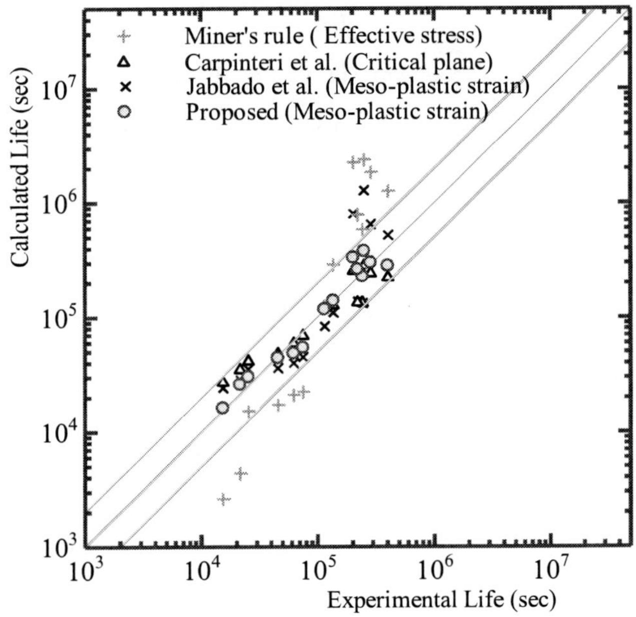

Figure 6. Comparison of calculated life versus experimental life for random loading fatigue tests.

(a)

Critical member

(b)

Secondary (local) stressed area

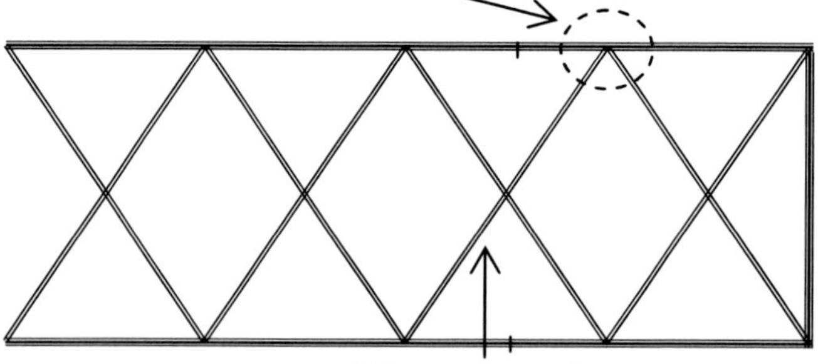

Primary stressed area

(c)

Figure 7. Considered connection: (a) General view of the bridge, (b) Close view of the considered connection and the critical member, (c) Schematic representation of the critical member and related areas for primary and local stresses.

5. Case Study: Fatigue Life Estimation of a Riveted Connection

Fatigue life evaluation of a riveted connection is discussed in this section. The evaluations are especially based on secondary stresses, which are generated around the riveted connection due to stress concentration effect of primary stress.

5.1. Considered Connection

One of the critical connections which were found from a detailed investigation of one of the longest railway bridges in Sri Lanka (Siriwardane et al., 2008) is selected for the life estimation. The selected riveted connection is shown in Figure 7.

5.2. Stress Analysis

The fatigue damage is evaluated based on the state of stress due to release of contact (tightness) of rivet while all the riveted locations have no clamping force. The clamping force is generally defined as compressive force in the plates which induced by the residual tensile force in the rivet. The residual tensile force in the rivets is occurred when the rivet gets shortened in length due to cooling after hot rivet is inserted in to the hole of plates in order to connect them and second head is formed from the protruding shank. Finally clamping force generates a tri-axial stress state in the connected plate. Since this study considers all riveted location have no clamping force (value of clamping force is zero), the connected members subjected only to biaxial stress state (Akesson, 1994). Therefore, a critical member without rivets can be considered to analyze biaxial state of stress (Figure 8 (a)). To make the continuity of stress field between global structure and the sub-structure (shown in Figure 8 (a)), it is required to use any interface between the two structures at every iterative step. Therefore, in this model, the primary tensile stress history (Figure 9) of the critical member, which has been obtained from measured strain histories at mid span of the member, is applied on the bottom face (*ab* of Figure 8 (a)) as an uniform pressure. The position of the *ab* boundary of the sub-model was determined considering the distribution of far field primary stress of the member. The four-node shell elements were used for the finite element analysis. The actual air gap restraint conditions were considered in the model to represent unilateral contact between rivet and plate. When the rivet is not properly in contact with the plate, the particular rivets tend to transfer less or zero amount of total load, and this leads to unexpected stress redistribution around the riveted area. The other rivets, which are carrying the load, are called as active rivets. In this study, fatigue lives were evaluated stepwise by reducing the contribution of active number of rivets in the connection. As a result, analyses were done considering five different cases such as all six rivets are active, five rivets are active, four rivets are active, three rivets are active and two rivets are active. For each case, the different numbers of arrangements of rivets were considered by changing the location of rivets and analyzed separately to choose critical arrangement (location) of rivets. The different numbers of arrangements were determined by the theory of combination in Mathematics. For all six rivets are active case, analysis was done based on only one arrangement of rivets (one combination). But for the other cases, six, fifteen, twenty and fifteen numbers of arraignments were considered respectively. Finally, the critical arrangements were selected based on maximum von Mises stress value. The obtained maximum stress contours are shown in Figure 8 for each considered case. This shows that stresses are operating well below the yield limit of the material and highly stressed locations (critical locations) are subjected to biaxial proportional state of stress. The principle stress variations versus the traffic sequence were obtained for all the cases. Since the considered member subjected to biaxial stress state, only two in plane principle stress components (principle stress 1 & 2) have values (principle stress perpendicular to plate is zero). The plots for the considered all cases are shown in Figure 10 to 14. Assuming that single day traffic sequence (time table) is repeating every day, one day time history is considered as the loading block in this study.

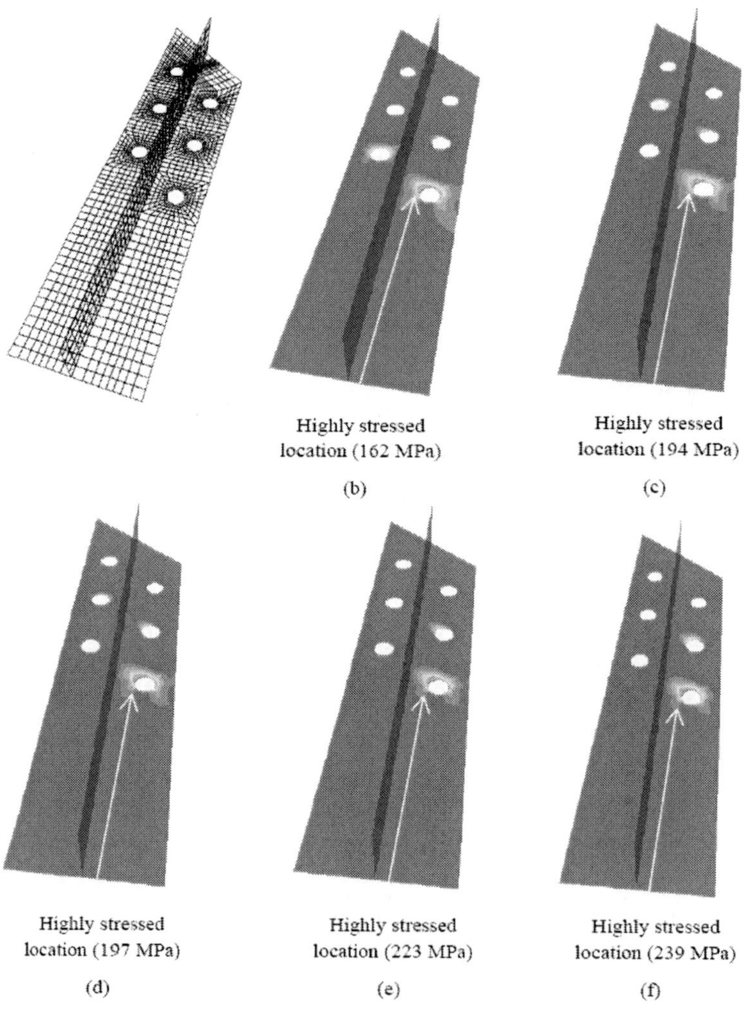

Figure 8. Stress analysis: (a) FE mesh, (b) Maximum von Mises stress contour when all six rivets are active, (c) Maximum von Mises stress contour when five rivets are active, (d) Maximum von Mises stress contour when four rivets are active, (e) Maximum von Mises stress contour when three rivets are active, (f) Maximum von Mises stress contour when two rivets are active.

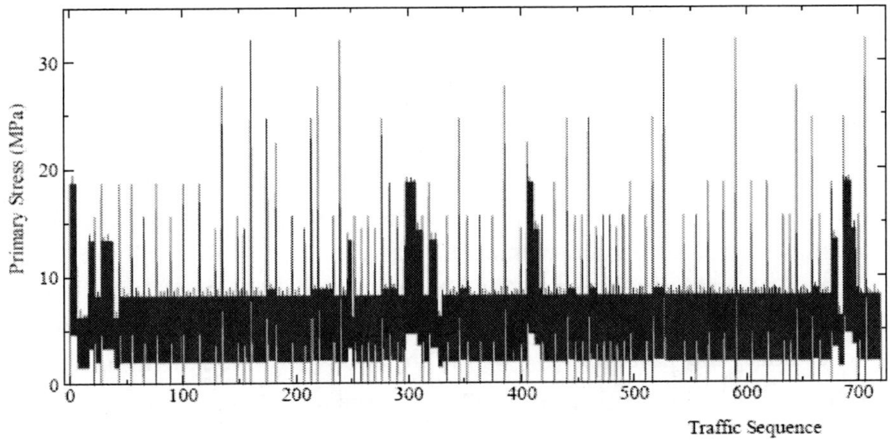

Figure 9. Primary stress variation of the member with traffic sequence per single day.

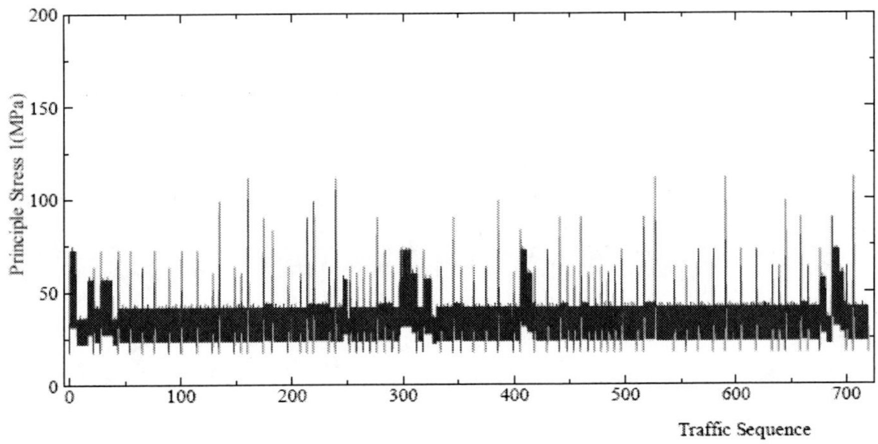

(a)

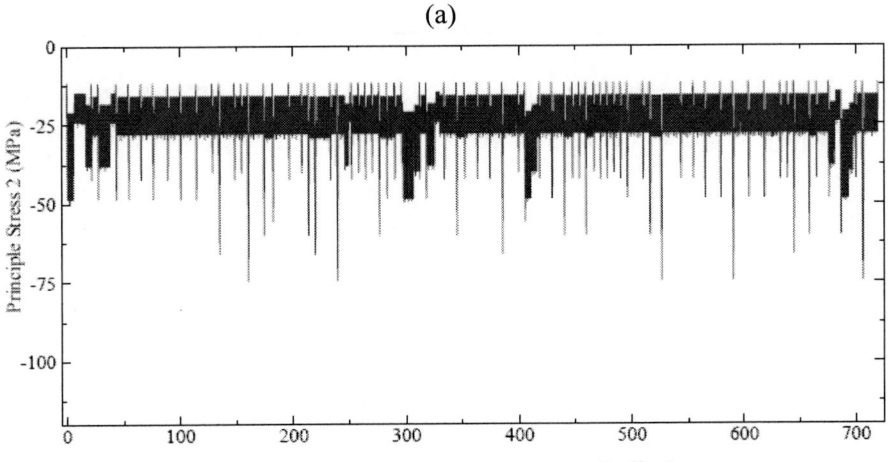

(b)

Figure 10. Principle stress variation with the traffic sequence per single day when all six rivets are active: (a). Principle stress 1, (b). Principle stress 2.

5.3. Determination of Model Parameters

The bridge material was Wrought iron and relevant design (mean minus two standard deviations) *S-N* curve, which represents low or no clamping force effect at the rivets, was obtained from the UK railway assessment code and previous paper (Railtrack, 2001; Cullimore, 1967). The design reversed tension- compression fatigue limit was determined from above *S-N* curve corresponding to 10^8 failure number of cycles. Then, general relations between the fatigue limits of particular material was used to calculate fully reversed torsion fatigue limit of Wrought iron material (Troshchenko et al., 1975). Using these data and following the guidelines given in section 3.4, a constrained minimization problem was performed to estimate the parameters $\alpha, \beta, \gamma, \xi$, c and A of Wrought iron. The calculated values are 1.0832, 15 MPa, 0.0473, 1.36, 12923 MPa and 613 respectively.

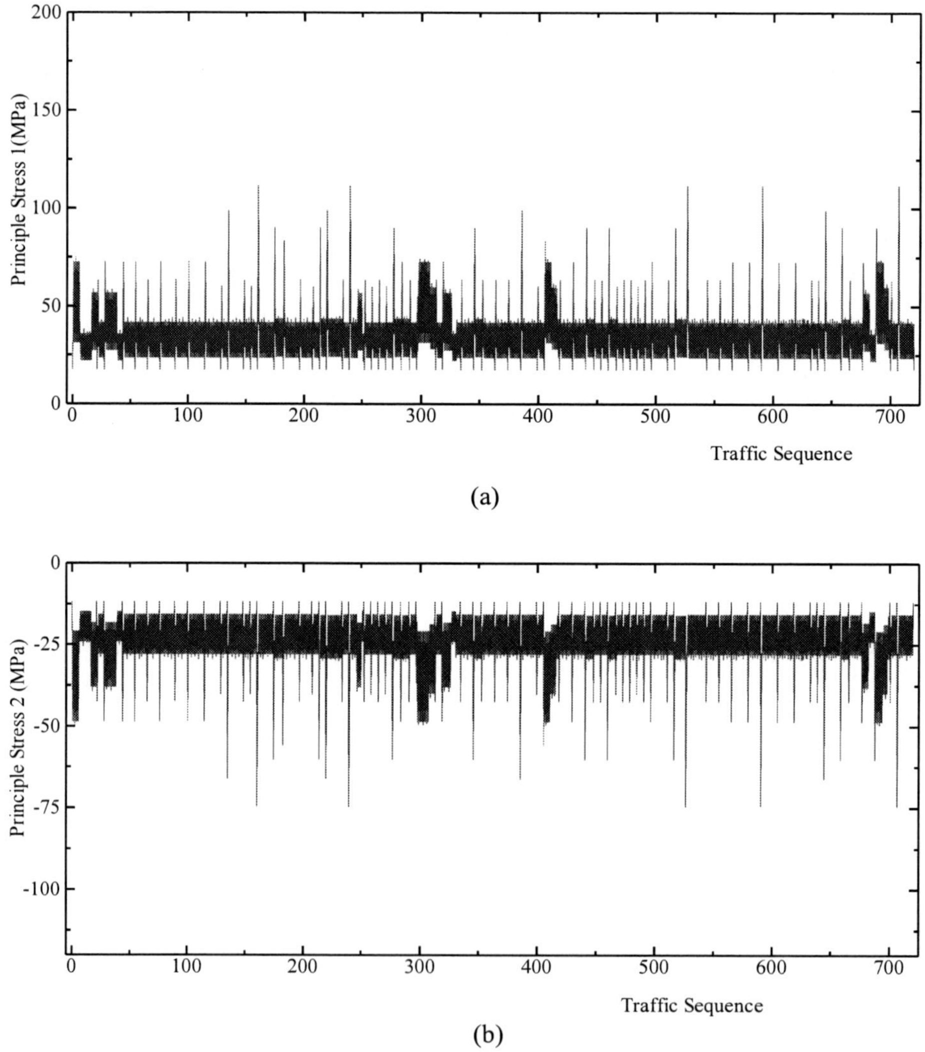

Figure 11. Principle stress variation with the traffic sequence per single day when five rivets are active: (a). Principle stress 1, (b). Principle stress 2.

5.4. Estimation of Accumulated Plastic Meso-Strain

The obtained principle stress variations are complex and also of irregular shape (Figure 10 to 14). These stresses should be reduced into a series of equivalent stabilized cycles (zero mean stress cycles) which give the same damage to the material. In order to achieve this goal, initially the famous rainflow cycle counting technique (Downing et al., 1982; Suresh, 1998; Liu, 2005) is used to identify the stress ranges and sequences of closed stress cycles (stress-strain hysteresis loops; Christ, 1996; Suresh, 1998). Then modified Goodman relation (Suresh, 1998) is used to transfer these counted cycle to mean stress zero equivalent stabilized cycle. These transferred principle stresses are plotted in Figure 15 to 19 for the all considered cases. These figures clearly exhibit that behavior of principle stresses are proportional (Figure 20). Then the macroscopic hydrostatic stress amplitudes p_a for each stabilized cycle were calculated using the both mean stress zero equivalent principle stress amplitudes (Figure 15 to 19) of each cycle. Then the radius of the smallest hypersphere which contains the entire history of the stabilized cycle k^*, was calculated from Eq. (11) with both mean stress zero equivalent principle stress amplitudes (Figure 15 to 19) of each cycle. Using the values of α, β and γ (determined in section 5.3) and previously calculated p_a with Eq. (12) the maximum and the minimum values of mesoscopic yield stresses $k_{\max}$ and $k_{\min}$ were calculated for each stabilized cycle. Finally, using the values of c (determined in section 5.3), k^*, $k_{\max}$ and $k_{\min}$,the accumulated plastic meso-strain per each stabilized cycle is calculated using Eq. (10) and plotted in Figure 21 & 22.

5.5. Fatigue Life Estimation

The proposed fatigue model in multiaxial fatigue (section 3) was utilized to predict the service life of the riveted connection. Using the parameters ξ and A which were determined in section 5.3, the mesoscopic law of fatigue life for constant amplitude loading was derived from Eq. (13) for Wrought iron (bridge material). Then the updated damage indicator (accumulation of damage) was calculated by following the flow chart shown in Figure 2. The variation of accumulated plastic meso-strain per each stabilized cycle versus time (Figure 21 & 22) was used for above calculation. The fatigue lives were determined when damage indicator reach to the value of unity. The predicted lives are shown in Table 5. The defined fatigue life describes the time duration from the date when considered feature of riveted connections appeared, to the date of fatigue failure. Further, it was considered that future sequence and density of rail passage is similar to the present period of operation. The predicted values were also compared with three previous models based estimations (similar approaches considered in section 4.1 and Table 2) as shown in Table 5.

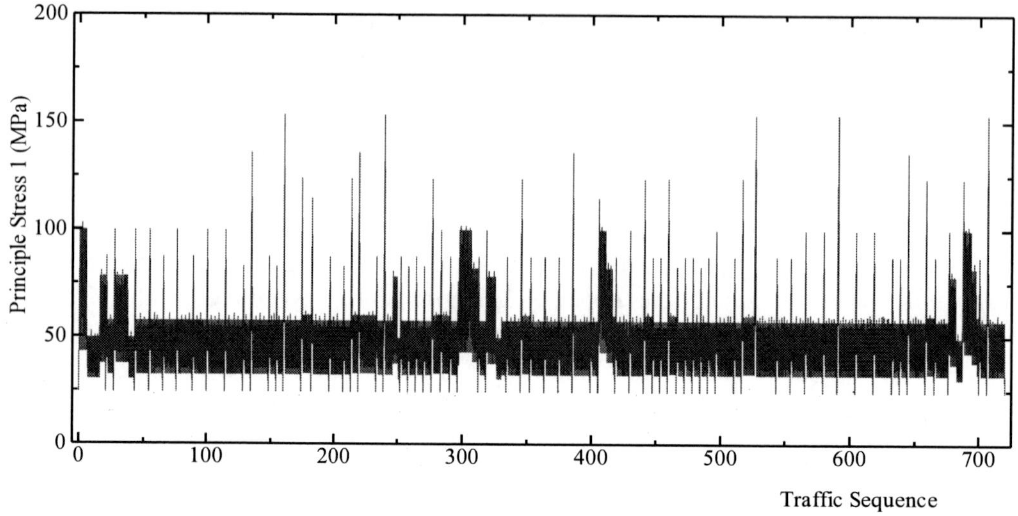

(a)

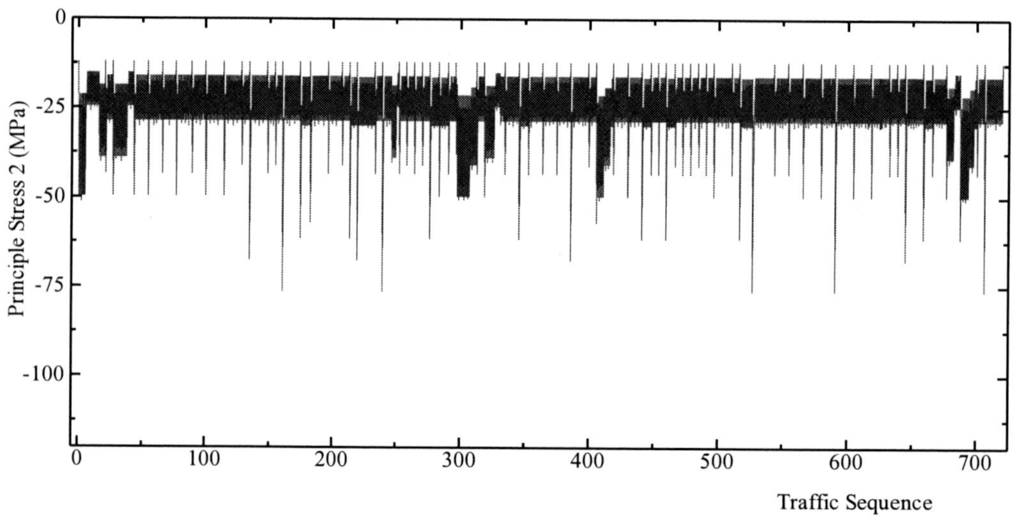

(b)

Figure 12. Principle stress variation with the traffic sequence per single day when four rivets are active: (a). Principle stress 1, (b). Principle stress 2.

Table 5. Comparison of calculated fatigue lives of riveted connection

Considered feature of the connection	**Fatigue life (months)**			
	Effective stress based approaches		**Mesoscopic plastic strain based approaches**	
	Miner's rule	**Sequential law**	**Jabbado et al.**	**Proposed model**
All six rivets are active	1957	1315	361	294
Five rivets are active	927	607	235	192
Four rivets are active	533	341	146	123
Three rivets are active	299	192	117	101
Two rivets are active	93	66	96	75

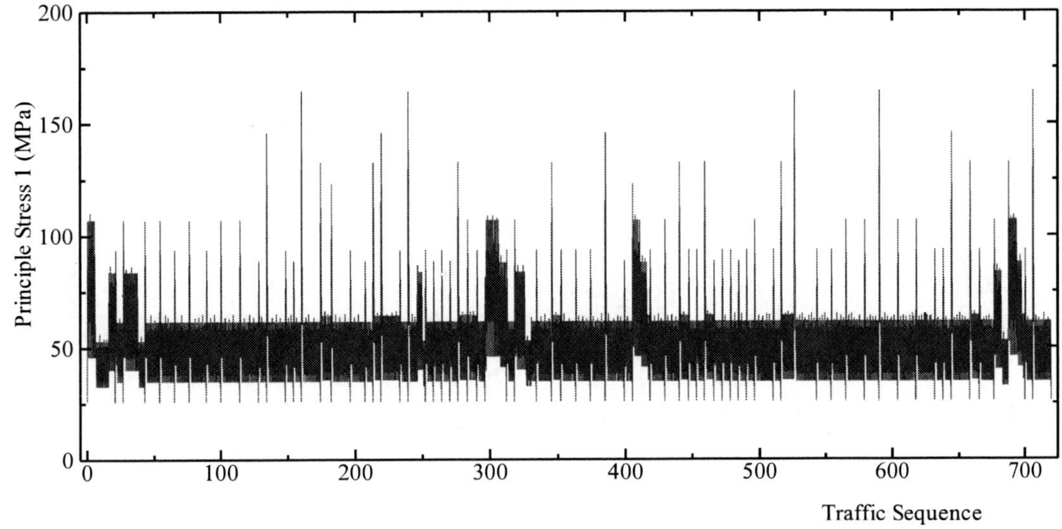

(a)

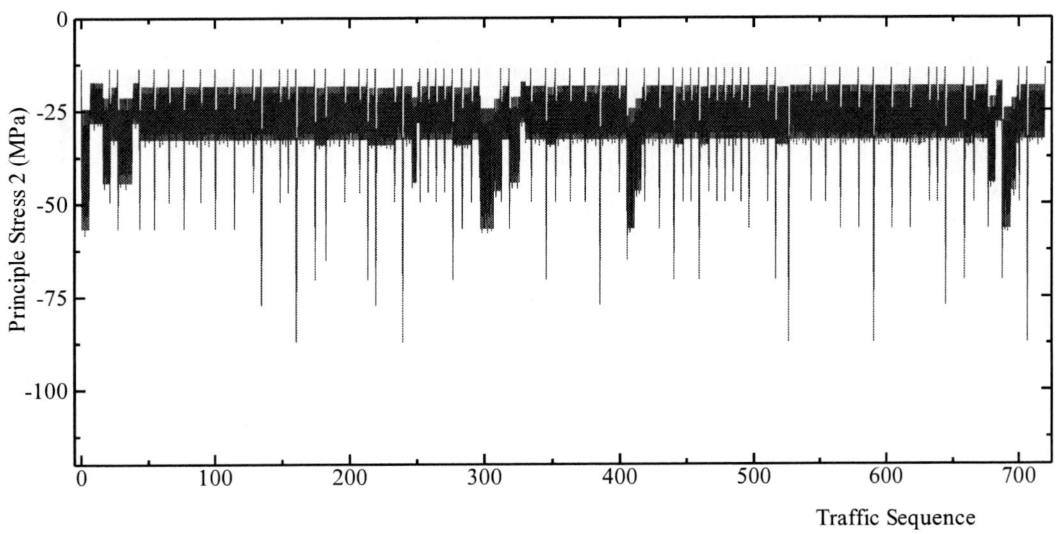

(b)

Figure 13. Principle stress variation with the traffic sequence per single day when three rivets are active: (a). Principle stress 1, (b). Principle stress 2.

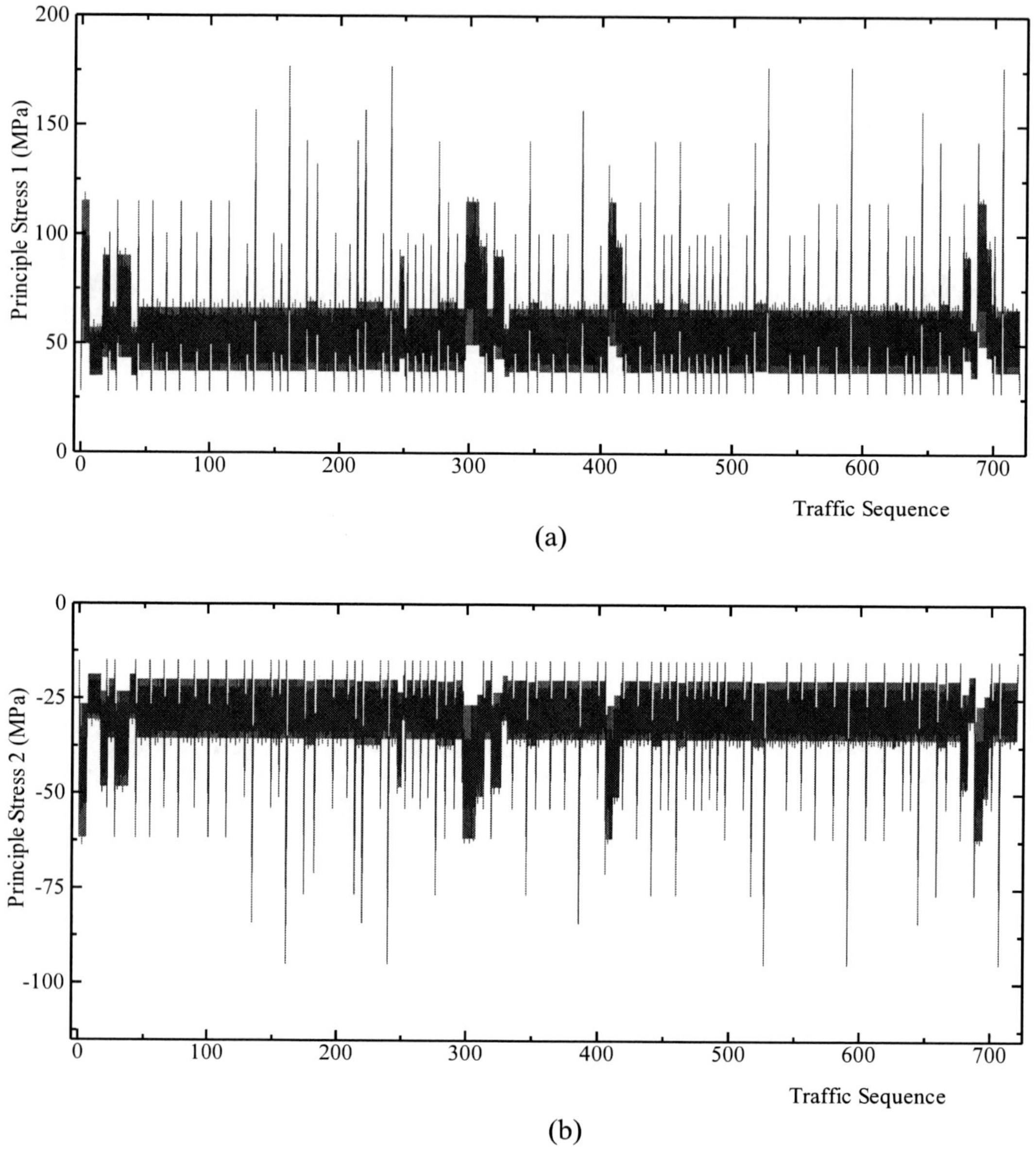

Figure 14. Principle stress variation with the traffic sequence per single day when two rivets are active: (a). Principle stress 1, (b). Principle stress 2.

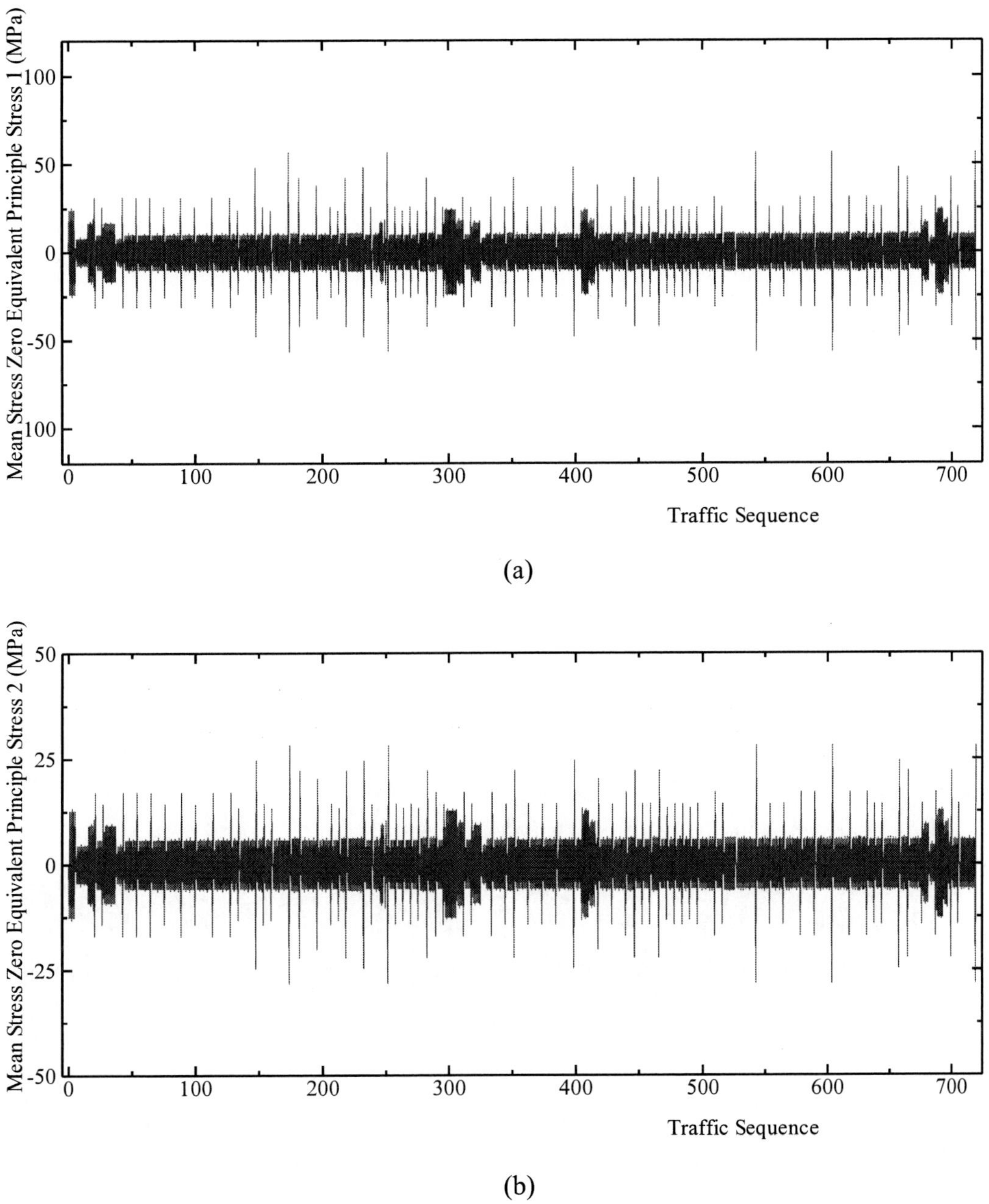

Figure 15. Mean zero equivalent principle stress variation with the traffic sequence per single day when all six rivets are active: (a). Principle stress 1, (b). Principle stress 2.

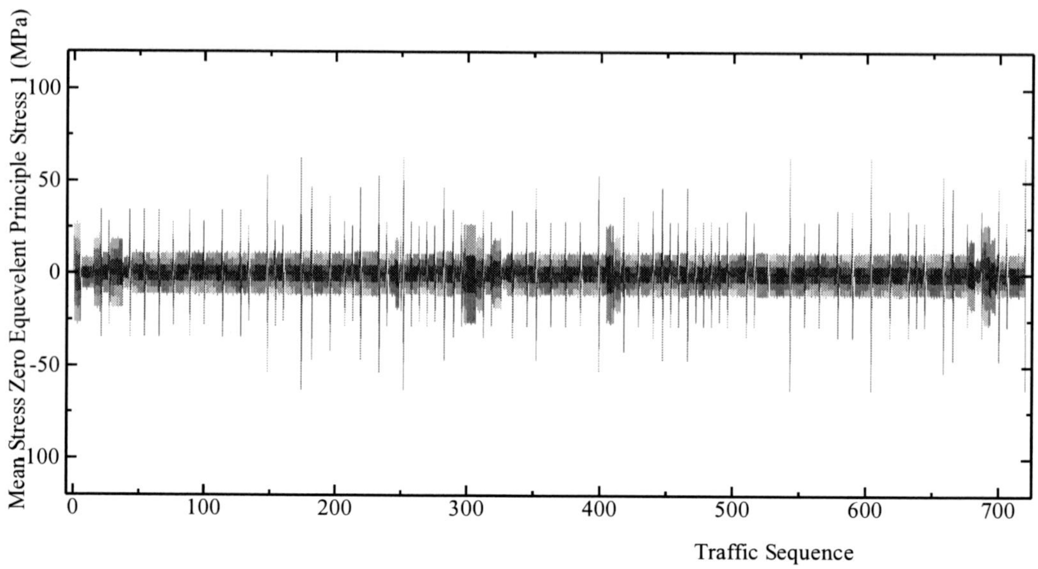

(a)

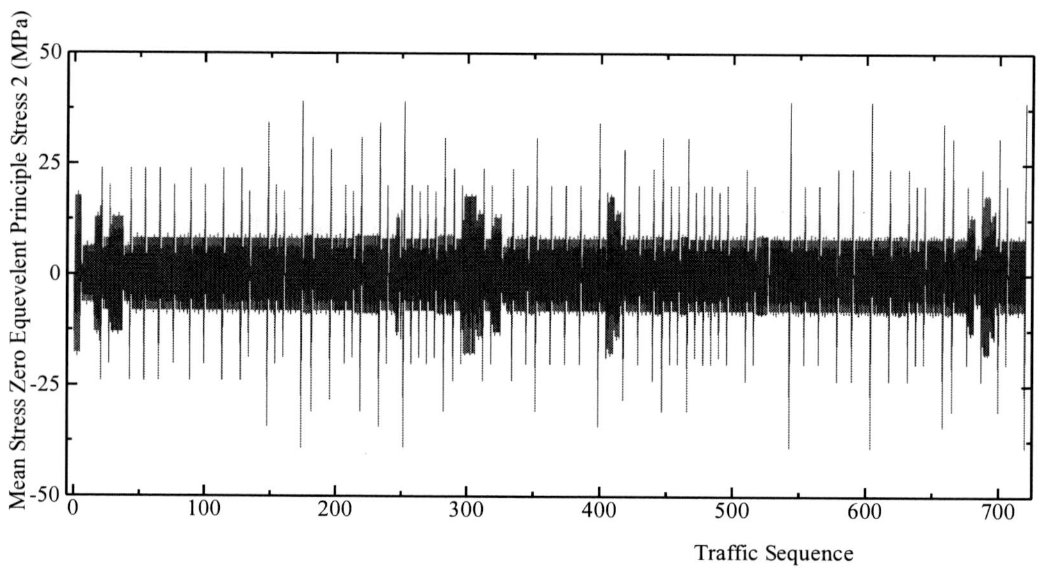

(b)

Figure 16. Mean zero equivalent principle stress variation with the traffic sequence per single day when five rivets are active: (a). Principle stress 1, (b). Principle stress 2.

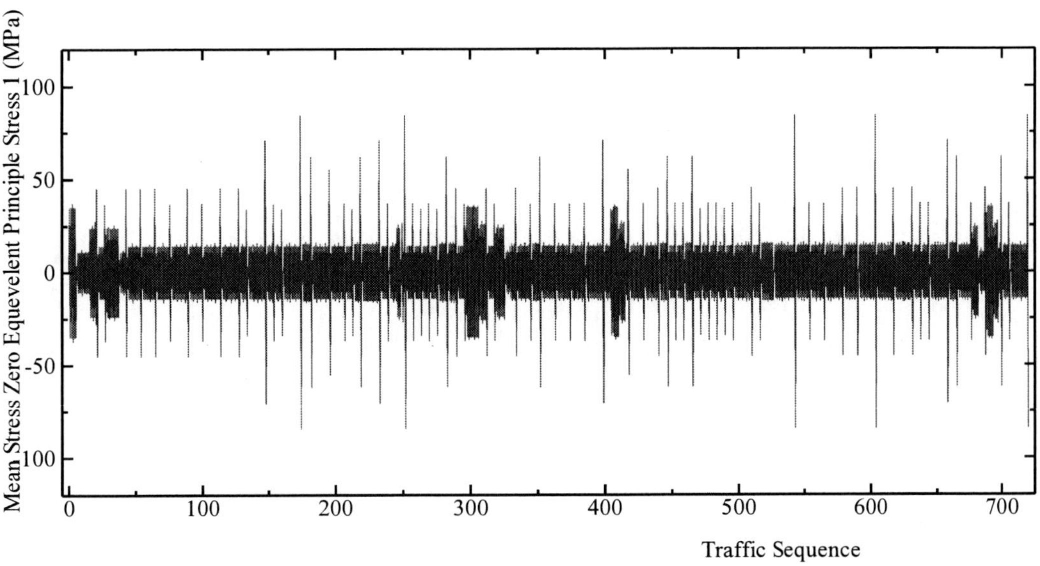

(a)

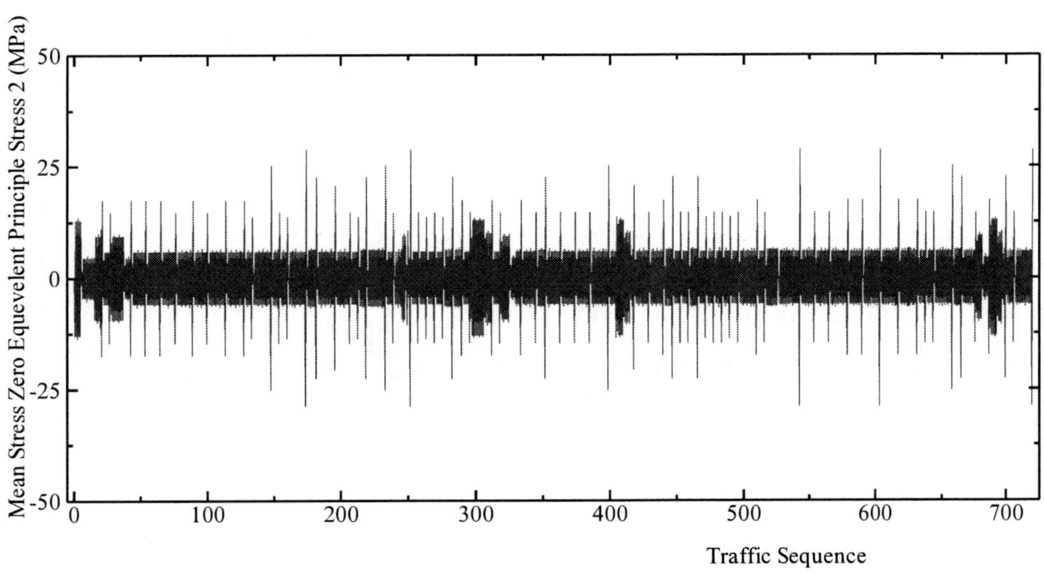

(b)

Figure 17. Mean zero equivalent principle stress variation with the traffic sequence per single day when four rivets are active: (a). Principle stress 1, (b). Principle stress 2.

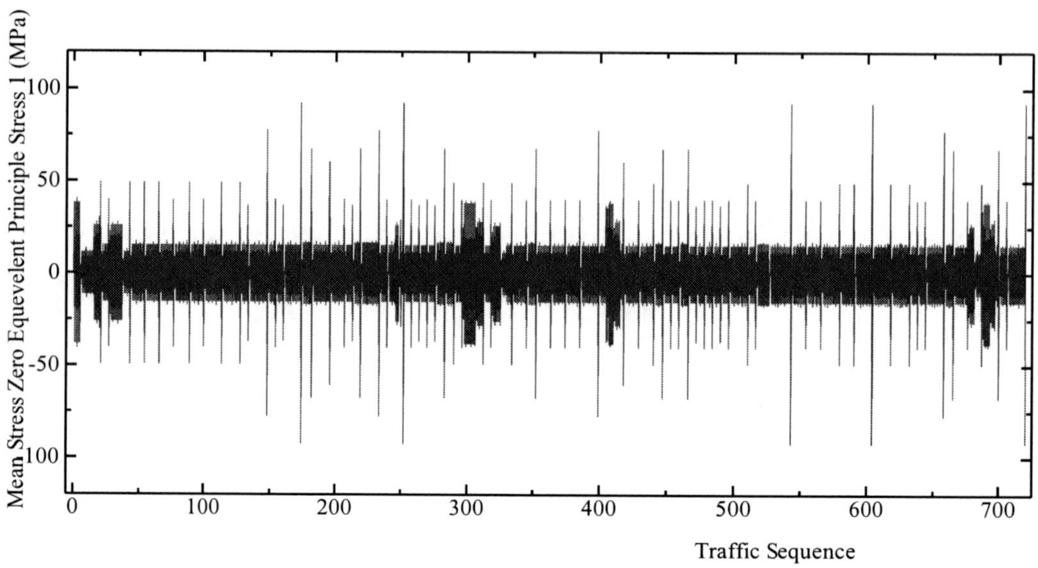

(a)

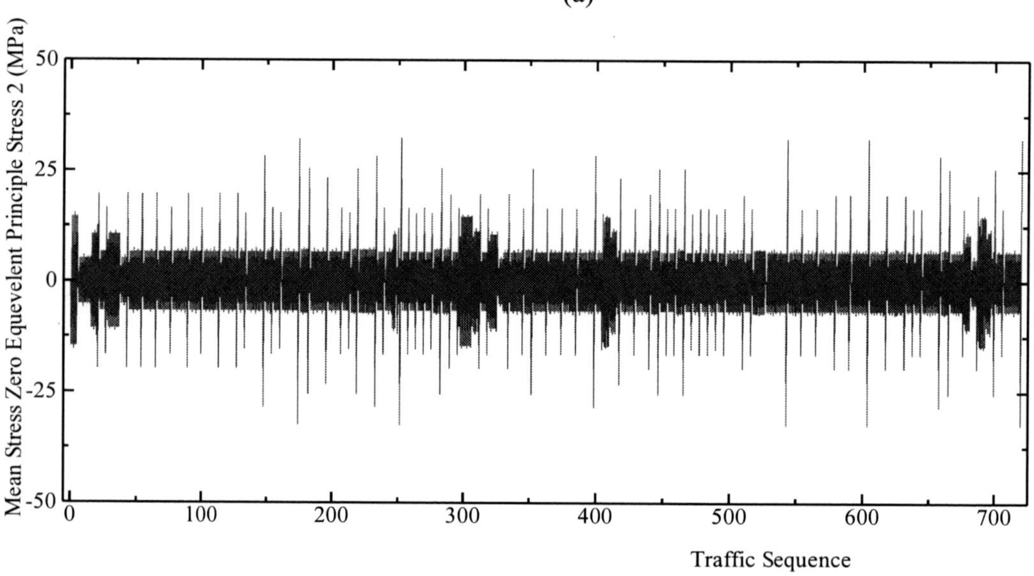

(b)

Figure 18. Mean zero equivalent principle stress variation with the traffic sequence per single day when three rivets are active: (a). Principle stress 1, (b). Principle stress 2.

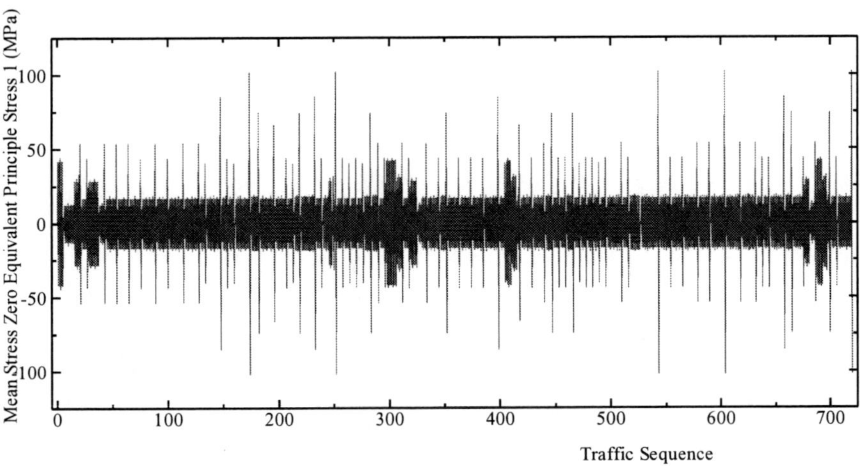

(a)

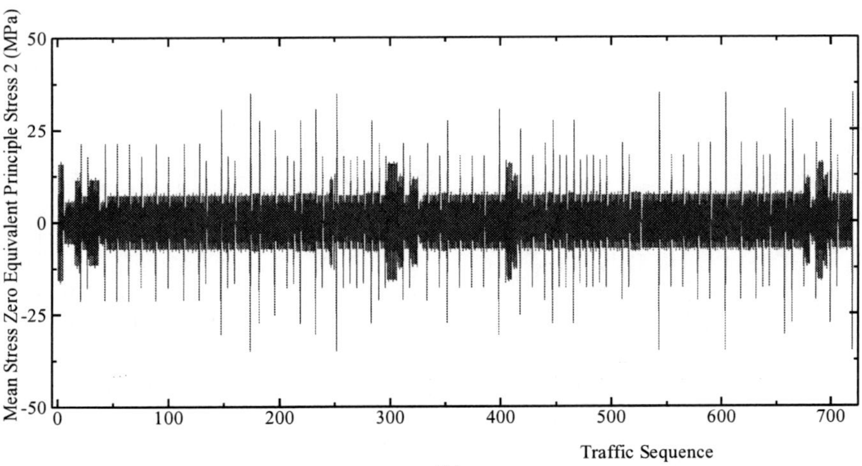

(b)

Figure 19. Mean zero equivalent principle stress variation with the traffic sequence per single day when two rivets are active: (a). Principle stress 1, (b). Principle stress 2.

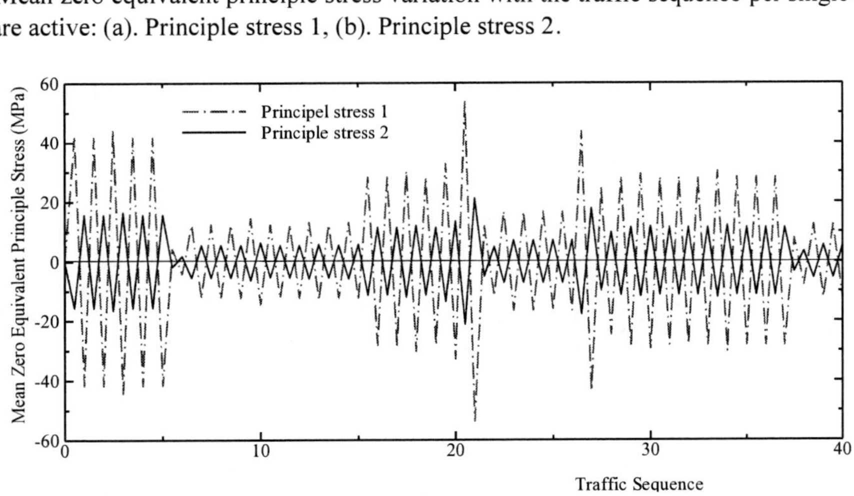

Figure 20. Enlarge fragment of the mean zero equivalent principle stresses when two rivets are active.

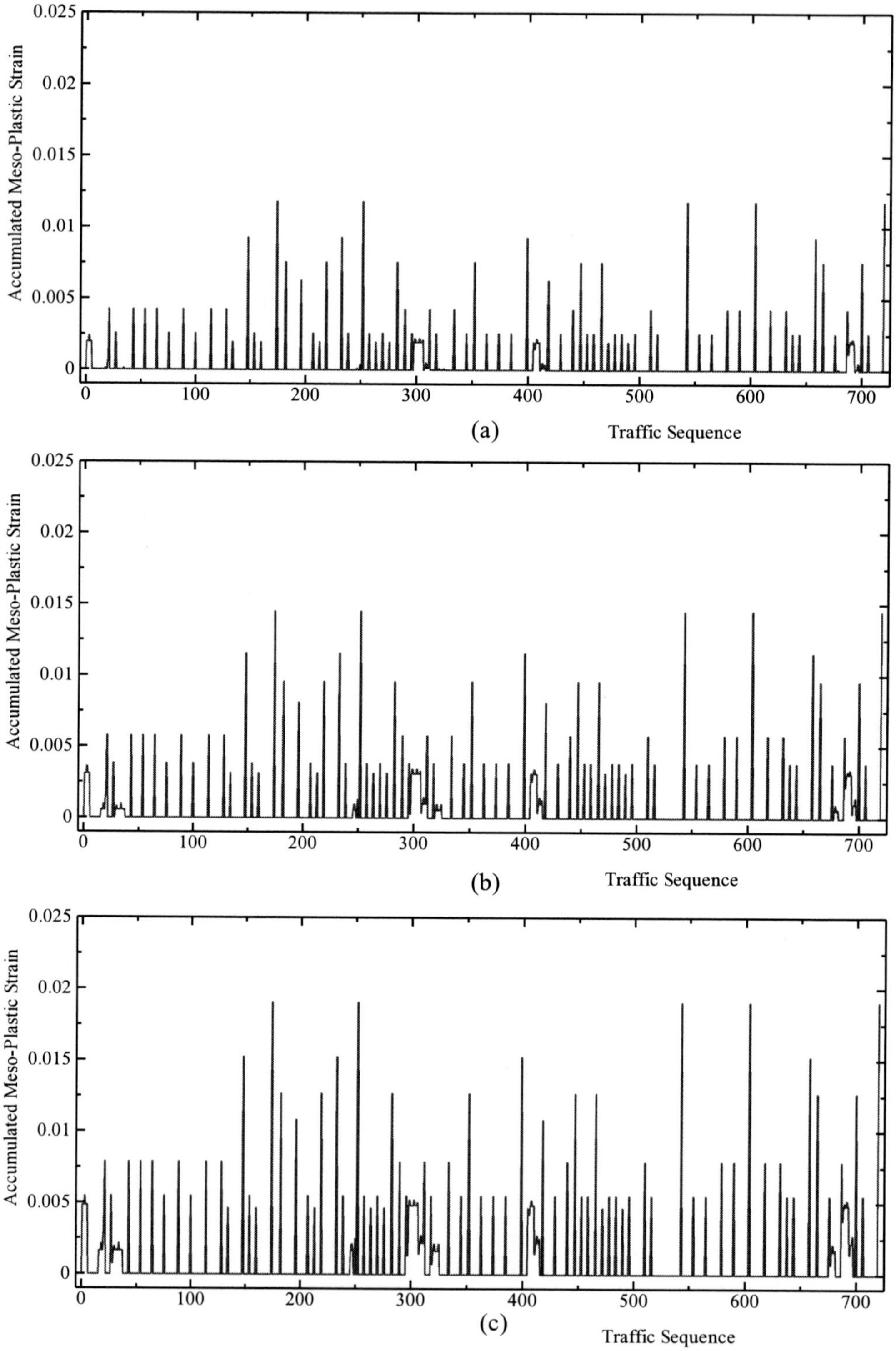

Figure 21. Accumulated meso-plastic strain variation with the traffic sequence per single day, (a) when all six rivets are active: (b). when five rivets are active, (c). when four rivets are active.

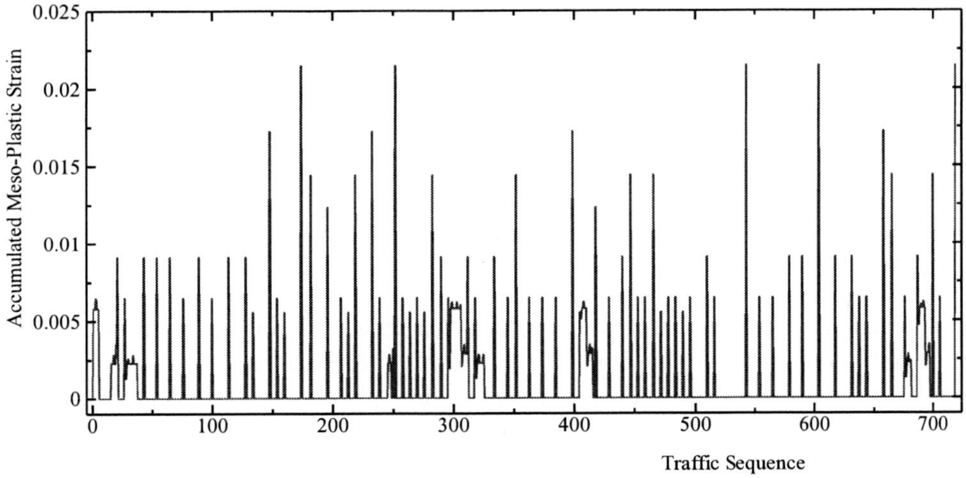

(a)

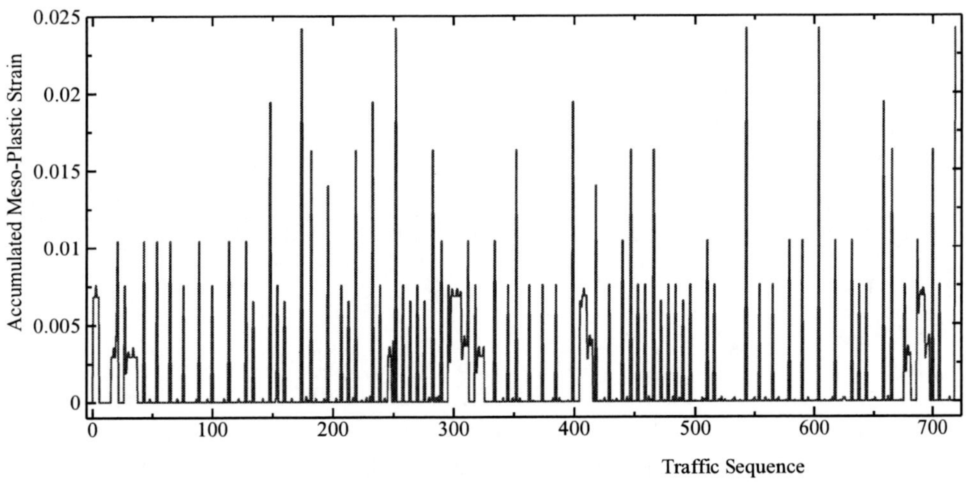

(b)

Figure 22. Accumulated meso-plastic strain variation with the traffic sequence per single day, (a) when three rivets are active: (b). when two rivets.

5.6. Comparison and Discussion of Results

The comparisons revealed that the proposed fatigue model based estimations deviate from previous fatigue theories based estimations. Further it can be observed that for all six rivets are active case, the effective stress based approaches predicted lives significantly deviate and overestimate from proposed model prediction than two rivets are active case. However most of field situations of existing connections are closer to case of all six rivets are active. Hence it reveals that there were possibilities of unexpected failures of connections in practice in history since previous fatigue models predictions are beyond the real life than proposed model (section 3). However, only based on this particular connection, it is possible to say that most of existing connections in field may also have such deviation or overestimation of remaining fatigue life because stress histories are different from connection to connection. But there may have some situations in practice which are similar to this case study. So this idea makes a suspect that this over prediction of life of previous approaches may have some level of contribution to happen some of bridge failures. Finally this case study confirms the importance of accurate fatigue models to estimate fatigue life of existing connections.

6. Conclusion

A new damage indicator based mesoscopic scale fatigue model was proposed to estimate the life of structural connections more preciously. A verification of the new model was conducted by comparing the estimated fatigue life with experimental life under variable amplitude and random multiaxial proportional loading. The proposed fatigue model was further utilized to estimate the remaining fatigue life of a riveted bridge connection.

Experimental verification of the proposed mutiaxial high cycle fatigue model shows that proposed model gives a better comparison to the experimental behavior than previous fatigue models. The case study (section 5) reveals that over prediction of life of previous approaches may have some level of contribution to happen some of bridge failures in past and hence confirm the importance of accurate fatigue models to estimate fatigue life of existing connections. These observations tend to conclude that the application of proposed fatigue model is advisable for the evaluation of fatigue life of structural connections under service loading. This model is applicable to most of the civil engineering structures since associated material parameter determination procedures are dependent on commonly available fatigue test data.

Acknowledgment

The authors wish to express their sincere gratitude to Prof. M.P Ranaweera, University of Peradiniya, Sri Lanka, Prof. A. Carpinteri and Prof. A. Spagnoli, University of Perma, Italy, team of experts who worked in the Sri Lankan Railway Bridge project, for their great advice in carrying out this research. Also the authors convey their kind gratitude to Prof. E. Macha and Prof. Z. Marciniak, Opole University of Technology, Poland, for providing us with the necessary loading histories and results of multiaxail fatigue tests.

References

Akesson, B. *Fatigue life of riveted railway bridges.* PhD Thesis, Sweden: Chalmers University of Technology, 1994.

Alampalli, S; Lund, R. Estimating fatigue life of bridge components using measured strains. *Journal of Bridge Engineering*, 2006, 11(6), 725-736.

Carpinteri, A; Spagnoli, A; Vantadori, S. A multiaxial fatigue criterion for random loading. *Fatigue & Fracture of Engineering Materials & Structures*, 2003, 26(5), 515-522.

Christ, HJ. Cyclic stress-strain response and microstructure. *ASM Hand book*, 1996, 19, 72-95.

Cullimore, M. SG. The fatigue strength of wrought iron after weathering in service. *The structural Engineer*, 1967, 45(5), 193-199.

Downing, SD; Socie, DF. Simple rainflow counting algorithms. *International Journal of Fatigue*, 1982, 4(1), 31-40.

Fisher, JW; Yen, BT; Wang, D. Fatigue and Fracture Evaluation for Rating Riveted Bridges." *NCHRP Report No 302*, Transportation Research Board, National Research Council, Washington D.C, 1994.

Imam, B; Righiniotis, TD; Chryssanthopoulos, MK. Fatigue assessment of riveted railway bridges. *International Journal of Steel Structures*, 2005, 5(5), 485-494.

Imam, BM; Righiniotis, TD; Chryssanthopoulos, MK. Numerical modeling of riveted railway bridge connections for fatigue evaluation. *Engineering Structures*, 2007, 29 (11), 3071-3081.

Jabbado, M; Maitournam, MH. A high cycle fatigue life model for variable amplitude multiaxial loading. *Fatigue & Fracture of Engineering Materials & Structures*, 2008, 31(1), 67-75.

Karolczuk, A; Macha, E. Fatigue fracture planes and expected principle stress directions under variable amplitude loading. *Fatigue & Fracture of Engineering Materials & Structures*, 2005, 28 (1-2), 99-106.

Kluger, K; Lagoda, T. Application of the Dang-Van criterion for life determination under uniaxial random tension-compression with different mean values. *Fatigue & Fracture of Engineering Materials & Structures*, 2004, 27(6), 505-512.

Li, ZX.; Chan, THT. Fatigue criteria for integrity assessment of long span steel bridge health monitoring. *Theoretical and Applied Fracture Mechanics*, 2006, 46(2), 114-127.

Liu, A. *Mechanics and mechanism of fracture.* USA: ASM Internationals, 2005.

Marciniak, Z; Rozumek, D; Macha, E. Fatigue lives of 18G2A and 10HNAP steels under variable amplitude and random non-proportional bending with torsion loading. *International Journal of Fatigue*, 2008, 30(5), 800-813.

Mesmacque, G; Garcia, S; Amrouche, A; Rubio-Gonzalez, C. Sequential law in multiaxial fatigue, a new damage indicator. *International Journal of Fatigue*, 2005, 27(4), 461-467.

Morel, F. A fatigue prediction method based on a mesoscopic approach in constant amplitude multiaxial loading. *Fatigue & Fracture of Engineering Meterials & Structures*, 1998, 21(3), 241-256.

Papadopoulos, IV. Multiaxial fatigue limit criterion of metals: A mesoscopic scale approach, High cycle metal fatigue in the context of mechanical design. *CISM Course and lecture No. 392*, Springer Verlag, Berlin, 1999, 89-143.

Railtrack. RT/CE/C/025. *Rail track line code of practice: The structural assessment of underbridges*, Railtrack, 2001.

Siriwardane, S; Ohga, M; Dissanayake, R; Taniwaki, K. Application of new damage indicator-based sequential law for remaining fatigue life estimation of railway bridges. *Journal of Constructional Steel Research,* 2008, 64(2), 228-237.

Suresh, S. *Fatigue of materials*; Second edition, Cambridge University Press, UK. 1998.

Troshchenko, VT.; Shestopal, LR. *Criteria of torsional fatigue strength of metals*. Institute of Strength of Materials, Academy of sciences of Ukrainian SSR, 1975, 2, 127-131.

In: Structural Steel and Castings
Editor: Lena M. Becker, pp. 185-206

ISBN: 978-1-61728-104-4

Chapter 5

SOLVENT BASED THERMOSET POLYMER CLAY NANOCOMPOSITE CASTINGS FOR GAS BARRIER APPLICATIONS

*V. Mittal**

Institute of Chemical and Bioengineering, Department of Chemistry and Bio-engineering, ETHZ Zurich, Zurich, Switzerland

Abstract

Thermosetting polymer nanocomposites castings based on polyurethane and epoxy polymers have been described. The goal was to develop these materials with high gas barrier properties for achieving their use in the packaging applications. The synthesis methodology for the generation of these castings and the analysis of subsequent parameters like thickness of the coatings, density of the composite etc. have been explained. The optimization of the curing parameters like curing time, temperature, crosslinker amount etc. was obtained by using differential scanning calorimetry. The filler used in the study have been modified with different surface modifications in order to quantify the effect of the interaction of the filler surface modification with the polymer matrices on the composite microstructure and properties. It was concluded that the exfoliated platelets contribute maximum to the barrier properties of the castings. Also, it was the interactions of the filler modifications with the polymer, rather than initial high basal plane spacing of the filler by exchanging long alkyl chains, which leads to exfoliation of the filler platelets and hence enhancement of the barrier properties.

Keywords: epoxy, castings, polyurethane, clay, delamination, surface modification, barrier properties, mechanical performance, intercalation, curing, excess, thermal degradation.

* E-mail address: vikas.mittal@chem.ethz.ch, Current Address: BASF SE, Polymer Research, 67056 Ludwigshafen, Germany. (Corresponding author)

1. Introduction

1.1. Polymer Nanocomposites

Pure polymer matrices need to be reinforced with fillers in order to improve their low mechanical, thermal and barrier properties. Conventional composites generally have the phase mixing on a microscopic scale. Consequently, these composites require the addition of large amounts of reinforcing fillers in order to achieve impressive property enhancements, which generally make them bulkier and opaque. Polymer-layered silicate nanocomposites, with phase mixing of the organic and inorganic phases at a nanometer level, have generated much research interest in the recent years due to the remarkable enhancements in the composite properties at very low volume fractions [1-8]. The inorganic filler i.e. layered silicate (aluminosilicates) clay particles, used for such composites must have at least one dimension less than 100 nm. Presence of the confined polymer chains in-between the nanometer thick delaminated elementary clay layers produces an altogether different morphology. Because of these tremendous interfacial contacts of the nanoparticles with the polymer matrix owing to their high surface area and platy structures, the generated properties are much more superior to the parent materials. The requirement of less volume fraction of filler as compared to the conventional composites also helps these nanocomposites in retaining their transparency and low density. Although the possibilities to generate nanocomposites were reported on more than one occasions, however, real research effort started with the advent of nanocomposites with polyamides in the early nineties developed by Toyota researchers [1-2]. In these composites, the authors proved to achieve tremendous gains in mechanical as well as thermal properties of polyamide composites by the dispersion of organically modified inorganic filler platelets at nanometer level. Since then, a large number of research studies have been reported dealing with a wide variety of polymer and inorganic filler systems and a number of different aspects of polymer nanocomposite properties have been studied in detail. The platy nature of the clay layers or platelets leads to high aspect ratio and the properties of the composites are observed to be dependant on the aspect ratio or geometrical shape of the platelets, their volume fraction as well their interactions with the organic phase at the interface. The improvement in the mechanical properties of the nanocomposites is generally reported, and the other properties like thermal resistance etc. are also automatically improved. The more sensitive properties like gas barrier properties however need to be studied separately as automatic enhancement of the gas barrier properties along with mechanical properties is not always achieved.

1.2. Inorganic Filler Materials

Montmorillonite from the family of 2:1 layered phyllosilicates is generally a material of choice as reinforcing filler because of its easy swelling in water and high surface area available. One such 2:1 aluminosilicate layer has two silica sheets condensed with an alumina sheet [9,10]. The resultant structure consists of alumina octahedral sheet sandwiched between two sheets of inward pointing silica tetrahedra [11]. These layers are stacked one above the other leading to regular gaps or interlayers in between. Since the type of cation occupying tetrahedral and octahedral sheet is limited more by ionic size and coordination than by valency, there is considerable scope for isomorphic replacement or substitution in these structures [11]. In montmorillonite, the common situation is the partial replacement of Al^{3+} with Mg^{2+} in the octahedral sheet. Due to these substitutions, the layers have a permanent net negative charge in their structural framework. This positive charge deficiency is balanced by sorption of extraneous inorganic alkali or alkaline earth metal cations, which are present in the interlayer as well as on the surfaces. Most of such inorganic minerals have high energetic hydrophilic surfaces, which make them incompatible with the hydrophobic polymer matrices. Because of the relatively weaker attractive forces holding the elementary clay layers in montmorillonite, these layers can be delaminated in water to give nano-sized platelets with very high aspect ratios, whose inorganic surface cations can then be exchanged with organic cations. An exchange of inorganic cations with organic cations renders the clay organophilic and hydrophobic and lowers the surface energy of the clay layers. It then becomes possible for the polymer chains to diffuse between the expanded clay interlayer and to delaminate the clay stacks to individual layers. Long chain alkyl ammonium salts are widely used for exchanging the inorganic cations to modify the clay surfaces [11-24]. Figure 1 is the representation of the surface modification process, where the exchange of surface cations ith the long chain ammonium ions has been shown.

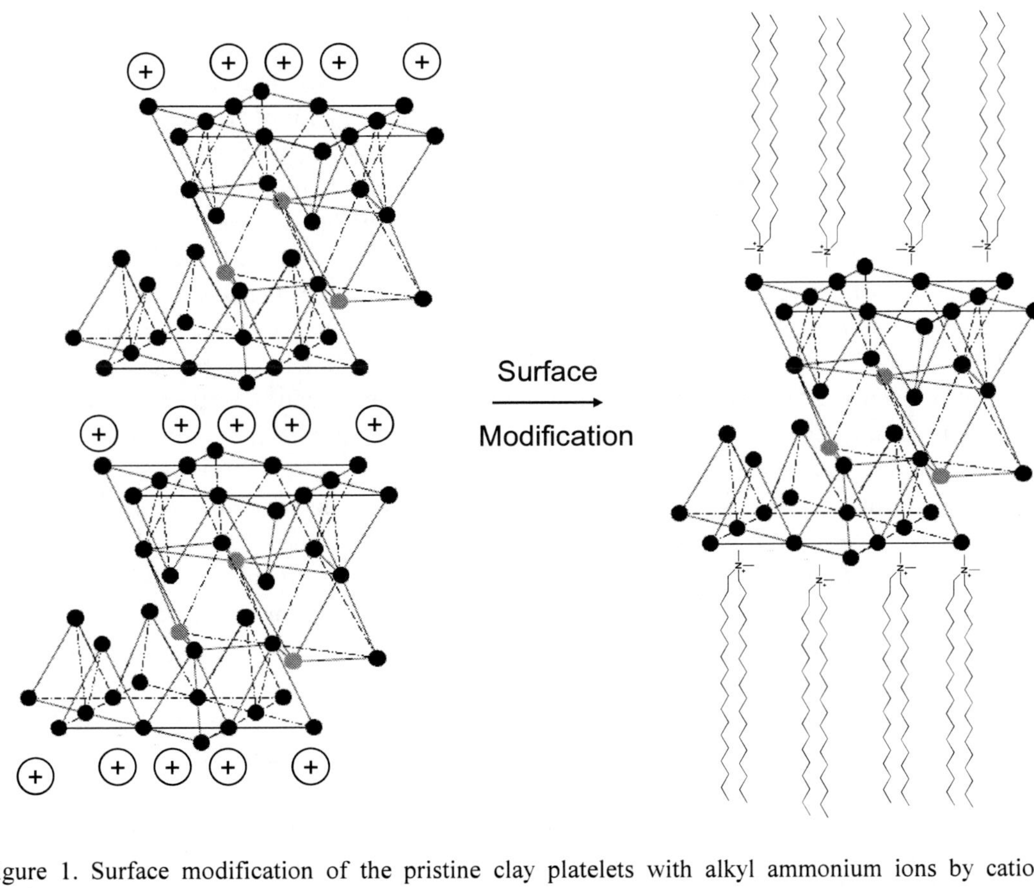

Figure 1. Surface modification of the pristine clay platelets with alkyl ammonium ions by cation exchange.

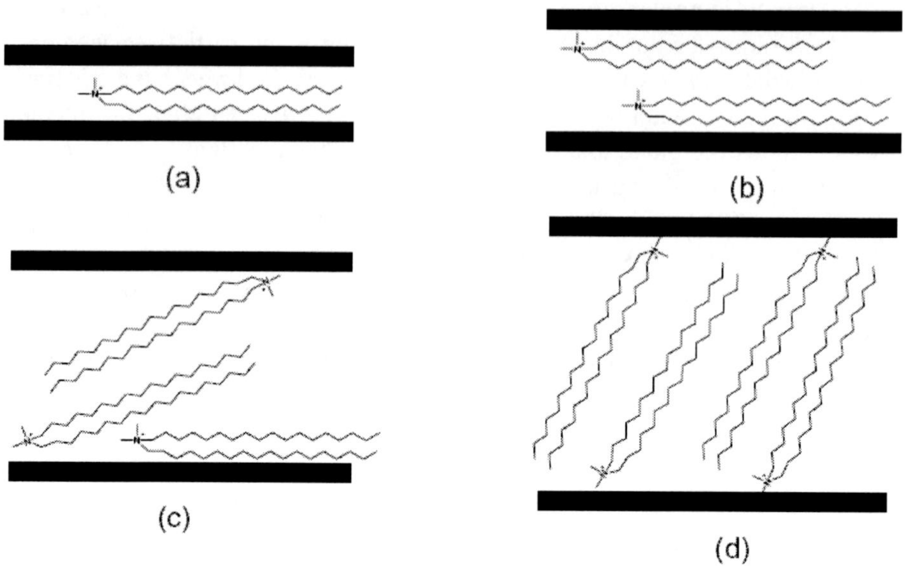

Figure 2. Various arrangements of the surface modification molecules on the clay surface, (a) monolayer, (b) bilayer, (c) pseudo trilayer and (d) paraffin structure.

Though montmorillonite has been the most commonly used aluminosilicate in these studies, however, owing to its low charge density (0.25-0.5 equiv.mol^{-1}), a larger area/cation is available on the surface which leads to a lower basal plane spacing in the clay after surface ion exchange with alkyl ammonium ions. On the other hand, minerals with high charge density (1 equiv.mol^{-1}) like mica and, hence, subsequently smaller area/cation, do not swell in water and, thus, do not allow the cation exchange. However, aluminosilicates with medium charge densities of 0.5-0.8 equiv.mol^{-1} (like vermiculite), offer a potential of partial swelling in water and cation exchange which can lead to much higher basal plane spacing in the modified mineral, if optimum ion exchange is achieved. Owing to the higher basal plane spacing, the electrostatic interactions holding the layers together can be expected to be lower than similar montmorillonite counterparts, thus, increasing the potential of better properties of the hybrid nanocomposites.

In order to describe the structure of the interlayer in organoclays, it is believed that to counter the negative charge on the surface, the cationic head group of the alkyl ammonium molecule preferentially resides at the surface layer, leaving the organic tail radiating away from the surface. Two parameters then define the equilibrium layer spacing in a given temperature range: the cation exchange capacity of the layered silicate, driving the packing of the chains and the chain length of organic tails [15]. According to X-ray diffraction (XRD) data, the organic chains have long been thought to lay either parallel to the silicate layer, forming mono or bilayers, or depending on the packing density and the chain length, to radiate away from the surface, forming pseudo trilayers or even tilted 'paraffinic' arrangement [16]. A more realistic description was proposed by Vaia et al. [17], based on FTIR experiments. By monitoring frequency shifts of the asymmetric CH_2 stretching and bending vibrations, it was reported that the intercalated chains exist in states with varying degrees of order. In general, as the interlayer packing density or the chain length decreases, the intercalated chains adopt a more disordered, liquid like structures resulting from an increase in the gauche/trans conformer ratio. When the available surface area per molecule is within a certain range, the chains are not completely disordered but retain some orientational order similar to that in the liquid crystalline state. Figure 2 shows the various possibilities of the arrangement of the alkyl ammonium chains inside the clay interlayers.

1.3. Nanocomposite Synthesis and Types of Nanocomposite Microstructures

The following are some of the techniques employed excessively to synthesize the polymer clay nanocomposites:

Monomer intercalation method: The layered silicate is swollen in liquid monomer. Polymerization of the monomer is then carried out in the interlayers of the clay mineral, resulting in an expanded interlayer distance and silicate layers that are homogenously dispersed on a nanometer level at the end of the polymerization.

Common solvent method or solution method: The layered silicate is delaminated to the individual layers in a solvent where the polymer is also soluble. This method is generally used for the thermosetting polymers. The polymer is then crosslinked, and the solvent is evaporated.

Melt intercalation method: This is by far the most commonly employed method to achieve the composites with thermoplastic matrices especially polyolefins. The polymer melt is mixed with the powder mineral and is mixed extensively so as to obtain the homogenous dispersion of the clay layers in the matrix. In the case of non-polar thermoplastic materials like polyethylene and polypropylene, the difference in the surface energies of the polymer and the clay always hinders the intercalation or exfoliation. A successful approach was suggested by Kato et al. [18], in which a PP oligomer grafted with maleic anhydride or hydroxyl groups was used along with PP as a compatibilizer, and an increased intercalation was reported.

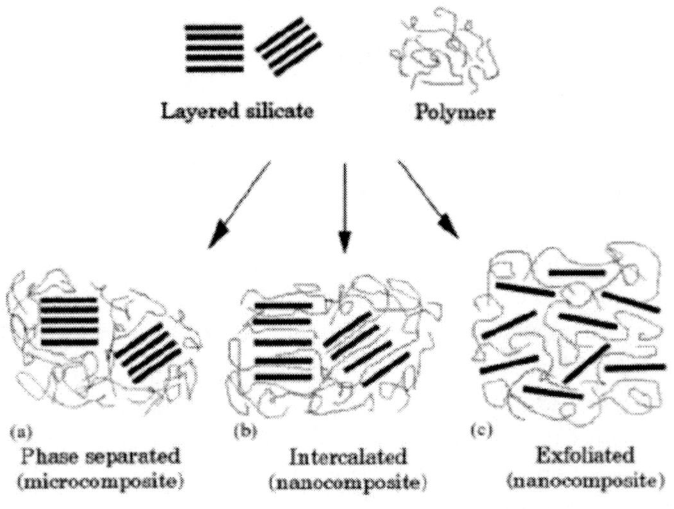

Figure 3. Representation of intercalated or exfoliated nanocomposites.

Depending on the nature of the components used (layered silicate, organic cation and polymer matrix) and the method of preparation, three main types of composites may be obtained as shown in Figure 3 [15]. Inability of the polymer to intercalate between the silicate sheets causes a phase separated composite (Figure 3a) to form, whose properties stay in the same range as traditional microcomposites. Two types of nanocomposites can be achieved beyond this classical family of composites. Intercalated structure (Figure 3b) in which a single (and sometimes more than one) extended polymer chain is intercalated between the silicate layers resulting in a well ordered multilayer morphology built up with alternating polymeric and inorganic layers. Complete and uniform dispersion of the silicate layers in a continuous polymer matrix leads to an exfoliated or delaminated structure (Figure 3c). Two complementary techniques are used in order to characterize these structures. X-ray diffraction allows the interlayer spacing to be determined in intercalated nanocomposites owing to the preserved repetitive multilayer structure. The intercalation of the polymer chains usually increases the interlayer spacing, in comparison with the spacing of the organoclay used, leading to a shift of the diffraction peak towards lower angle values (angle and layer spacing values being related through the Bragg's relation: $\lambda = 2d \sin \theta$, where λ corresponds to the wavelength of the X-ray radiation used in the diffraction experiment, d the spacing between diffractional lattice planes and θ is the measured diffraction angle or glancing angle). As far as exfoliated structure is concerned, no more diffraction peaks are visible in the XRD diffractograms either because of a much too large spacing between the layers (i.e. exceeding 8 nm in the case of ordered exfoliated structure) or because the nanocomposite does not present ordering anymore. But this idealized classification can be erroneous owing to the sensitivity of the X-ray signal to sample preparation, orientation, impurities and time of scanning. It is the exfoliated morphology of the composites which is of interest as it leads to maximum enhancement in the composites properties. The aspect ratio of the platelets in exfoliated state is very high and it helps to generate better organic-inorganic interfacial interactions, stress transfer, heat transfer and barrier to the permeant molecules as the filler platelets do not allows the penetrant molecules to pass through. The intercalated platelets on the other hand have no increment in the aspect ratio and thus are not as effective as the exfoliated platelets.

1.4. Thermoset Nanocomposite Castings

Thermoset polymer castings find uses in a wide range of applications, packaging application being one of them. The castings for packaging application need to be highly resistant to gas permeation. Apart from that, they should also be mechanically strong. Polymer nanocomposite castings by the incorporation of the high aspect ratio organically modified filler platelets and by properly compatibilizing the organic and inorganic components can lead to enhancements in their gas barrier as well as mechanical properties. As mentioned above, the filler need to be exfoliated in the polymer matrix as the intercalated filler platelets are not very effective in improving especially the gas barrier performance.

Polyurethane and epoxy are the two most important thermoset polymer systems [19-24] extensively employed in castings. Commercial development of the polyurethanes was achieved from the work of German chemists attempting to circumvent the Du Pont patents on nylon 66. Bayer et al. were able to produce fibre-forming polymers by reacting aliphatic di-isocyanates and aliphatic diols (glycols). Subsequent work resulted in the production of useful products by using polymeric hydroxyl-containing compounds such as polyesters to give rubbers, foams, coatings and adhesives. By 1980 global consumption was of the order of 3 x 10^6 tonnes per annum. In the year 2000, global production was of the order of 9 x 10^6 tonnes per annum.

A wide range of polyurethane-type products has become available in recent years for casting applications. These include simple solutions of linear polyurethanes, two-pot alkyd-isocyanate and polyether-isocyanate systems and a variety of prepolymer and adduct systems. The coatings vary considerably in hardness and flexibility and find use mainly because of their toughness, abrasion resistance and flexibility. Uses include metal finishes in chemical plant, wood finishes for boats and sports equipment, finishes for rubber goods, adhesives coatings for the packaging laminates and rain erosion- resistant coatings for aircraft. Many isocyanates have good adhesive properties.

In the early stage of their development, the epoxy resins were used almost entirely for surface coating and developments in this field are to a large extent due to the works of S. O. Greenlee. By the beginning of the 1980s world capacity for epoxide resins reached about 6 x 10^5 tonnes per annum. But with a global consumption of about 10 million tonnes per annum for thermosetting plastics, epoxide resins had a share of about 3%. By the late 1990s, however, the world market for epoxide resins had risen to about 7.5 x 10^5 tonnes per annum. About half of epoxide resin production is used for surface casting applications, with the rest divided approximately equally between electronic applications (particularly for printed circuit boards and encapsulation), the building sector and miscellaneous uses.

The epoxide resins of the glycidyl ether type are usually characterized by parameters like resin viscosity (of liquid resin), epoxide equivalent, hydroxyl equivalent, average molecular weight (and molecular weight distribution), melting point (of solid resin) and heat distortion temperature (deflection temperature under load) of cured resin. Resins must, however, be compared using identical hardeners and curing conditions. Amine hardeners cross-link epoxide resins either by a catalytic mechanism or by bridging across epoxy molecules. In general the primary and secondary amines act as reactive hardeners whilst the tertiary amines are catalytic. Being cross-linked, the epoxy resins do not dissolve without decomposition but can be swollen by liquids of similar solubility parameter to the cured resin. The chemical resistance is as much dependent on the hardener as on the resin since these two determine the nature of the linkages formed. The epoxide resins are used in a large number of fields, including in surface coatings, in adhesives, in potting and encapsulation of electronic components, in tooling, for laminates in flooring and to a small extent in moulding powders and in road surfacing.

Owing to their immense commercial importance and potential for expanding use in gas barrier applications in packaging industry, the current chapter focuses on the polyurethane and epoxy thermoset systems for the synthesis of nanocomposites. Also, the gas barrier properties are focused as these properties have been generally neglected in the literature, the nanocomposites have been often synthesized to achieve improvements in the mechanical performance and it is assumed that the enhancement in the mechanical performance brings automatic improvement in the barrier performance, which is not necessarily true owing to the significant dependence of the barrier properties on the interface development between the organic and inorganic phases. In the packaging industry, conventionally the thermoset polymers like polyurethanes and epoxy are generally used as adhesives. They are used to adhere together various polymer films in the packaging laminates and do not contribute towards other properties of the laminate. Thus, it is of interest to incorporate filler platelets in these polymers to generate thermoset polyurethane or epoxy nanocomposites which apart from being adhesives can also contribute to the barrier as well mechanical performance of the laminates. Common solvent or solution method is the commonly used method for the generation of thermoset polymer nanocomposite castings. The prepolymer is generally dissolved in a solvent and to this solution, dispersion of filler in the same solvent is added. The addition of crosslinker followed by crosslinking and removal of solvent leads to the generation of polymer nanocomposites. The microstructure and the properties of the castings are significantly affected by the synthesis methodology used for the generation of the nanocomposites. The rate of crosslinking, time and temperature of crosslinking, amount of crosslinker etc. all have significance on the final properties. These parameters are required to be optimally designed especially taking into account the effect of presence of clay in the system so that the polymerization and crosslinking rate in the interlayers is not overtaken be the polymerization and crosslinking of the polymer taking place outside the interlayers. The forthcoming sections focus on the synthesis of the thermoset nanocomposite castings, the optimization of curing parameters as well as gas barrier properties of generated castings.

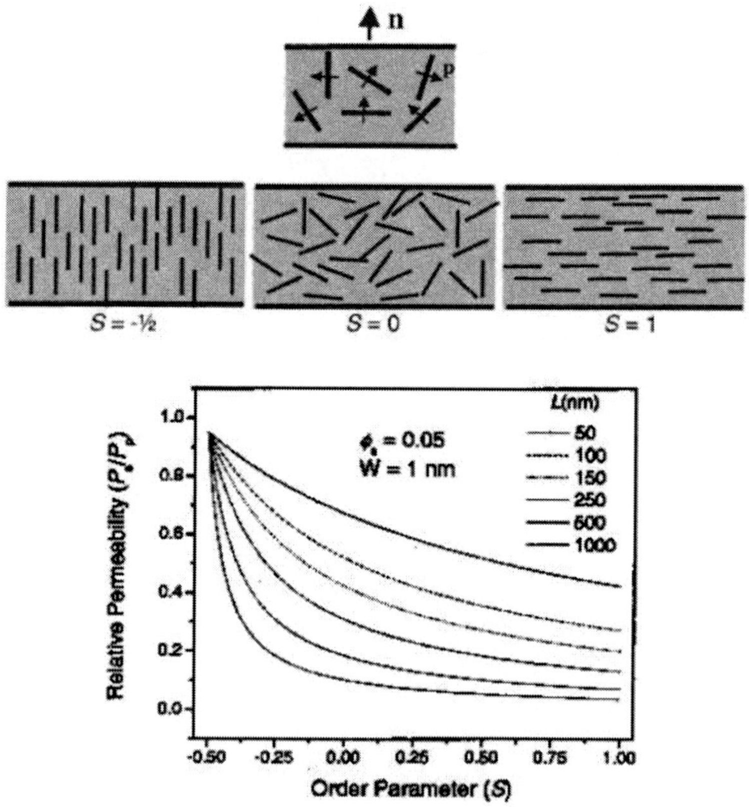

Figure 4. Representation of platelet orientation on the permeability through nanocomposites. Reproduced from reference 25 with permission from American Chemical Society.

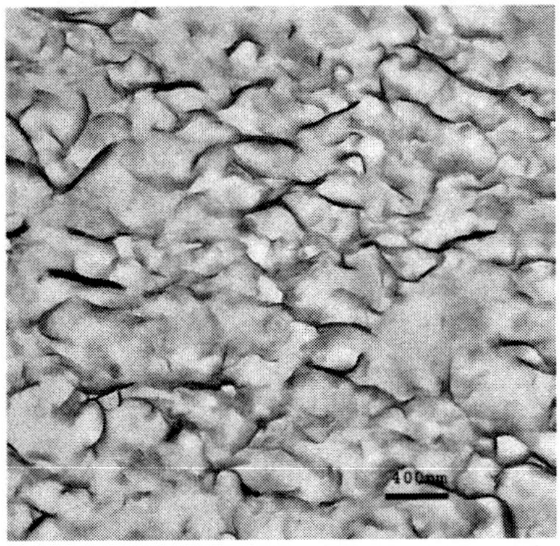

Figure 5. TEM micrograph of the epoxy nanocomposite indicating the misalignment, bending and folding of filler platelets.

It has been mentioned above that the exfoliated platelets contribute maximum toward the barrier performance of nanocomposites. However, the alignment of the platelets in the polymer matrix is also of importance as the misaligned platelets are not as effective as the aligned ones as shown in Figure 4 [25]. For the same aspect ratio of the platelets, the aligned platelets were observed to have higher extent of decrease in permeation as compared to aligned ones. The castings generally generate completely misaligned, bent and folded platelets as the platelets do not undergo any shear induced alignment in the case of solvent based castings owing to low viscosity. Such alignment is more possible in the case of melt mixing of the high polymer weight polymer with the clay in the absence of any solvent. Figure 5 shows the transmission electron micrograph of epoxy nanocomposite castings where the complete misalignment of the filler platelets is clearly visible (black lines correspond to the cross-section of the alumino-silicate clay platelets.

2. Synthesis of Nanocomposite Castings

For the synthesis of nanocomposite castings, suitable substrate foils like polypropylene or polyamide were chosen. The substrates were required as no free standing casting could be generated owing to the adhesive nature of the polymer systems.

The substrates were chosen in such a way that the permeation of either oxygen or water vapor through these foils was very high as compared to the castings so that there was no hindrance in the measurement of the permeation properties of the castings. The first step in the synthesis process involved the swelling of the organically modified montmorillonite in a solvent (tetrahydrofuran, THF or dimethylformamide, DMF). The swelling was assisted by the sonication for 10 minutes in intervals while keeping the amplitude of sonication lower as well as cooling the suspension so as to avoid any mechanical damage to the platelets and thermal damage to the surface modification ionically bound to the platelets. The next step was the addition of polyurethane prepolymer or epoxy resin in the suspension and to better intercalate the prepolymer molecules in the filler interlayers, sonication was used. The curing agent was then added. The amount of the curing agent is very critical as it affects the microstructure and properties of the composites. In the case of polyurethane nanocomposites 0.2 g of crosslinker was added for 6 g of prepolymer, whereas in case of epoxy nanocomposites, the crosslinker amine to epoxy mole ratio of 0.3:1 was maintained. The optimization of the parameters for the nanocomposite synthesis is also reported in the next section. The polyurethane prepolymer was a ca. 60% solution of a polyester urethane with iso-cyanate end groups in ethyl acetate, which also contained 1-3% of diphenylmethane-4,4'-diisocyanate. The cross-linker used for polyurethane system consisted of a mixture of polyetherglycol carrying amino groups (ca. 80-90%) and 3-(aminopropyl)triethoxysilane (ca. 10-20%). In case of epoxy nanocomposites, bisphenol A diglycidyl ether (4,4'-isopropylidene-diphenol diglycidyl ether) with an epoxide equivalent weight 172-176, was used as epoxy resin and tetraethylenepentamine (TEPA) was the amine crosslinker. The montmorillonites modified by various surface modifications were used to synthesize nanocomposite castings and the filler content was also varied in order to quantify the effect of increasing filler fraction on gas barrier properties. The castings were drawn with a bar coater and thickness of the wet castings was also controllable. Figure 6 shows the casting assembly with a substrate foil and a bar coater placed on casting plate which had vacuum underneath to properly place the substrate foil. After drawing castings on the substrate foils, they were dried at room temperature and atmospheric pressure followed by room temperature and reduced pressure. The polyurethane castings were cured at 40°C overnight under reduced pressure followed by further curing at 60°C for 4 hours. The epoxy castings were cured at 70°C overnight followed by post curing at 90°C for further 4 hours.

The composite properties are often related to the inorganic volume filler fraction in the nanocomposite. It can be calculated by dividing the product of composite density (ρ_c) and mass fraction (m_m)of the non- modified montmorillonite in the composite with the density (ρ_m, 2.6 g.cm3) of the pristine unmodified montmorillonite [27,28]. The alumino-silicate mass fraction in the composite can be obtained by knowing the correct mass fraction of the organic monolayer on the filler (m_{ml}). Tis value can be obtained from thermogravimetric analysis of the organically modified montmorillonite (OM) by subtracting the mass loss due to moisture desorption and that due to dehydroxylation of the Na-montmorillonite from the total mass loss [26]. The density of the composite (ρ_c) was calculated from that of the unfilled polyurethane (PU), determined as described below and that of the OM (ρ_{OM}) as follows:

$$\rho_c = \frac{1}{\dfrac{m_{OM}}{\rho_{OM}} + \dfrac{m_{PU}}{\rho_{PU}}}$$

Here m_{OM} is the mass fraction of the OM and m_{PU} is the mass fraction of the polyurethane. Similarly, it can also be tailored for epoxy nanocomposites. The density of the OM (ρ_{OM}) was calculated similarly as composite density from that of the monolayer (assumed to be equivalent to that of octadecanol ρ_{ml} = 0.81 g.cm^3) and that of Na montmorillonite.

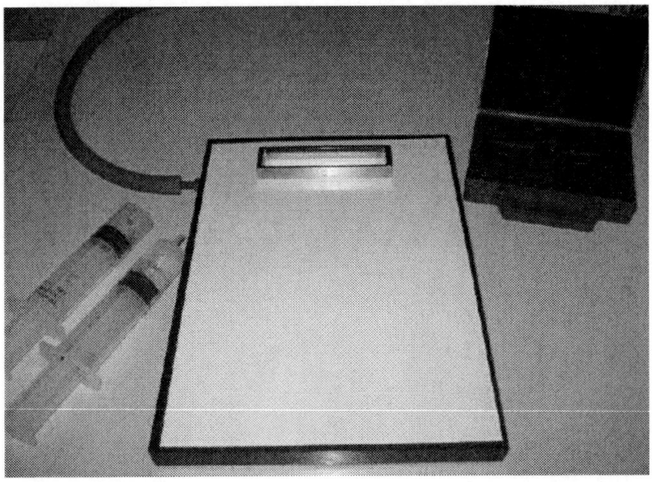

Figure 6. Assembly showing the coating plate carrying the substrate foil and the bar coater.

The density of the substrate foils (PP and PA) and pure polyurethane and epoxy was determined by weighing them in air and in ethanol using an analytical balance (Mettler AE 200) and a homemade device similar to the Mettler density kit ME-33360. the density was calculated by dividing the product of density of ethanol ρ_{Et} sample mass in air M with (M- M_1), where M_1 is sample mass in ethanol. The thickness of the substrate film (t) was determined by weighing a piece of defined area. For accuracy, the average of five measurements was taken. The thickness of the casting (t_{coat}) was similarly determined as follows:

$$t_{coat} = \frac{M_1 - M_2}{\rho_c * area}$$

where M_1 is the mass of the coated film, M_2 is the mass of a substrate film with the same area and ρ_c is the density of the composite.

The oxygen transmission rate through PP foils coated with unfilled polyurethane, epoxy and with the nanocomposites was measured at 23 °C, 1 atm and 0% RH. The oxygen transmission rate was normalized with respect to the film thickness. The water vapor transmission rate through PA foils coated with unfilled polyurethane, epoxy and with the nanocomposites was measured at 23 °C, 1 atm and 100% RH. The transmission rate was also normalized with respect to the film thickness. The transmission rate through the nanocomposite film was calculated from that of the coated PP or PA foil as follows:

$$\frac{t_{tot}}{T_{tot}} = \frac{t_1}{T_1} + \frac{t_2}{T_2}$$

where t_i are the film thicknesses and T_i are the normalized transmission rates [29].

An average of four measurements for each sample is reported.

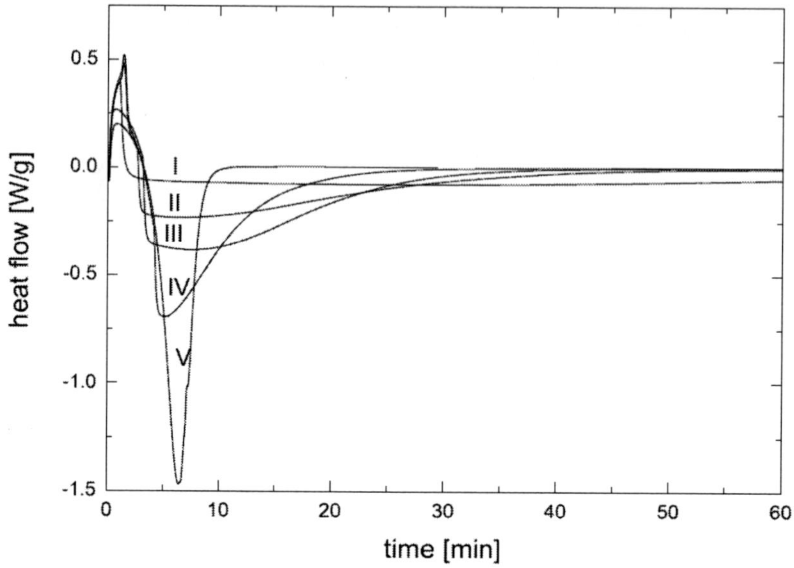

Figure 7. DSC thermograms showing the curing of epoxy at different temperatures at a fixed amine to epoxy mole ratio of 1:1; I: 40°C, II: 55°C, III: 60°C, IV: 70°C and V: 100°C.

3. Optimization of Curing Parameters

The example of epoxy system is presented here. The curing parameters such as time, temperature and mole ratio of the curing agent to epoxy were required to be optimized in order to have the curing process which provides time for the intercalation of epoxy and amine in the clay interlayers, but is also quick enough to avoid excessive extragallery polymerization [73]. Figure 7 shows the DSC plots of epoxy curing runs at different temperatures of 40°C, 55°C, 60°C, 70°C and 100°C, while keeping the amine to epoxy mole ratio constant at 1:1. The trials runs at different amine to epoxy mole ratios, while keeping the temperature constant at 70°C are shown in Figure 8. From these results, curing temperature of 70°C and epoxy to amine ratio of 0.3:1 were chosen as optimal values as it provided enough curing time necessary for exfoliation, along with maintaining thermal stability of the substrate foils at the curing temperature.

Approximately 45 minutes were required to achieve maximum curing in this chosen system. Effect of presence of filler on the overall curing process has also been described in Figure 9. Presence of dioctadecyldimethylammonium modified clay (2C18-OM) led to the advancement of the curing significantly as shown by curve II as compared to curve I without the presence of filler. It is because of the acidic nature of the ammonium ions which have been reported to catalyze the epoxy-amine polymerization reaction [22, 30-32]. Although with quaternary ammonium ions the extent of intergallery catalysis is reported to be lower as compared to primary, secondary and tertiary counterparts, but still it can be significant. Curve III indicates the curing process carried out at 70°C at a reduced amine to epoxy mole ratio of 0.2:1. The slowing down of the curing process is clearly visible. The influence of curing temperature on the filler exfoliation was also analyzed through x-ray as shown in Figure 10. Using an amine to epoxy mole ratio of 0.3:1, the curing at 70°C led to decrease in the peak intensity as compared to the same material cured at 60°C, indicating optimal curing rate and time are the most important criterion. Thus, a balance of curing time, curing temperature, amount of crosslinker etc. is required to achieve filler exfoliation in the polymer.

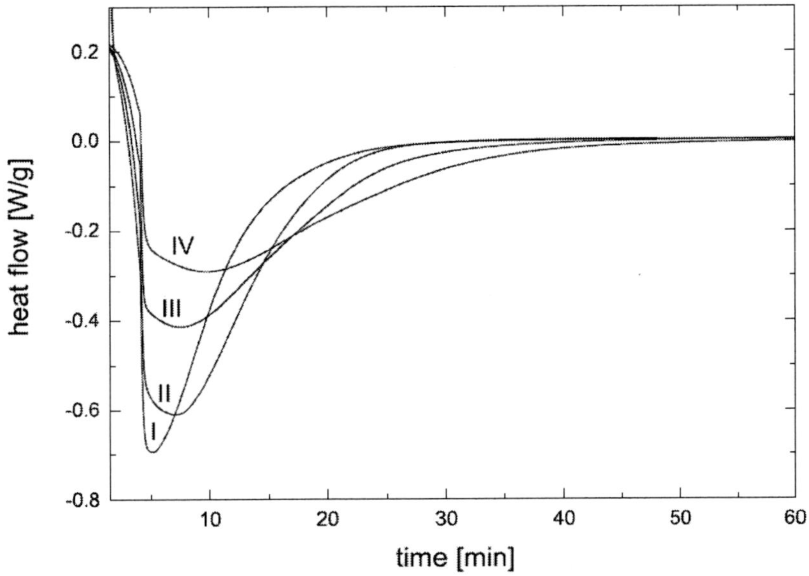

Figure 8. Curing of epoxy at 70°C using different amine to epoxy mole ratios; I: 1:1, II: 0.6:1, III: 0.4:1 and IV: 0.3:1.

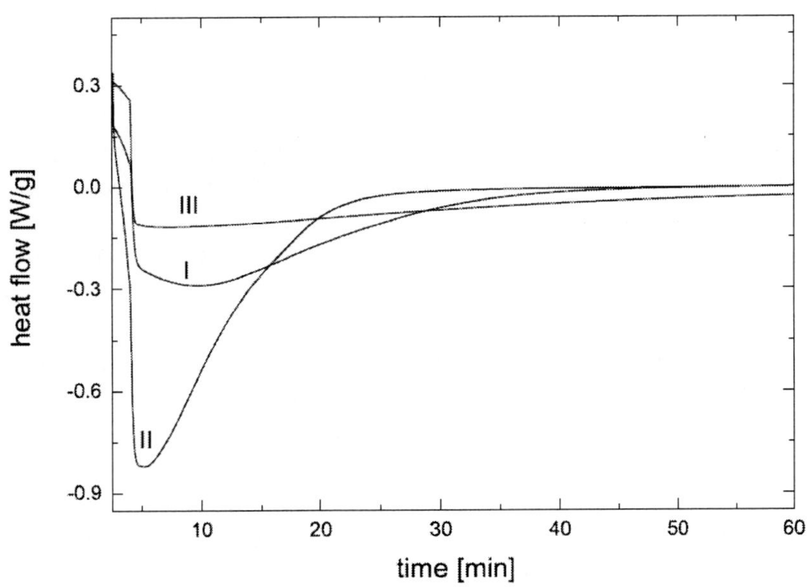

Figure 9. DSC thermograms indicating the effect of filler on curing at 70°C; I: no filler, amine to epoxy mole ratio 0.3:1, II: 3.5 vol% 2C18-OM, amine to epoxy mole ratio 0.3:1 and III: 3.5 vol% 2C18-OM, amine to epoxy mole ratio 0.2:1.

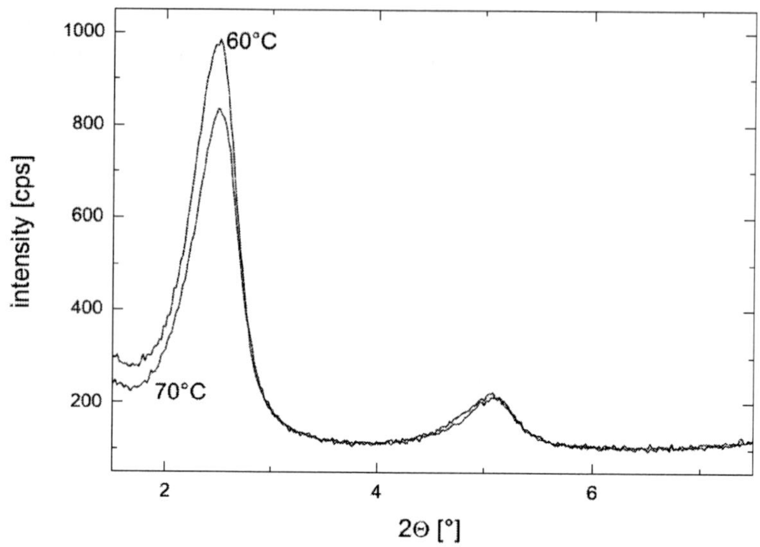

Figure 10. X-ray diffractogram showing the effect of temperature on composite microstructure (3.5 vol% 2C18-OM filler).

4. Gas Barrier Properties of Castings

For polyurethane nanocomposites, three modified montmorillonite with trade names Nanofil 804, Nanofil 32 and Nanofil 15 were chosen. The chemical structure of the organic cations in Nanofil 804 was bis(2-hydroxyethyl) methyl hydrogenated tallow ammonium; for Nanofil 32 was alkylbenzyldimethylammonium (benzalkonium; alkyl being mostly dodecyl chains) and for Nanofil 15 was dimethyl dihydrogenated tallow ammonium, where tallow represents a mixture of long alkyl homologues with octadecyl the most prominent component. The different modifications were chosen in order to quantify the effect of interactions of the surface modification molecules with the polymer matrix on the microstructure and properties of nanocomposite castings. Figure 11 shows the chemical structures of the cations exchanged on the montmorillonite surface. The modification in Nanofil 804 contained two hydroxyl groups in the structure and these groups were expected to have better match of polarity with the polyurethane matrix. The chemical reaction of these groups with the isocyanate groups in the prepolymer during curing was also hoped. The chemical reaction of the surface modification with the polymer leads to adhering of the polymer chains more strongly with the inorganic filler and the generated composite can have better performance than the system where the filler and polymer are not chemically bound to each other. The modification in the case of Nanofil 32 had a benzyl group in the structure apart from a long chain. The benzyl groups are also expected to have netter van der Waals interactions with the polymer which can also lead to better dispersion of the filler in the polymer matrix. The modification in the case of Nanofil 15 filler had two long alkyl chains, thus, this modification represented the most hydrophobic modification. This modification leads to highest basal plane spacing among all the modified montmorillonites.

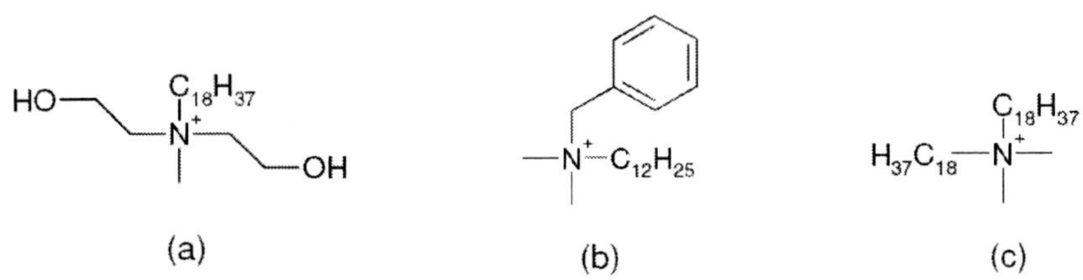

Figure 11. Chemical structures of the cations exchanged on the montmorillonite surface (a) Nanofil 804, (b) Nanofil 32 and (c) Nanofil 15.

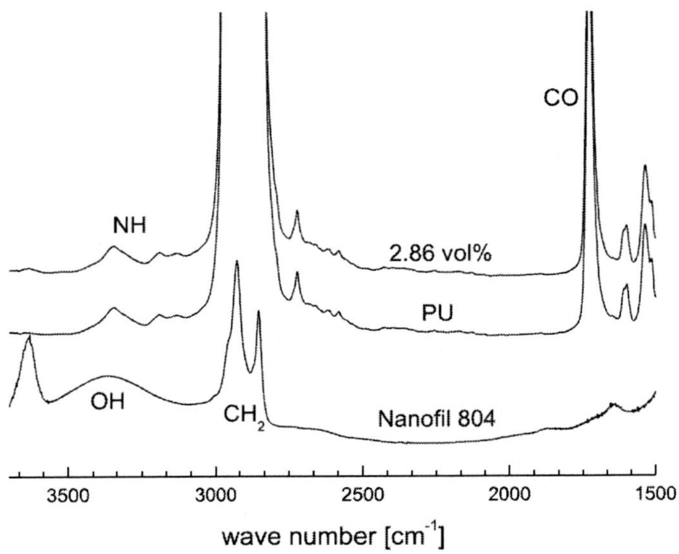

Figure 12. IR spectra of Nanofil 804 and its 2.86 vol % polyurethane-nanocomposite compared to that of the unfilled polyurethane matrix. Reproduced from reference 27 with permission from American Chemical Society.

As mentioned above that by using two hydroxyl groups in the modification, it was expected that these groups could chemically react with the isocyanate groups in the prepolymer thus tethering the polymer chemically with the filler which can have significant effect on composite morphology in terms of increased polymer intercalation on the filler interlays followed by filler exfoliation an enhancement of barrier properties [33,34]. The IR analysis on the modified filler, pure polymer and 2.86 vol% polyurethane composite was carried out in order to gain further insights into the possibility of chemical tethering of polymer and Nanofil 804 filler platelets with each other as shown in Figure 12 [27]. A broad absorption band at 3362 cm^{-1} was observed in the modified clay which can be related to the hydroxyl groups of the modification. On reaction of these groups with the polymer, this absorption band should disappear in the polymer nanocomposites. However, the analysis was complicated by the NH stretching absorption band in the polyurethane at 3345 cm^{-1} as the band was broad and partially superimposed with OH absorption band. Another complication arose from the residual OH groups of the cross-linker, which did not react during curing reaction. These groups could also be observed in the IR analysis as a shoulder on the NH absorption band in the unfilled polymer. Though there is a shoulder 3360 cm^{-1} in the nanocomposites spectrum, but owing to the above observations it can not be safely related to the OH groups from the filler modification. Thus, it was not possible to confirm the chemical reaction between the two phases. However, the other means to obtain an indirect insight was through comparing microstructure and properties. The composites with Nanofil 804 filler had similar basal plane spacing as composites containing Nanofil 32 filler. The oxygen permeation reduction through the composites containing Nanofil 804 was also similar to the Nanofil 32 filled composites (see table 1). As the organic modification in the case of Nanofil 32 has no possibility to chemically react with the prepolymer, it thus indicates that most probably no chemical reaction of the hydroxyl groups in Nanofil 804 with the prepolymer took place. The OH group can become unavailable for the chemical reaction owing to their deactivation caused by their hydrogen bonding with the filler surface. The presence of long alkyl chains in the surface modification can also mask or shield them from the reaction sites. However, it is only an indirect observation as the microstructure and properties of the composites depend on the interaction of the surface modification with the polymer and modification in Nanofil 32 may also have interacted very well physically thus equalizing the effect of chemical bonding of modification of Nanofil 804 filler with the prepolymer.

The gas barrier properties of the polyurethane nanocomposites are reported in Table 1 and 2. The composites with Nanofil 804 and Nanofil 32 showed a decrease in the oxygen permeation as a function of filler volume fraction. A decrease of roughly 35% in oxygen permeation was observed at 5 vol% of the filler. The addition of even 5 vol% of the filler retained the transparency of the castings, the loss of which can hinder the use of these materials for packaging applications. The oxygen permeation through the Nanofil 15 filled composites increased as a function of filler volume fraction. As earlier mentioned that this filler had completely non polar modification, and owing to incompatibility between the non polar surface modification and polar polymer, the oxygen barrier performance was deteriorated. The water vapor permeation is fundamentally different from oxygen permeation as both the permeants differ in the nature of their interaction with the polymer. The water vapor permeation in the composites shown in Table 2 was more a function of hydrophobicity of filler, the more hydrophobic the filler, the better resistance to water vapor permeation.

In the case of epoxy nanocomposites, a wider range of surface modification was synthesized in order to study their effect on the filler exfoliation and composite properties. The fillers were also self-modified in this case as the commercially modified montmorillonites have been observed to contain an excess of surface modification molecules. These excess surface modification molecules can have detrimental effect on the composite properties as these molecules can specifically interact with the prepolymer molecules thus affecting the stability of the whole system. In a recent study [38], the epoxy composites prepared with commercially modified montmorillonites were observed to have significant deterioration in the oxygen permeation resistance, the same modified cation when self exchanged on the surface of montmorillonites followed by extensive cleaning to wash off the excess surface modification, the oxygen permeation properties of the resulting nanocomposites were significantly improved.

Table 1. Oxygen transmission rate through the pure polyurethane and its nanocomposites with varying amounts of filler content. Reproduced from reference 27 with permission from American Chemical Society

V_m, inorganic	oxygen transmission rate ($mm.cm^3/m^2.day$)		
vol fraction	Nanofil 804	Nanofil 32	Nanofil 15
0.0000	6.3	6.3	6.3
0.0141	5.0	5.2	7.4
0.0213	4.8	4.7	8.6
0.0285	4.6	4.3	8.6
0.0358	4.6	4.5	9.0
0.0431	4.5	4.3	9.2
0.0542	4.2	4.0	–

Table 2. Water vapor transmission rate through the pure polyurethane and its nanocomposites with varying amounts of filler content. Reproduced from reference 27 with permission from American Chemical Society

V_m, inorganic	water vapor transmission rate ($mm.g/m^2.day$)		
vol fraction	Nanofil 804	Nanofil 32	Nanofil 15
0.0000	1.9	1.9	1.9
0.0106	1.4	1.2	1.2
0.0213	1.2	1.1	1.1
0.0321	1.1	1.0	0.9
0.0431	1.0	0.9	0.8
0.0542	0.9	0.7	0.6

The surface modifications exchanged on the Cloisite montmorillonite substrate were benzyldibutyl(2-hydroxyethyl)ammonium chloride (Bz1OH), benzylbis(2-hydroxyethyl) butylammonium chloride (Bz2OH), benzyltriethanolammonium chloride (Bz3OH), and benzyl(2-hydroxyethyl) methyloctadecylammonium chloride (BzC18OH), benzyldimethylhexadecylammonium chloride (BzC16) and dioctadecyldimethylammonium chloride (2C18). The exchange reaction was monitored by high resolution thermogravimetric analysis. The high resolution thermogravimetric analysis allows the detection of the presence of any excess surface modification as the excess modification molecules are not ionically bound to the filler surface and are only physically trapped in the filler interlayers. This makes these excess molecules thermally less stable than the ionically bound molecules. Thus, the presence or absence of low temperature degradation in the thermogravimetric analysis indicates the presence or absence of excess surface modification. The thermogravimetric analysis is also helpful in quantifying the amount to faction exchange on the filler surface. From the amount weight loss, the amount of surface exchange can be calculated. Thus, if the surface exchange is not optimal, the filler can be re-exchanged to improve the amount of organic surface coverage. Thus, high resolution thermogravimetric analysis not only ensures the cleanliness of the surface but also the extent of organophilization. The thermogravimetric analysis of the modified clays used in the study is shown in Figure 13. The presence of only one degradation peak in the temperature range of 200-300°C indicates that the excess surface modification molecules could be successfully washed off. It should be noted that the low temperature degradation of the excess surface modification molecules is also detrimental to the interfacial interactions and composite properties when thermoplastic matrices are used, where the filler is often mixed with the polymer at high temperatures.

The X-ray diffractograms of the modified fillers as well as their 3.5 vol% epoxy nanocomposites are depicted in Figure 14. The fillers treated with modifications with longer alkyl chains had much higher basal plane spacing than the fillers modified with Bz1OH, Bz2OH and Bz3OH modifications. In the composites also, the increase in the basal plane spacing of the filler was more in the case of BzC18OH, BzC16 and 2C18 modified fillers. However, as the X-ray diffraction signals are qualitative in nature owing to their dependency on the sample preparation, filler orientation, defects in the crystal structure etc, therefore, microstructure of the composites is also needed to be studied by measuring the composite properties. Effect of scanning time in the case of X-ray diffraction was also found to have a significant effect on the intensity. Same composite film scanned for increasing amounts of time per step had corresponding increase in the intensity as shown in Figure 15. Thus the scanning time should be high enough so as to avoid misinterpreting the diffractograms. Also, the diffraction peak in the X-ray analysis does not rule out the presence of exfoliated platelets. Similarly, the absence of any diffraction peak in the analysis can still miss some intercalated tactoids of filler. To quantify if the basal plane spacing has some correlation with the gas barrier properties, the relative oxygen permeation through the nanocomposites was plotted against the filler basal plane spacing as shown in Figure 16 [28]. As one can observe that an increase in the basal plane spacing was observed with increasing relative permeability which is in contrast to all expectations. It thus indicates that the intercalated platelets do not have impact on the oxygen permeation of the composites, but it is only the exfoliated platelets which are effective in improving the barrier performance of the castings. The relative oxygen permeation of the Bz1OH, Bz2OH and Bz3OH containing fillers was observed to be much better than the other systems (Figure 16) even if the basal plane spacing of the fillers was initially low. On the other hand, the fillers with much higher initial basal plane spacing had much poorer barrier performance, indicating that in the case of polar polymer, it is not the initial basal plane spacing, but the interactions of the filler modification with the polymer which have more influence on the composite properties.

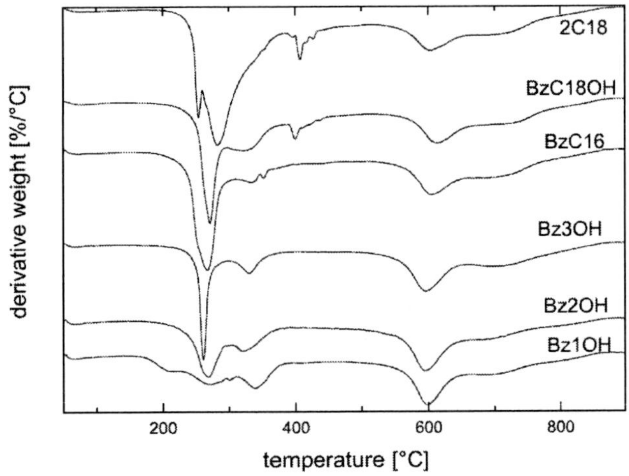

Figure 13. TGA thermograms of the modified fillers used for the synthesis of epoxy nanocomposites.

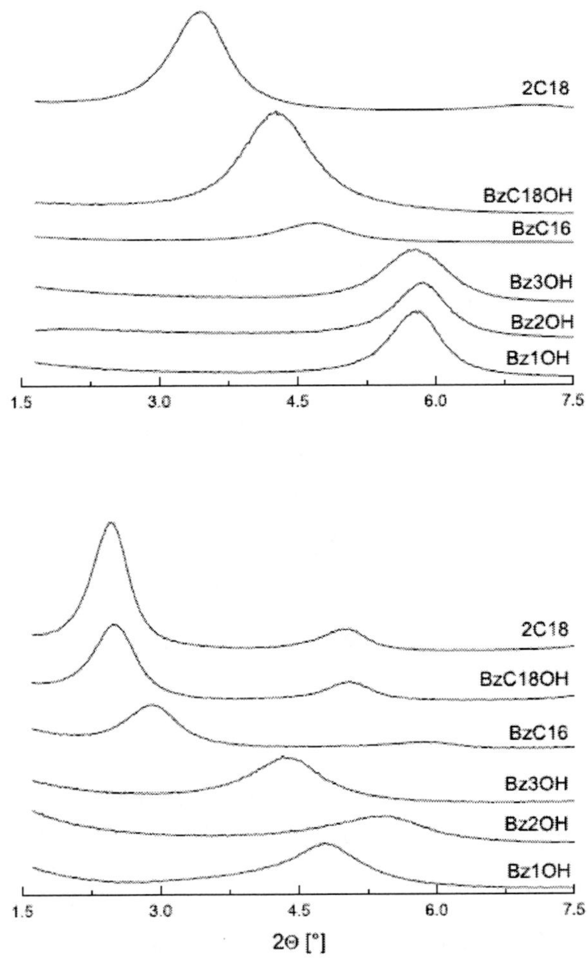

Figure 14. X-ray diffractograms of (a) modified fillers (top) and (b) their 3.5 vol% epoxy nanocomposites (bottom).

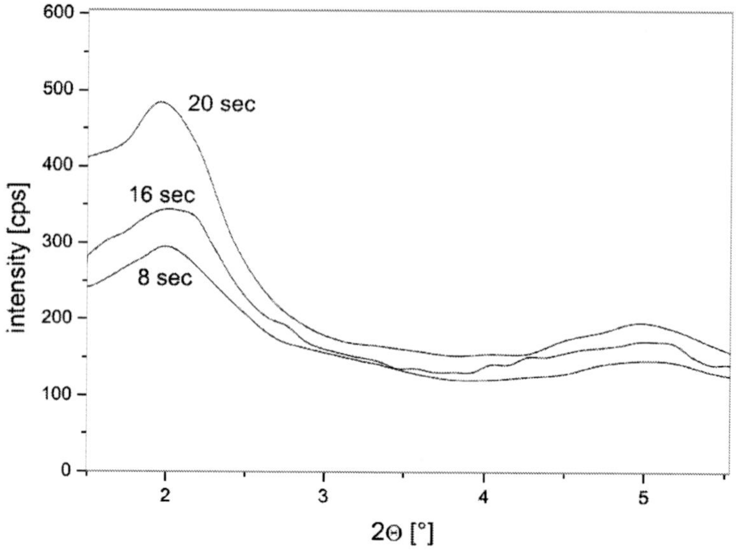

Figure 15. Effect of scanning time on the intensity of the diffractogram measured with 2 vol% BzC16-modified montmorillonite epoxy nanocomposite.

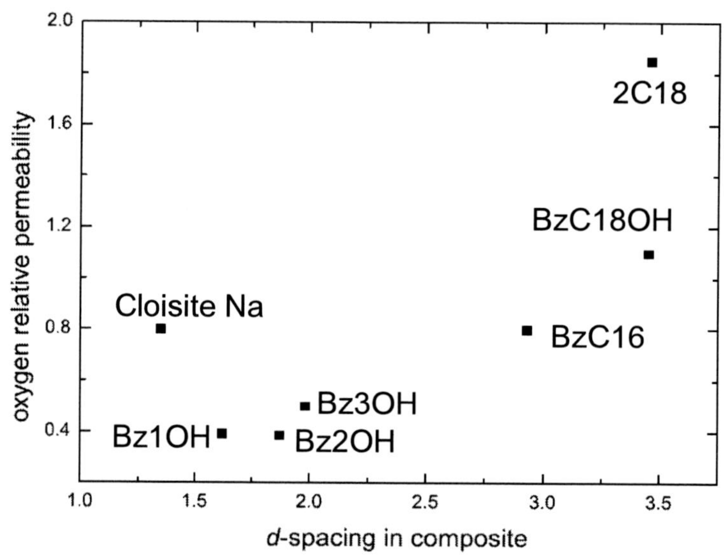

Figure 16. Oxygen relative permeability of the epoxy nanocomposites as a function of filler basal plane spacing in composite. [28]

Effect of the filler volume fraction on the gas barrier properties was also studied. For this, BzC16 and Bz1OH modified montmorillonites were chosen as fillers. Table 3 reports the filler density, weight fraction if the filler as well as composite density for these experiments. The composite density was not significantly increased even at 5 vol% loading of the filler indicating that the composites retained not only the optical transparency, but also low weight. Table 4 shows the oxygen and water vapor permeation of the epoxy nanocomposites as a function of filler volume fraction. As earlier observed Bz1OH composites had much better improvement in the oxygen barrier performance owing to much better interactions with the epoxy polymer. On the other hand, BzC16 composites had only 20% reduction on the oxygen permeation at 5 vol% of the filler as compared t pure polymer. The water vapor permeation, as earlier reported, was not affected much by different surface modification and was a function of hydrophobicity of the filler.

Table 3. Density of the organo-montmorillonites and their composites as a function of filler volume fraction

OM	modified filler density [g.cm^{-3}]	filler weight fraction [%]	composite density [g.cm^{-3}]
Bz1OH (1 vol%)	1.93	2.67	1.19
Bz1OH (2 vol%)	1.93	5.20	1.20
Bz1OH (3.5 vol%)	1.93	8.96	1.22
Bz1OH (5 vol%)	1.93	12.06	1.24
BzC16 (1 vol%)	1.69	2.85	1.19
BzC16 (2 vol%)	1.69	5.59	1.20
BzC16 (3.5 vol%)	1.69	9.51	1.21
BzC16 (5 vol%)	1.69	13.20	1.23

Table 4. Oxygen and water vapor permeation through the epoxy nanocomposites as a function of filler volume fraction. Reproduced from reference 28 with permission from American Chemical Society

composite	oxygen permeability coefficient [$cm^3.\mu m/(m^2.d.mmHg)$]	water vapor trans. rate [$g.\mu m/(m^2.d.mmHg)$]
neat epoxy	2.0	10
Bz1OH (1 vol%)	1.4	8.9
Bz1OH (2 vol%)	0.91	8.0
Bz1OH (3.5 vol%)	0.77	6.7
Bz1OH (5 vol%)	0.55	5.7
BzC16 (1 vol%)	1.9	7.9
BzC16 (2 vol%)	1.7	6.5
BzC16 (3.5 vol%)	1.6	5.3
BzC16 (5 vol%)	1.6	4.9

References

[1] Yano, K; Usuki, A; Okada, A; Kurauchi, T; Kamigaito, O. *J. Polym. Sci., Part A: Polym. Chem.*, 1993, 31, 2493.

[2] Kojima, Y; Fukumori, K; Usuki, A; Okada, A; Kurauchi, T. *J. Mater. Sci. Lett*, 1993, 12, 889.

[3] Lan, T; Kaviratna, PD; Pinnavaia, T. J. *Chem. Mater*, 1994, 6, 573.

[4] Lan, T; Pinnavaia, T. J. *Chem. Mater*, 1994, 6, 2216.

[5] Burnside, SD; Giannelis, EP. *Chem. Mater*, 1995, 7, 1597.

[6] Wang, Z; Pinnavaia, T. J. *Chem. Mater*, 1998, 10, 3769.

[7] Messersmith, PB; Giannelis, EP. *J. Polym. Sci., Part A: Polym. Chem.*, 1995, 33, 1047.

[8] Yano, K; Usuki, A; Okada, A. *J. Polym. Sci., Part A: Polym. Chem.*, 1997, 35, 2289.

[9] Bailey, SW. In Book *Reviews in Mineralogy*, SW; Bailey, Ed; Virginia Polytechnic Institute and State University, Blacksburg, 1984.

[10] Bailey, SW. In Book *Crystal Structures of Clay Minerals and their X-ray Identification*; GW; Brindley, G; Brown, Eds; Mineralogical Society: London, 1980.

[11] Theng, BKG. *The Chemistry of Clay-Organic Reactions*, Adam Hilger: London, 1974.

[12] Lagaly, G. In Book *Developments in Ionic Polymers*; AD; Wilson, HJ; Prosser, Eds; Elsevier Applied Science Publishers: London/New York, 1986.

[13] Pinnavaia, T. J. *Science*, 1983, 220, 365.

[14] Giannelis, EP. *Adv. Mater,* 1996, 8, 29.

[15] Alexandre, M; Dubois, P. *Mater. Sci. Engg. R: Reports*, 2000, 28, 1.

[16] Lagaly, G. *Solid State Ionics*, 1986, 22, 43.

[17] Vaia, RA; Teukolsky, RK; Giannelis, EP. *Chem. Mater.* 1994, 6, 1017.

[18] Kato, M; Usuki, A; Okada, A. *J. Appl. Polym. Sci.*, 1997, 66, 1781.

[19] Brydson, JA. *Plastic Materials*, Newnes-Butterworths, London, Third ed; 1975.

[20] Lee, S. *The Polyurethane Book;* Wiley: New York, 2002.

[21] Wirpsza, Z. *Polyurethanes Chemistry, Technology and Applications*, Ellis Horwood: New York, 1993.

[22] May, CA. *Epoxy Resins Chemistry and Technology*, Dekker: New York, Second ed; 1988.

[23] Lee, H; Neville, K. *Handbook of Epoxy Resins*, McGraw-Hill: New York, 1967.

[24] Ellis, B. *Chemistry and Technology of Epoxy Resins*, Blackie Academic & Professional: London, 1993.

[25] Bharadwaj, RK. *Macromolecules*, 2001, 34, 9189.

[26] Osman, MA; Ploetze, M; Suter, UW. *J. Mater. Chem.*, 2003, 13, 2359.

[27] Osman, MA; Mittal, V; Morbidelli, M; Suter, UW. *Macromolecules*, 2003, 36, 9851.

[28] Osman, MA; Mittal, V; Morbidelli, M; Suter, UW. *Macromolecules*, 2004, 37, 7250.

[29] Crank, J. *The Mathematics of Diffusion;* Clarendon Press: Oxford, Second ed; 1998.

[30] Lan, T; Kaviratna, PD; Pinnavaia, T. J. *Chem. Mater*, 1995, 7, 2144.

[31] Kamon, T; Furakaw, H. *Adv. Polym. Sci.*, 1986, 80, 177.

[32] Barton, JM. *Adv. Polym. Sci.*, 1985, 72, 120.

[33] Tien, YI; Wei, KH. *Macromolecules*, 2001, 34, 9045.

[34] Tien, YI; Wei, KH. *J. Appl. Polym. Sci.*, 2002, 86, 1741.

[35] Mittal, V. *J. Mater. Sci.*, 2008, 43, 4972.

In: Structural Steel and Castings
Editor: Lena M. Becker, pp. 207-223
ISBN: 978-1-61728-104-4

Chapter 6

GAS EMISSION FROM MOLDS WITH SYNTHETIC ADDITIVES DURING METAL CASTING

***G. Ya. Gerasimov*[1] *and M. Yu. Pogosbekyan*[2]**
[1]Institute of Mechanics, Moscow State University, Moscow, Russia
[2]Moscow Road-Transport Institute, Moscow, Russia

Abstract

This chapter analyses the main features of gas emission from molds during metal casting. A set of experiments on determination of the parameters of the gas emission from molding sands was carried out. The thickness of the casting, the thickness of the mold wall, the process time, and the initial temperature of the molten metal were varied. The investigated molding sands were prepared from quartz sand and synthetic binders: M19-62 (urea resin), FF-1C (furan resin), and PK-104 (phenol-formaldehyde resin). It was shown that the highest rate of thermal decomposition has urea resin. A mathematical model of gas emission from molds in the stage of solidification and cooling of metal was constructed. The model includes heat and mass transfer in the mold's capillary-porous medium containing various synthetic resins as a binder, the kinetics of the resin's thermal decomposition, and secondary reactions between products formed. It was shown that the model gives a fairly good description of the process. The kinetic parameters of the thermal decomposition of considered resins were determined on the base of fitting of calculated results to experimental data.

Introduction

The synthetic resins widely used in the foundry industry to prepare casting molds are one of the main sources of pollution of foundries with toxic gases [1]. During the casting process, the molten metal causes thermal decomposition of the carbonaceous additives in the mold, which results in the formation of potentially hazardous gases that are emitted to the atmosphere [2]. The estimation of the degree of pollution and the choice of possible arrangements for its lowering make it possible to improve the working conditions of operational personnel and reduce the level of environmental pollution.

Synthetic resins used in foundry are high-molecular organic compounds whose thermal properties are depended on the chemical composition of monomers involved and on the type of chemical bonds between them [3]. Investigations in this field are mainly associated with study of depolymerization mechanisms of resins at high temperatures [4-12]. The most commonly used organic binders in molds are synthetic resins of cold solidification that are products of polycondensation of formaldehyde with urea, phenol, and furfuryl alcohol [13]. Among them urea-formaldehyde (urea) resins have the lowest thermal stability, the highest saturation with nitrogen, and a relatively low binding ability, but at the same time they are cheapest and readily available. Urea-formaldehyde-furan (urea-furan) resins are the easiest in handling and provide a high quality of molds. Therefore, they exceed in the volume of applications other classes of acid-hardening resins in the foundry industry. Phenol-formaldehyde-furan (phenol-furan) resins are distinguished by a higher thermal stability, which explains their use in producing of steel and high-quality grey iron casts. Phenol-formaldehyde (phenol) resins also have a high thermal stability and are cheaper than phenol-furan resins, but they exhibit a higher toxicity because of emission of phenol vapours at high temperatures.

The mechanism of the thermal decomposition (pyrolysis) of resins entering into molding sands as organic binders differs from the corresponding mechanism obtained under laboratory conditions because of the presence of secondary chemical reactions between products formed. Therefore, the numerical investigation of gas emission from molds during metal casting must include the entire set of physical and chemical processes occurring in the mold in its contact with liquid metal. In the present work, an attempt was made to construct the most complete mathematical model of gas emission from molds in the stage of solidification and cooling of metal. The model includes heat and mass transfer in the mold's capillary-porous medium containing various resins as a binder, the kinetics of the resin's thermal decomposition, and secondary reactions between products formed. To verify the model a set of experiments on determination of the parameters of gas emission from molding sands was carried out. The results of calculations may provide the basis for prediction of the level of foundry pollution with molding gases.

Experimental

Measurements of the parameters of gas emission from molding sands during solidification of liquid metal were carried out on the experimental device, which incorporates the heat insulated upper half-mold including a cavity for casting with liquid metal and the lower half-mold including a heat insulated cylinder filled with the molding sand under investigation. The sand was fixed at the bottom using a grid, below which there is a cavity where the molding gases emitted from the sand are mixed with the carrier gas (e.g., nitrogen or argon). A flow sensor precalibrated for a given gas monitored the flow rate of the carrier gas. The current time was counted out from the moment of the beginning of the process of filling up the molding cavity with liquid metal.

The device is equipped with an experimental data processing system. A humidity detector for the molding gases and a block of heat sensors distributed throughout the volume of the investigated molding sand in addition to the carrier gas flow sensor are linked up to this system. In a particular case, a flow sensor of undiluted molding gases and a molten metal heat sensor are included into the system. Signals from all sensors are fixed and treated in on-line regime with help of computer.

Table 1. Main characteristics of investigated resins

Resin	Elemental composition, mass %				Thermodestruction products	
	C	H	O	N	C_s	V, nm^3/kg
M19-62	39.0	6.0	25.0	30.0	0.22	1.26
FF-1S	72.0	6.4	21.6	-	0.40	0.94
PK-104	78.0	6.1	15.2	0.7	0.50	0.89

The rate of molding gases emission was calculated from the flow rate of the carrier gas and the concentration of molding gases in the sampled gaseous mixture. The chemical composition of the molding sand was determined by the chromatographic and gas-analytic methods. The main measured parameter was the volume of the emitted molding gas $Q = Q(t)$ reduced to normal conditions. The relative coefficient of gas emission $a = Q/t^{1/2}$, which is the most generalized characteristic of gas-generating properties of any molding sands, was computed from the Q values.

The investigated molding sands were subjected to liquid grey iron that was inserted into molding cavity of the experimental device. The casting thickness δ and the thickness of the mold wall h were varied over the range of 8 – 28 mm and 15 – 100 mm, respectively. The temperature of the liquid grey iron was varied from 1550 to 1700 K. The molding sands were prepared from 1K02B quartz sand and synthetic resins: M19-62 (urea resin), FF-1S (furan resin), and PK-104 (phenol-formaldehyde resin).

The elemental composition of the investigated resins and the characteristics of their thermal decomposition are given in Table 1. The mass fraction of solid carbon C_s and the volume V of gas released from the resin were measured for the case of flash heating of the resin sample to 1700 K. In the absence of experimental data, an approximate estimation of these quantities can be obtained on the basis of the elemental composition of the resin at the assumption that the complete decomposition of the resin results in the formation of solid carbon and a gas consisting of H_2, CO, and N_2. In this case, $C_s = 0.01$ ([C] – [O]), $V = 0.224$ ([H]/2 + [O]/16 + [N]/28), where [X] is a mass percentage of element X in the resin.

Description of Model Approaches

Numerical investigation of various physical phenomena occurring during casting process is the most efficient approach to solution of the arising problems. It gives an insight into the details of fluid flow, heat transfer, and solidification of liquid metal [14-17]. The thermal decomposition of organic binders in the molding sand and the emission of molding gases to the atmosphere is the complex process that depends in the general way on thermal and mass-transfer history of the contacting systems (mold/liquid metal/ environment). Therefore, the construction of the complete mathematical model, which includes the heat and mass transfer in capillary-porous medium of the mold, the kinetics of the thermal decomposition of synthetic additives, and secondary reactions between pyrolysis products, allows to estimate the degree of pollution under metal casting and to choice possible arrangements for its lowering.

Thermal Decomposition of Resin

The process of synthetic resins decomposition under heating conditions is described by numerous chemical reactions. Nevertheless, all resins can be subdivided into two groups depending on the character of chemical bonds breaking [3]. The first group includes resins, whose depolymerization is determined by the breaking of bonds in the main chain with formation of low-molecular volatile products. This group includes urea-formaldehyde resins widely used in molds production [13]. The second group includes resins prone to intramolecular abstraction of corresponding functional groups followed by reactions of cyclization and condensation that give rise to formation of considerable quantity of carbonised non-volatile products. This group includes phenol-formaldehyde and furan resins. The thermal decomposition of resins in both cases can proceed by a radical, ionic, or molecular mechanism in a chain or nonchain manner [5].

Global Kinetic Model

In spite of the considerable progress made in understanding of pyrolysis mechanisms for different resins, modeling of the thermal decomposition process is confined, as a rule, to the simplest kinetic models describing the conversion of the initial polymer as a whole without concrete definition of the process products (see, e.g., [18, 19]). Such global kinetic models consider the yield of thermal decomposition products into gas phase by the simple kinetic equation of the form:

$$dX/dt = -kX, \qquad k = A\exp(-E/RT), \qquad X\big|_{t=0} = X_0, \tag{1}$$

where X is concentration of the gas forming fraction in resin, kmole/(kg of resin); k is the rate constant of the resin thermal decomposition; A and E are kinetic parameters of the process; R is universal gas constant.

The analytical solution of Eq. (1) in the one-dimensional approach as applied to the heating conditions of the molding sand in the experimental device gives the concentration field $X(x, t)$ inside the sand:

$$X(x,t) = X_0 \exp\left\{ -A\int_0^t \exp\left[-E/RT(x,t)\right]dt \right\}. \tag{2}$$

The volume Q of gaseous decomposition products emitted from the molding sand into the gas phase in time t is determined by the formula:

$$Q(t) = 22.4\rho_{mix}\varphi_{mix}h\left[X_0 - h^{-1}\int_0^h X(x,t)dx\right], \tag{3}$$

where ρ_{mix} and φ_{mix} are the density and the mass fraction of resin in the molding sand; h is the thickness of the mold wall. The initial concentration of the gas-forming fraction X_0 in the resin can be expressed in terms of the V value: $X_0 = 0.0446\ V$. The solution of the heat conduction equation in the one-dimensional approach as applied to the heating conditions of the molding sand in the experimental device (see below) permits to determine the temperature field $T(x, t)$, with help of which the concentration field $X(x, t)$ and the volume of emitted gaseous products Q are calculated by formulas (2) and (3).

Model of Functional Groups

This kinetic approach is applicable mainly to resins of the second group (see above), which includes phenol-formaldehyde and furan resins. In the first stage of the decomposition process for such resins ($T < 600$ K), the polymer changes are small and the quantity of decomposition products is negligible. The destruction rate grows sharply, when the process temperature reaches 600 – 900 K. In this stage of the process, the complete depolymerization of the resin is not achieved, and there can be only a random cleavage of polymer chains with formation of carboxyl (-COOH), carbonyl >C=O), and methyl ($-CH_3$) functional groups, whose abstraction leads to formation of light gas components: CO_2, CO, and CH_4. At the same time, the separation of large molecular fragments forming oligomer fraction (OF) takes place with its yield in gas phase. On the basis of the average molecular weight of the fraction ($\mu_{OF} = 340$), one can believe that the fraction consists of the fragments of the resin's three-dimensional structure, i.e., oligomers containing three or four condensed benzene rings on average [20]. In the final stage of the process ($T > 900$ K), the polymer undergoes deeper changes, which are accompanied by the transformation of aromatic structures into condensed carbon [21].

In accordance with the above-described mechanism of the process, the resin of the second group may be represented as a condensed aromatic structure containing different functional groups (carboxyl, carbonyl, methyl, and others) in the form of substitutes. The rapid heating of the resin leads to the rapture of bonds attaching the functional groups to aromatic rings with formation of light gaseous components. The thermal cleavage of polymer chains gives the yield of oligomer fraction and carbonization of residue. The atomic hydrogen releasing from aromatic structures goes on stabilization of free radicals and formation of H_2.

In accordance with the considered scheme of the resin thermal depolymerization, the yield of components into gas phase can be describe by the model of functional groups [22]. In this model the system of kinetic equations has the form:

$$dY_i / dt = -k_i Y_i, \qquad k_i = A_i \exp(-E_i / RT), \qquad Y_i\big|_{t=0} = Y_{i0}, \tag{4}$$

where Y_i is the concentration of the ith functional group in the resin, kg/(kg of initial resin); A_i and E_i are kinetic parameters of the ith functional group yield.

The kinetic parameters of the model (A_i, E_i, and Y_{i0}) as applied to the pyrolysis of phenol-formaldehyde resin was obtained in [23] by fitting the calculated curves describing the dynamics of the process under isothermal conditions to experimental data [20]. Table 2 gives the values of these parameters for main functional groups of the resin participating in the formation of the gas phase. The concentration of condensed carbon (carbon residue) is $Y_C = 0.5$. For want of experimental data, the activation energy of the reaction of H_2 yield was taken equal to the corresponding value for thermal decomposition of coal organic mass [22]. It should be noted that in practical situations the gas phase contains a much larger number of components, in particular, acetylene, ethylene, acetone, propanol, benzene, and others [20]. Some of them are the products of thermal destruction of the resin; others are formed in secondary chemical reactions. The molar concentrations of these components are low, and they haven't considerable influence on the dynamics of the process.

Secondary Chemical Reactions

Intense decomposition of phenol-formaldehyde type resin in the molding sand occurs at temperatures of the order of 1000 – 1200 K [7]. The volatile decomposition products migrate then in the capillary-porous medium of the mold undergoing different secondary reactions, mainly, cracking, condensation, and polymerization of the OF with deposition of a certain amount of carbon. The cracking of the OF vapor is determined first of all by the rupture of methylene bonds between benzene rings. Therefore, in the simplest case one can take the mixture of phenol and methane for modeling of the subsequent transformation of the OF disintegration products.

Sufficiently detailed kinetic mechanisms of combustion of aromatic hydrocarbons including phenol are available at present [24]. The high temperature transformation of phenol in a neutral medium may be represented by a chain of reactions of peroxide radical C_6H_5O formation followed by its transformation to cyclopentadienyl radical *cy*-C_5H_5, the opening of the cycle, and the formation of propargyl radical C_3H_3 [25]:

$$C_6H_5OH \xrightarrow{-H} C_6H_5O \xrightarrow{-CO} cy - C_5H_5 \rightarrow C_5H_5 \rightarrow C_3H_3 + C_2H_2. \tag{5}$$

The recombination of propargyl radicals is the main channel of benzene formation [26]. The structural growth of aromatic molecules and their condensation to soot particles is a complex chemical process [27]. The simplest model of the process includes the formation of aromatic radicals A_k• followed by the addition of acetylene molecules:

$$A_k + H \xrightarrow{-H_2} A_k \bullet \xrightarrow{C_2H_2} A_kC_2H_2 \bullet \xrightarrow{C_2H_2} A_{k+1}. \tag{6}$$

Table 2. Kinetic parameters of functional groups yield for phenol-formaldehyde resin.

Group	Y_{i0}	A_i, s^{-1}	E_i, kJ/mole
OF	0.289	4	58.24
CO	0.140	2×10^2	99.81
CH_4	0.040	4×10^4	124.81
H_2	0.028	6×10^6	191.34
CO_2	0.012	1×10^2	74.86

In the present work, the kinetic mechanism of oxidation of benzene and its derivatives [28] was used to model the high temperature transformation of phenol in the medium of volatile products of phenol-formaldehyde resin destruction. The mechanism consists of 130 elementary reactions and was supplemented with the reactions entering in the chains (5) and (6) whose parameters are given in Table 3.

Table 3. Kinetic mechanism of high temperature transformation of phenol.

Reaction	log *A*	*n*	*E*, kJ/mole
$OF \rightarrow C_6H_5OH + CH_4$	13.00	-	272.14
$C_6H_5OH \rightarrow C_6H_5O + H$	16.43	-	371.96
$C_6H_5O \rightarrow cy\text{-}C_5H_5 + CO$	11.87	-	183.63
$cy\text{-}C_5H_5 \rightarrow C_5H_5$	13.87	-	139.42
$C_5H_5 \rightarrow C_3H_3 + C_2H_2$	14.00	-	71.18
$C_3H_3 + C_3H_3 \rightarrow A_1$	12.70	-	-
$C_3H_3 + H \rightarrow C_3H_4$	13.30	-	-
$C_6H_5 \rightarrow C_4H_3 + C_2H_2$	15.20	-	343.32
$C_4H_3 + H_2 \rightarrow C_4H_4 + H$	10.20	-	74.15
$C_4H_4 \rightarrow C_2H_2 + C_2H_2$	15.06	-	362.24
$A_k\bullet + C_2H_2 \rightarrow A_kC_2H_2\bullet$	38.85	-8.02	68.66
$A_kC_2H_2\bullet + C_2H_2 \rightarrow A_{k+1} + H$	16.20	-1.33	22.61
$A_k\bullet + C_4H_4 \rightarrow A_{k+1} + H$	33.52	-5.70	106.76
$A_k \rightarrow$ products	7.82	-	200.00

Rate constant: $k = AT^n\exp(-E/RT)$ $(cm^3\ mole^{-1})^{m-1}\ s^{-1}$, m is the reaction order.

The volatile products of the resin pyrolysis move then in the pore space of the mold to its external boundary arriving the region of low temperatures. In this region, the condensation and repolymerization of the unreacted part of the OF occur with the formation of carbon residue and the volatile repolymerization products. The detailed mathematical description of this process is quite difficult. In the simplest case, one can take it into account by direct addition of CO and H_2 to the gas phase and C to condensed carbon. The amount of the addition is determined on an analysis of the OF element composition.

Processes of Heat and Mass Transfer

During the process of the metal solidification in the casting mold, the amount of the solid phase increases, whereas the amount of the liquid phase decreases. There are three zones in solidifying casting: a solidified layer, a two-phase zone (liquid metal and metal crystals), and a liquid layer. The crystallization rate and, accordingly, the increase in the solidified metal thickness are determined by the thermal regime of the mold, which also influences on the final quality of the casting (porosity, grain structure, blowholes, and so on).

Formation of Temperature Field

Let us consider the model problem of the casting formation in a dry mold in the case of one-side heat emission through the mold wall. It is assumed that the temperature of a liquid metal in the process of its crystallization is constant and the overheating of the metal is small. Since the concentration of resin in the sand is low, it is assumed that the thermal effect of pyrolysis reactions, the convective transfer of heat by the gas components, and the influence of the change in the sand mass have small effect on the formation of a temperature field. In this case the heat conductivity equation in a one-dimensional formulation has the form:

$$\rho_{mix} c_{mix} \partial T / \partial t = \lambda_{mix} \partial^2 T / \partial x^2, \tag{7}$$

where ρ_{mix}, c_{mix}, and λ_{mix} are density, heat capacity, and heat conductivity of the molding sand. The equation for determination of the thickness of the solidified metal layer ξ in the mold and the boundary condition on the metal-mold interface are as follows:

$$\rho_m H_m d\xi / dt = -\lambda_{mix} \partial T / \partial x\big|_{x=0}, \qquad T(t,0) = T_{m0}, \tag{8}$$

where ρ_m and H_m are the density and heat of crystallization of the metal. The equation describing the change of metal temperature T_m at cooling of the solidified metal and the boundary condition on the interface are written in the form:

$$\rho_m c_m \delta dT_m / dt = \lambda_{mix} \partial T / \partial x\big|_{x=0}, \qquad T(t,0) = T_m, \tag{9}$$

where δ is casting thickness. The corresponding boundary condition at the external boundary of the mold is defined by the relation:

$$-\lambda_{mix} \partial T / \partial x\big|_{x=h} = \alpha [T(h,t) - T_a] + \varepsilon\sigma [T^4(h,t) - T_a^4], \tag{10}$$

where α is heat transfer coefficient, ε is the emissivity of the mold surface, σ is Stephan-Boltzmann constant, h is mold wall thickness, and T_a is the ambient temperature. The initial conditions are as follows: $\xi(0) = 0$ and $T(x, 0) = T_a$. The density, heat capacity, and heat conductivity of the molding sand for the prescribed composition and degree of compacting are determined experimentally and have the values: $\rho_{mix} = 1600$ kg/m^3, $c_{mix} = 1200$ J/(kg·K), $\lambda_{mix} = 0.7$ W/(m·s). The properties of the grey iron are as follows: $\rho_m = 6900$ kg/m^3, $H_m = 240$ kJ/kg, $c_m = 540$ J/(kg·K).

Filtration of Pyrolysis Products

The filtration of volatile products of the resin pyrolysis in the capillary-porous medium of the mold is described by the continuity equation [29]:

$$\partial(\rho \varepsilon_{mix} y_i) / \partial t + \partial(y_i N)/\partial x = \rho_{mix} \varphi_{mix} G_i, \tag{11}$$

$$G_i(x,t) = Y_{i0} A_i \exp\{-E_i / RT(x,t) - A_i \int_0^t \exp[-E_i / RT(x,t)] dt\},$$

where ε_{mix} is the porosity of the molding sand, y_i is the mass fraction of the ith gas component, N is the mass flux of the volatiles in the molding sand, and G_i is the rate of yield of the ith functional group from the resin. The expression for $G_i(x, t)$ is obtained by integration of the kinetic equation (4). Since the mass-mean gas velocity in the medium, $v = N/(\rho\varepsilon_{mix})$, is much smaller than the sound velocity, one can accept the quasi-stationary regime for the process description and omit the first term in Eq. (11).

The mass fluxes $N_i = y_i N$ entering in Eq. (11) are generally dependent on gradients of the concentrations y_i and pressure p [29]. The influence of components diffusion on N_i in considered case is insignificant, and the mass flux of pyrolysis products due the pressure difference plays a dominant role. Therefore, the pressure p and density ρ of the gas in the medium are determined from equations:

$$N(x,t) = -\rho\varepsilon_{mix}(r^2/8\eta)\partial p/\partial x, \qquad \rho(x,t) = (p/RT)(\sum_i y_i/\mu_i)^{-1}, \tag{12}$$

where r is the average pore radius in the molding sand, μ_i is the molecular weight of the ith gas component, and η is the gas viscosity. Boundary conditions for Eqs. (11) – (12) have the form: $N_i(0, t) = 0$, $p(h, t) = p_a$, where p_a is the ambient pressure.

Results and Discussion

A set of experiments was performed for determination of the gas release from mold at the stage of solidification of liquid metal and cooling of the casting in a wide range of variation of the process parameters. These data were used for calculation of the kinetic parameters of the thermal decomposition of investigated resins. The thickness of the casting, the thickness of the mold wall, the process time, and the temperature of the liquid metal were varied. The kinetic parameters of the resins thermal decomposition were obtained by fitting of calculated values of Q to corresponding experimental data with the use of the least square technique, where Q is the volume of gaseous decomposition products emitted from the molding sand into the gas phase. The results of the approximation are given in Table 4.

Table 4. Kinetic parameters of thermal decomposition of resins.

Resin	X_0	A_i, s^{-1}	E_i, kJ/mole
M19-62	0.0562	4.0	50.0
FF-1S	0.0421	1.3	66.5
PK-104	0.0399	2.0	66.5

The calculations show that a dominant role in the dynamics of gas release from the molding sand plays the formation of temperature field in the mold. Figure 1 represents temperature curves calculated for the case of the average values of the casting thickness δ and the thickness of the mold wall h. One can see that the metal solidifies and the temperature of the mold layer adjacent to the casting (left boundary of the mold with linear coordinate $x = 0$) is invariable, when the process time is less than 400 s. After this time, the casting begins to cool down, which leads to a reduction of the left boundary temperature. The heat transfer coefficient α in Eq. (10) for calculation of mold heat exchange with the environment (right boundary of the mold with linear coordinate $x = h$) is taken equal 20 W/(m^2·K) [30].

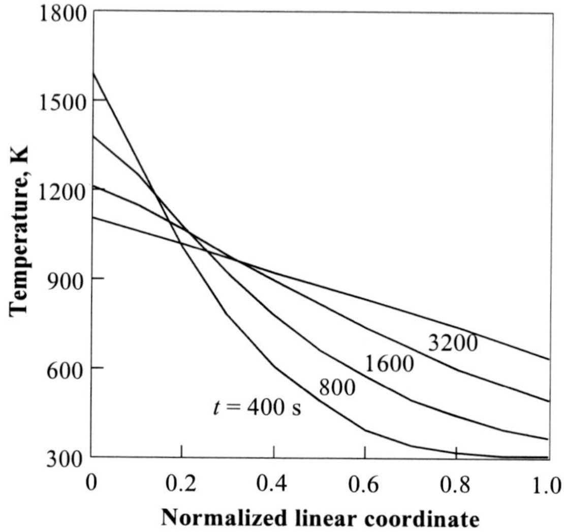

Figure 1. Calculated distribution of the temperature in the mold: h = 50 mm, δ = 15 mm, T_{m0} = 1600 K, T_a = 300 K.

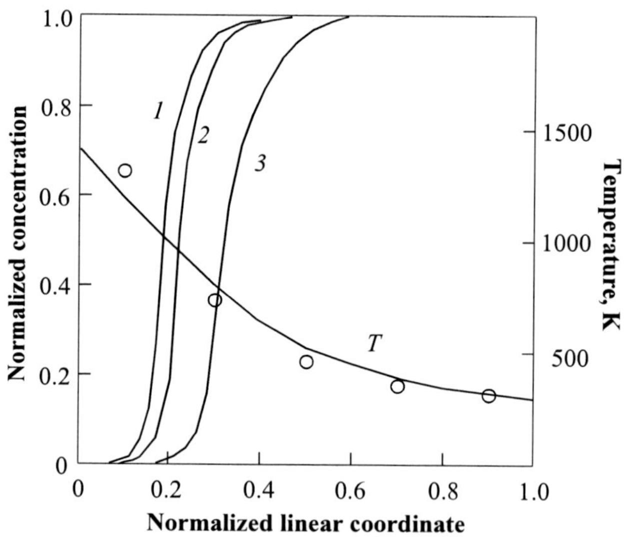

Figure 2. Calculated distributions of the concentration $X(x, t)$ of the gas forming fraction and the temperature $T(x, t)$ inside the mold for furan (1), phenol (2), and urea (3) resins: h = 100 mm, δ = 28 mm, t = 1900 s, T_{m0} = 1600 K. Points are the measured data.

The pressure in the zone adjacent to the casting calculated for average pore radius r = 5 μm sharply increases to p = 1.4 atm at process time $t = 10^{-2}$ s and then decreases to p = 1.03 atm, which is consistent with experimental data [31]. It should be noted that the influence of pressure on the thermal decomposition of various resins is insignificant. This process parameter plays the dominant role only at formation of the regime of the volatiles filtration in the mold's capillary-porous medium.

Figure 2 represents the calculated and measured field $T(x, t)$ inside the mold, as well as the concentration curves $X(x, t)$ for investigated types of resins calculated with use of recovered kinetic parameters (see data of Table 4). Analysis of the figure shows that urea resin has the highest thermal decomposition rate other things being equal. The decomposition process for this resin begins at a temperature of the order of 600 K, whereas the temperature of intensive decomposition for furan and phenol resins is about 800 K. The least resistance of urea resin to temperature loads is also illustrated in Figure 3, where the dependence of the gases volume flow W from the mold surface on the process time is shown.

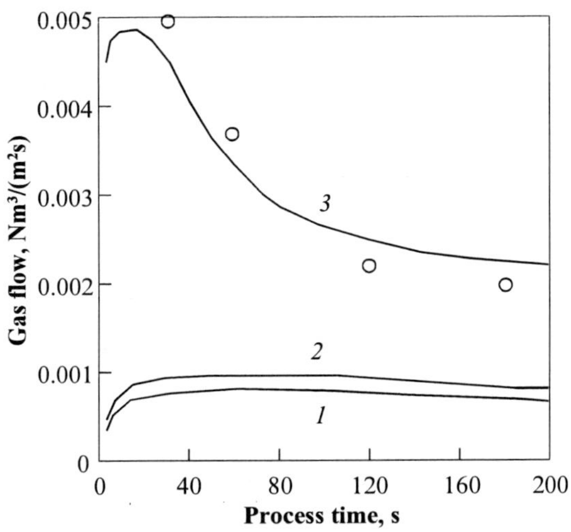

Figure 3. The gases volume flow W from the mold surface versus the process time for furan (1), phenol (2), and urea (3) resins: h = 50 mm, δ = 28 mm, φ_{mix} = 0.04, T_{m0} = 1600 K. Points are the measured data.

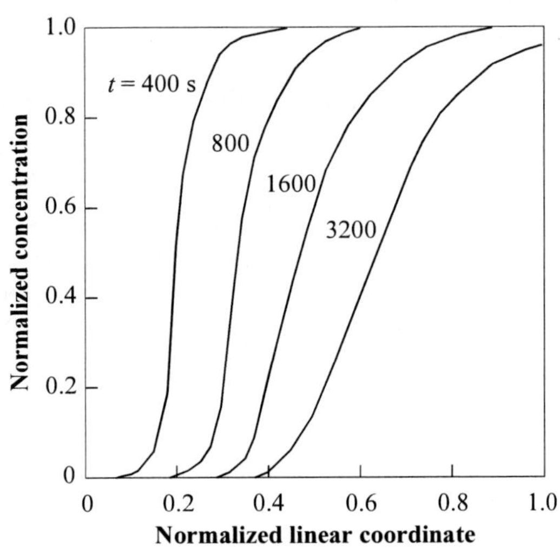

Figure 4. Calculated distribution of the oligomer fraction concentration in the mold with phenol-formaldehyde resin: h = 50 mm, δ = 15 mm, T_{m0} = 1600 K, T_a = 300 K.

The pyrolysis of the resin on the initial stage of the process occurs in a narrow zone of the molding sand. Its width is $\Delta x/h < 0.1$ and the temperature is higher than 1200 K. In the course of the process, the pyrolysis zone approaches the external boundary of the mold and its width increases, whereas the temperature decreases. This is clear from Figure 4, where the concentrations curves of oligomer fraction $Y_{OF}(x, t)$ for phenol-formaldehyde resin are shown. At the instant of time $t = 800$ s, the intensive OF yield from the molding sand occurs in the interval $0.15 < x/h < 0.25$, where the sand temperature T varies from 1200 K for $x/h = 0.15$ to 900 K for $x/h = 0.25$. At the instant of time $t = 1600$ s, the corresponding interval is equal to $0.5 < x/h < 0.8$, and the average temperature in it decreases to 700 K.

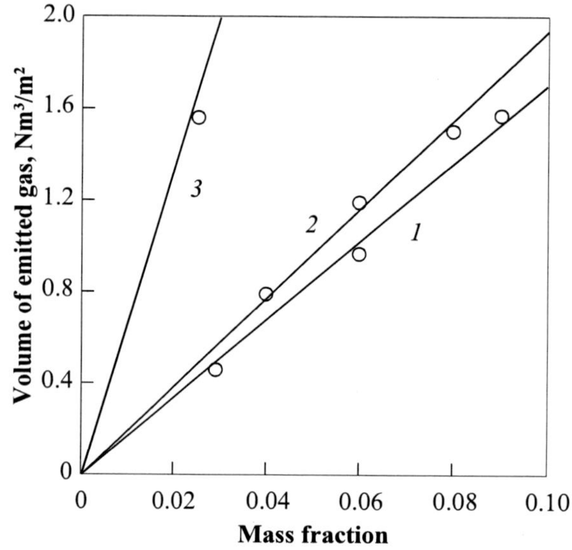

Figure 5. Volume of emitted gaseous decomposition products Q versus the initial content of resin φ_{mix} in the molding sand for furan (1), phenol (2), and urea (3) resins: $h = 100$ mm, $\delta = 28$ mm, $t = 1900$ s, $T_{m0} = 1600$ K. Points are the measured data.

The measured data for the volume Q of gaseous decomposition products emitted from the molding sand into the gas phase obtained at variation of different process parameters (mass fraction φ_{mix} of resin in the molding sand, casting thickness δ, thickness of the mold wall h, and initial temperature of liquid metal T_{m0}) and their comparison with calculated data are given in Figures 5–7. In general, there is a good agreement between received results for all considered variants. It should be noted that the experimental values of Q increase somewhat less rapidly than the calculated ones with increasing thickness of the mold wall (see Figure 6). This can be explained by the influence of the sorption processes in the molding sand that are not taken into account in the calculation.

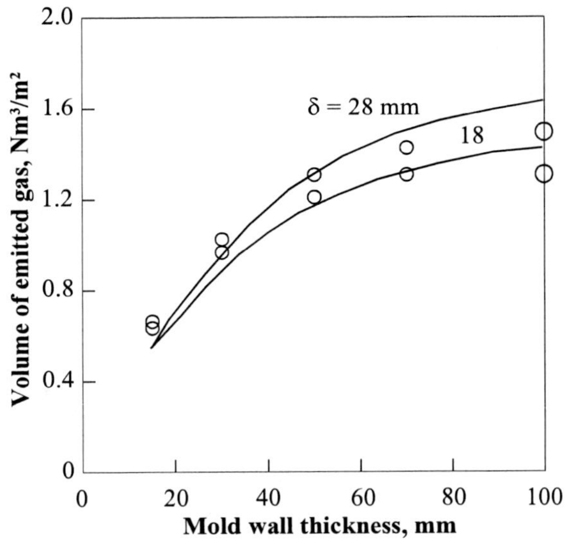

Figure 6. Volume of emitted gaseous decomposition products Q for urea resin versus the thickness of the mold wall h: $\varphi_{mix} = 0.025$, $T_{m0} = 1600$ K, the process time is varied from $t = 300$ s at $h = 15$ mm to $t = 1800$ s at $h = 100$ mm. Points are the measured data.

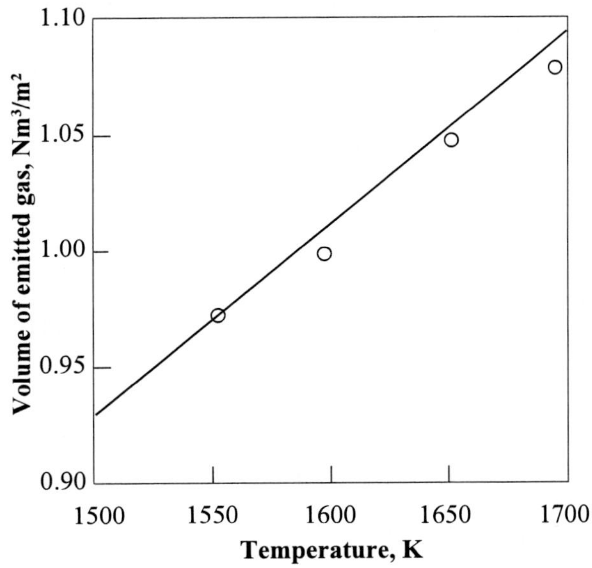

Figure 7. Volume of emitted gaseous decomposition products Q for phenol resin versus the initial temperature T_{m0} of liquid metal: $h = 500$ mm, $\delta = 18$ mm, $t = 1080$ s, $\varphi_{mix} = 0.06$. Points are the measured data.

The measured chemical composition of dry molding gase averaged over the time and the thickness of castings and mold walls is given in Table 5. It is seen that this composition does not depend strongly on the type of resins used in the present work to prepare molding sands. Besides the components given in the table the molding gas contains phenol (0.5 vol.% for PK-104 resin and 0.3 vol.% for FF-1S resin), formaldehyde (0.2 vol.% for M19-62 resin), ammonia (4 vol.% for M19-62 resin), hydrocarbons, and other toxic components. It was also determined that the content of hydrogen and carbon oxide in the molding gas decreases and the content of carbon dioxide increases with increasing process time (see Figure 8).

The calculated temperature and concentration fields in the mold show that the volatiles in the initial stage of the casting process arrive in the region of high temperatures. The estimation of the mass-mean gas velocity in the mold's capillary-porous medium for the porosity of the molding sand $\varepsilon_{mix} = 0.3$ gives the residence time of the volatiles in this region of the order of 1 s. Figure 9 shows the behaviour of the oligomer fraction of volatiles at thermal decomposition of phenol-formaldehyde resin as a function of the residence time for different temperatures. One can see that the cracking of the OF vapour leading to formation of light gas components (phenol and methane in this case) begins at temperatures above 1000 K.

The kinetic curves of aromatic components of volatiles in the case of phenol-formaldehyde resin decomposition in high temperature region are given in Figure 10. The figure analysis shows that the OF concentration decreases sharply with the increase of gas temperature. The disintegration of OF molecules leads to phenol formation followed by its transformation in benzene and subsequent growth of aromatic structure in accordance with kinetic scheme (6). On the one hand, this reaction sequence is simplified kinetic mechanism of polycyclic aromatic hydrocarbons formation. On the other hand, it adequately represents the main steps in the growth of the aromatic structure: hydrogen abstraction in the interaction of the aromatic molecule with hydrogen atoms, and the subsequent addition of acetylene. The decomposition of the aromatic molecules occurs due to their interaction with O atoms and O_2 molecules. Thus, the initial stage of the foundry process, which characterized by high temperature in pyrolysis zone, is responsible for formation of aromatic (benzene, phenol) and polycyclic aromatic hydrocarbons, whose concentration in the molding gas may be sufficiently high.

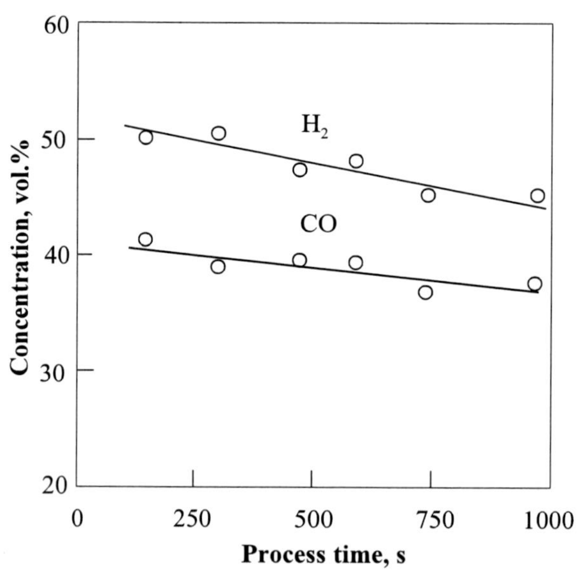

Figure 8. Concentration of components in the molding gas versus the process time t for urea resin: h = 30 mm, δ = 18 mm, φ_{mix} = 0.025, T_{m0} = 1600 K. Points are the measured data.

Table 5. Averaged composition of molding gas

Resin	H_2	CO	CO_2	CH_4	O_2	N_2	Others
M19-62	46.7	37.7	2.7	2.4	0.4	4.6	5.5
FF-1S	49.0	35.0	1.2	2.8	0.5	4.0	7.5
PK-104	49.6	36.4	0.5	7.7	0.3	2.8	2.9

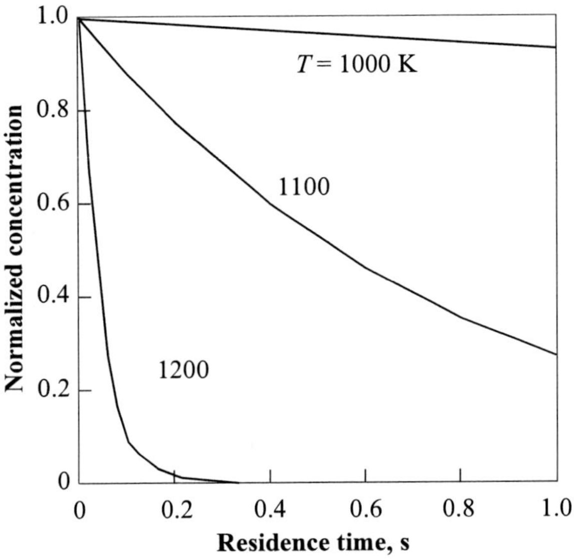

Figure 9. Concentration of the oligomer fraction of volatiles as a function of residence time in the high-temperature region.

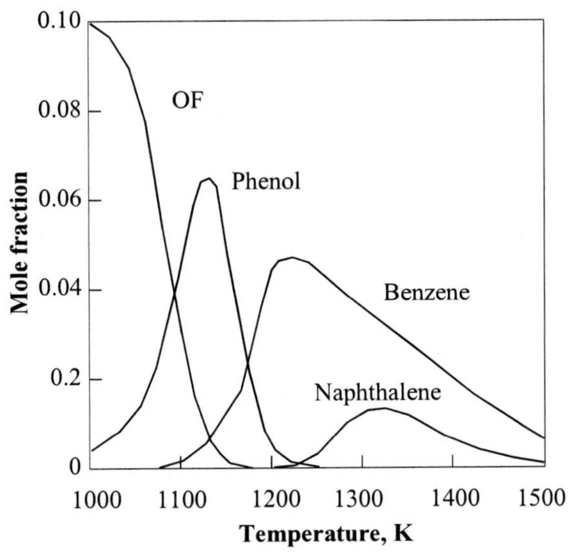

Figure 10. Temperature dependence of the concentrations of volatile aromatic hydrocarbons in high temperature region of the mold: h = 50 mm, δ = 15 mm, T_{m0} = 1600 K, T_a = 300 K. The residence time at given temperature is equal 1 s.

The results of measurements of the molding gases volume Q for different values of the process parameters show that the intensity of gas formation increases with the resin content in the molding sand φ_{mix} other things being equal. This increase in calculated data is proportional to φ_{mix}. Experimental values of Q grow somewhat slower. This fact can be explained, first, by the small increase in the porosity of the sand and, accordingly, by the decrease in its density with growth of φ_{mix}. Second, the increase in the endothermal effect of the resin thermal decomposition leads to a slowing down of the molding sand warming up.

Conclusion

The purpose of this study is the analyses of the main features of gas emission from molds during metal casting. A set of experiments was performed for determination of the process parameters at the stage of solidification of liquid metal and cooling of the casting. These data were used for calculation of the kinetic parameters of the thermal decomposition of the synthetic resins widely used in the foundry industry to prepare casting molds. The mathematical model of gas emission from molds was constructed. The model includes the heating of the molding sand by liquid metal, the thermal destruction of the resin, secondary reactions between products formed, and their filtration in the mold's capillary-porous medium. It was shown that the model gives a fairly good description of the process. The received results may provide the basis for prediction of the level of foundry pollution with molding gases.

References

[1] Boldin, AN; Davydov, NI; Zhukovski, SS. *Casting Molding Materials. Molding, Core Sand Mixtures and Coatings*, Mashinistroenie: Moscow, 2006.
[2] Dungan, RS; Reeves, JB. *Thermochim. Acta*, 2007, 460, 60-66.
[3] Grassie, N; Scott, G. *Polymer Degradation and Stabilization*, University Press: Cambridge, 1985.
[4] Allen, NS; Edge, M. *Fundamentals of Polymer Degradation and Stabilization*, Elsevier: London, 1992.
[5] Cohen, Y; Aizenshtat, Z. *J. Analyt. Appl. Pyrolysis*, 1992, 22, 153-178.
[6] Lytle, CA; Bertsch, W; McKinley, M. *J. Analyt. Appl. Pyrolysis*, 1998, 45, 121-131.
[7] Lin, JM; Ma, CC. *Polym. Degrad. Stab*, 2000, 69, 229-235.
[8] Sobera, M; Hetper, *J. Chromatogr*, 2003, A993, 131-135.
[9] He, G; Yan, N. *Polymer*, 2004, 45, 6813-6822.
[10] Yuan, L; Liang, GZ; Xie, JQ. et al. *Polym. Degrad. Stab*, 2006, 91, 2300-2306.
[11] Bahramian, AR; Kokabi, M; Famili, MHN; et al. *Polymer*, 2006, 47, 3661-3673.
[12] Yangfei, C; Zhiqin, C; Shaoyi, X; Hongbo, L. *Thermochim. Acta*, 2008, 476, 39-43.
[13] Zhukovski, SS. *Molding Materials and Technology of the Casting Mold*, Mashinistroenie: Moscow, 1993.
[14] Sulaiman, S; Hamouda, AMS. *J. Mater. Process. Technol.*, 2001, 113, 245-250.
[15] Janik, M; Dyja, H. *J. Mater. Process. Technol.*, 2004, 157-158, 177-182.
[16] Xie, Y; Yu, H; Ruan, X; et al. *J. Mater. Process. Technol.*, 2008, 199, 49-55.

[17] Kermanpur, A; Mahmoudi, S; Hajipour, A. *J. Mater. Process. Technol.*, 2008, 206, 62-68.
[18] Perng, LH; Tsai, CJ; Ling, YC. *Polymer*, 1999, 40, 7321-7329.
[19] Yang, KK; Wang XL; Wang, YZ. *Eur. Polym. J*, 2003, 39, 1567-1574.
[20] Madorsky, SL. *Thermal Degradation of Organic Polymes*, John Wiley & Sons: New York, 1964.
[21] Lenghaus, K; Qiao, GGH; Solomon, DH. *Polymer*, 2001, 42, 7523-7529.
[22] Solomon, PR; Serio, MA; Suuberg, EM. *Prog. Energy Combust. Sci.*, 1992, 18, 133-220.
[23] Gerasimov, GYa; Pogosbekyan, YuM. *J. Eng. Phys. Thermophys.*, 2007, 80, 545-554.
[24] Simie, JM. *Prog. Energy Combust. Sci.*, 2003, 29, 599-634.
[25] Richter, H; Howard, JB. *Phys. Chem. Chem. Phys.*, 2002, 4, 2038-2055.
[26] Fernandes, RX; Hippler, H; Olzmann, M. *Proc. Combust. Inst.*, 2005, 30, 1033-1038.
[27] Frenklach, M. *Phys. Chem. Chem. Phys.*, 2002, 4, 2028-2037.
[28] Emdee, JL; Brezinsky, K; Glassman, I. *J. Phys. Chem.*, 1992, 96, 2151-2161.
[29] Jackson, R. *Transport in Porous Catalyst*, Elsvir: New York, 1977.
[30] Kreith, F; Black, WZ. *Basic Heat Transfer*, New York: Harper & Raw, Publishers, 1980.
[31] Medvedev, YaI. *Gaseous Processes in Mold*, Mashinistroenie: Moscow, 1980.

In: Structural Steel and Castings
Editor: Lena M. Becker, pp. 225-258
ISBN: 978-1-61728-104-4

Chapter 7

THE DEVELOPMENT OF MHD APPLICATION RESEARCHES IN ELECTROMAGNETIC CONTINUOUS CASTING

Ben-Wen Li*

Key Laboratory of National Education Ministry for Electromagnetic Processing of Materials, POB 314, Northeastern University, Shenyang 110004, China

Abstract

The continuous casting technology appeared almost more than 40 years. The most attractive advantage of continuous casting machines is the shortest process from the melt liquid metal to the near—end—products. Further the energy consuming can be remarkably reduced. With the development of socio—economic, the qualitative requirements of metal products are getting stricter. Thanks to the deep scientific researches of magnetohydrodynamics (MHD), the applications of MHD are enjoying more and more extensions. For example, the electromagnetic continuous casting (EMCC) is the most spotlighted issue during the past double decades.

Magnetic fields can provide energy into media without any contact, and the energy supplement can be controlled locally and directionally. According to the types of magnetic fields produced and applied in the continuous casting process, the applications of electromagnetic fields in EMCC can be classified into the following categories (a) Electromagnetic braking (EMBr), (b) Electromagnetic soft contact (EMSC), (c) Electromagnetic stirring (EMS).

Based on the latest results of EMCC, a comprehensive review is given. The experiments, numerical simulations, together with some theoretical analysis and industrial applications are presented and discussed in detail.

Introduction

As mentioned by Asai[1] in the 5th international symposium on electromagnetic processing of materials, 2006, Sendai, Japan, ISIJ, the utilization of the Lornetz force can be

* E-mail address: heatli@hotmail.com; heatli@epm.neu.edu.cn.

traced back to 1923 and 1932 for electromagnetic levitation and mixing, respectively. Due to the great contribution of Alfvén for his establishment of MHD in 1940s, gradually the applications of MHD theory and knowledge in metal and materials industries have become flourish and fruitful. The concept, electromagnetic processing of materials, which abbreviated to EPM, has be established in IUTAM conference which was hold in Cambridge, England, 1982[2]; and the first symposium on EPM was hold in Nagoya, Japan, 1994. After that, the symposium on EPM had gone on its steady step till the sixth one, which was hold in Dresden, Germany, 2009. Concerned with modern EPM, not only the primary two branches, metallurgy and MHD, have gone more deeply, micro—MHD[3] or μ—MHD for example[4], but also many others, like materials, metallography and solidification, manufacture and fabrication, fluid flow and heat transfer, mathematical model and numerical simulation, etc., are involved and their interactions.

With the fast development of metallurgy and materials industry, the striving for high quality and high product is getting more and more possible and practical. Beginning from 1980's, the widely applications of MHD in metallurgy and materials industry have made and are making great progress in this area. Resultantly, the researches of EMCC are still going and spreading into more wide range and more intensive sense. With the help of electromagnetic techniques, the continuous casting process can produce near—net—shape productions with higher quality and productivity, further to save more energy; and the process itself can be very efficient and environment friendly.

Figure 1 shows an example of popular continuous casting process machine. According to different function requirement, the electromagnetic force can be applied locally and directionally on different compositive parts of the machine, say, the ladle, the tundish, the mold, and even the end strand, as shown partly in Refs. [5, 6]. This is the most typical case of EPM.

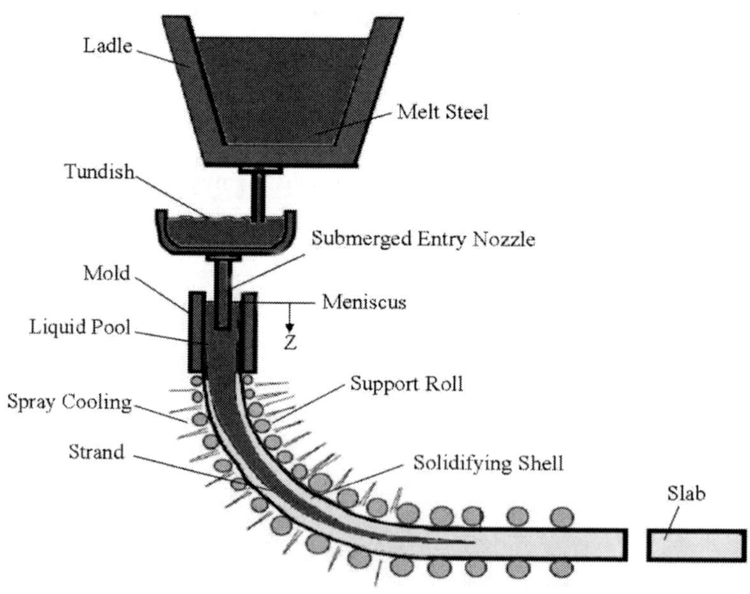

Figure 1. Whole view of continuous casting process.

The development of some main techniques mentioned in abstract will be reviewed and discussed. It should be noted that, in many early references, the abbreviation, EMCC or EMC, usually mean "electromagnetic continuous casting" and imply the application of soft contact with high frequency electromagnetic fields. However, according to the author's understanding, the EMCC should be a unitary concept, in which all kind of techniques concerned with the applications of MHD theory may be included. Thereafter, the EMSC will be used specially for "electromagnetic soft contact". In addition, the EMBR or EMBr are all abbreviated from "electromagnetic braking", and EMS from "electromagnetic stirring". Here in this writing, the strictly distinguishes between ferrous and nonferrous materials, typically steel and aluminum for example, will not been stated when talking about electromagnetic techniques. In the following sections, the mechanisms of EMSC, EMBr, and EMS, together with their characteristics and applications will be described, respectively. The combinations of above techniques and some newly appeared technologies are mentioned. Apart from the experimental researches in these aspects, the challenges and advances of numerical simulations are presented specially. Finally, some numerical research works concerned with the author in our laboratory are given, and the possible prospects arise during the description.

Electromagnetic Soft Contact (EMSC)

The description of the mechanism or principle of EMSC can be found in many references [7-12]. Figure 2 shows the phenomenon of initial solidification of the mold with or without EMSC. When a solenoid coil is surrounded horizontally on the top of the mold, the imposed high frequency alternating current in the coil will generate a magnetic field B and an induced current density J, which is in opposite phase to the imposed current, on the surface of molten metal. The interaction of J and B results in the Lorentz force, which directs inward the center of the mold. This force can also be understood as electromagnetic pinch.

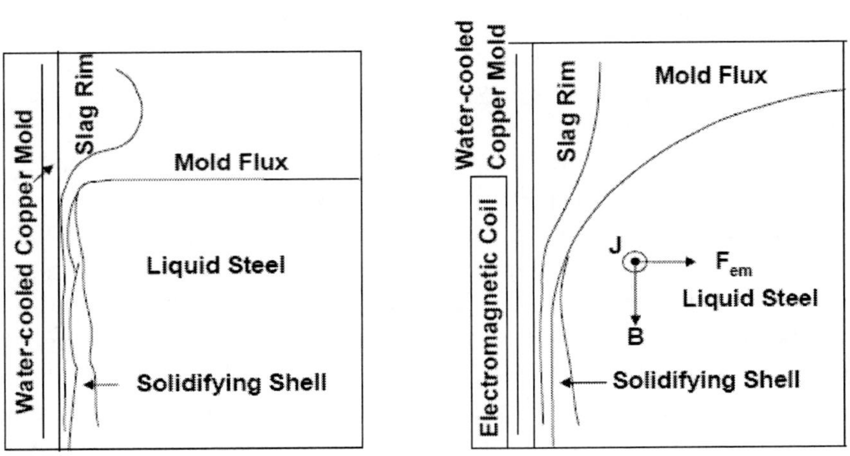

Figure 2. Phenomenon of initial solidification in the mold: without (left) and with (right) EMSC.

The electromagnetic force can partly or fully balance the hydrostatic pressure on the top of the mold, deform the shape of meniscus and even consequently the initial solidified shell, and prevent the metal from touching the slag rim. Then the meniscus shape is no longer almost the right angle as it is in conventional continuous casting mold, and a lager gap between the metal and the slag rim exists to allow more mold flux filling in. The increased mold flux consuming will obviously reduce the friction resistance between the shell and the mold, and the deformation of the meniscus can prevent the hook marks, which further develop into oscillation marks associated with surface cracks during the casting process. In addition, the shell deformation of medium—carbon steels caused by delta—gamma transformation can be effectively suppressed[10, 11]. As mentioned in Ref. [10], the oscillation marks can provide sites of entrapment of inclusions and bubbles, which cause surface defects of the products. On the other hand, the enlarged gap will reduce the heat extraction from the initial shell, and this leads to "mild cooling"[10]; and the induced heating due to the high frequency current in the metal can lag the level of initial solidification in the mold and prevent the meniscus stagnation of freezing.

In summary, compared with conventional continuous casting, the EMSC can efficiently eliminate surface defects, like oscillation marks or longitude cracks, roughness, pin holes due to inclusions or bubbles.

The very beginning researches and applications of EMSC may come from the functional concepts like, shaping, cold crucible, confinement[12, 13], levitation, in some nonferrous metals or alloys production. In 1985, Viviès and Ricou[14] investigated the fluid flow phenomena in a induction aluminum—molten furnace using high frequency alternating current; afterward, Li and Evans[15] made numerical and experimental researches on aluminum electromagnetic casting, Furui *et al.*[7] and Bermúdez *et al.*[16] also. In 2003, Park *et al.*[17] presented slit mold with EMSC for aluminum billet, and their results indicated that not only the surface and inner quality can be improved, but although the casting speed when using the slit mold with EMSC. Numerical simulations were performed for EMSC using cold crucible[18-20] and for magnetic semilevitation[21]. Electromagnetic force can change the microstructure of alloys and further improve the properties of the products during the EMSC continuous casting[22-24]. The topic of electromagnetic force changing the structure of solidification will be stated much more in EMS. Zhu *et al.*[25] used a combination of electromagnetic and centrifugal forces to act on solidified Al—7%Si alloy, and observed the microstructure and macrosegregation. They found that the electromagnetic force can produce convection in the solidification front.

Compared with aluminum, the steel has high specific weight, low electric conductivity, and the casting speed of steel is faster than that of aluminum[9]. In order to balance the hydrostatic pressure, much higher magnetic pressure is needed. These reasons make the EMSC for large scale commercial steel continuous casting unimplemented up to now. However, the researches on EMSC for steel continuous casting have been and still being active.

In 1992, Nakata and Etay[26] numerically studied the meniscus shape under alternating magnetic field, they concluded that the height of meniscus will becomes large as the magnetic flux density and the frequency of the magnetic field increases. Whereas in 2006, Deng *et al.*[27] made experimental researches on meniscus behavior using low molten Pb—Sn—Bi alloy in a round billet mold. This time, their results indicate that the lower the frequency, the higher the meniscus height, and with increasing of frequency, the free surface is more flattened and the meniscus becomes more stable. Finally they concluded that the optimal frequency for that round billet mold with a diameter of 100mm and 12 slits is 20kHz. In succession, Zhang *et al.*[28] experimentally measured the distribution of magnetic flux density in EMSC rectangular mold, they concluded again, with the same electric power input, the higher the frequency, the lower the flux density. In fact, early in 2003, He *et al.*[29] have finished experiments on EMSC using lower molten Sn—Pb alloy. Their results had shown that, compared with conventional continuous casting, the surface of billet becomes much smoother with EMSC. Recently, Deng *et al.*[30] numerically analyzed the effect of structure parameters on power and magnetic field in EMSC round billet mold of diameter 178mm, and they found that, compared with full-segment type mold, about 62% power dissipated in the mold when using partial—segment type mold. Finally they gave their optimal values for number, width, and length of slit are 24—32, 0.5—1.0mm, and 160mm respectively for this EMSC round billet mold. In Ref. [31], another research of effects of electric power frequency, current, and conductivity of billet on the electromagnetic pressure on the surface of the billet were numerically and experimentally performed.

From 1990's on, numerical simulation has played very important role in the researches of EMCC. There is another section on this topic later. Kageyama and Evans[32] computed the dynamic behavior of melts under EMSC using a modified hybrid method, Qian *et al.*[33] computed the electromagnetic fields in a EMSC rectangular mold using a modified finite—difference time domain (FDTD) method for the very first time (detail will be given late), Deng *et al.*[34] computed the magnetic fields and analyzed their characteristics for EMSC rectangular mold using finite element method, Zhang *et al.*[35] did the similar work but for a round billet mold. More complex numerical simulations may be referred to Refs. [36, 37], in which multi—physical phenomena were coupled.

The large progress of commercial applications of EMSC may come from Kim *et al.*[38], Park *et al.*[39], Kim *et al.*[40], and Park *et al.*[9]. First, the experiments of EMSC for steel continuous casting billet were carried out using low molten tin[38], then the experiments for steel at the laboratory scale were made[39], and at the same time a level meter technique was developed[40].

As mentioned and the results indicated in Ref. [30], one largest obstacle for the commercial application of EMSC to steel continuous casting is the energy losses in the mold. Nevertheless, Jacobson *et al.*[41] presented a improved mold for EMSC steel continuous casting in 2003, and they declared that a reduction of total active power of about 65% might be achieved according to their 1/3 scale experiments. It sounds very nice information for the commercial application of EMSC to steel continuous casting.

The new improvements or innovations to EMSC have appeared. They are, for example, the intermittent high frequency magnetic field in EMSC[42], the ultra—high—frequency up to 100kHz EMSC[43], the AMMF (Amplitude—modulated magnetic field) coupled with mold oscillation in EMSC[44]. The EMSC technique can also be used in some other industries, powder metallurgy[45] for example.

Electromagnetic Braking (EMBr)

With the increasing of continuous casting speed, the outlet velocity of molten metal from the submerged entry nozzle (SEN) has to be increased intensively. The liquid metal, in which the nonmetallic inclusions may be included, impinges on the initial solidification shell on the wall of the mold. The superheated liquid metal stream can remelt or even break the shell, and enhance the capturing of inclusions by the shell. On the other hand, the bifurcated molten metal streams from SEN will produce four recirculating flow regions within the mold, two in the upper, two in the lower part of the mold, respectively. The upper recirculating flow can strike on the free surface and leads to surface oscillation and slag entrapment into the metal; while the lower recirculating flow can bring inclusions, slag, or gas bubbles deep into the mold, and make it difficult for them to float up to the top mold flux. Certainly the inclusions, slag, or gas bubbles in the slab will lead to strict defects for the final products, like cracks, pin holes, etc. Now thanks to the EMBr technique, the flow pattern in the mold can be efficiently controlled, and further to improve the quality of the final product.

The mechanism of EMBr can be found in many references, for example in Refs. [1, 2, 11, 12, 46-48]. Very recently, Tian[49] described the principle of EMBr detailedly in her doctorial dissertation using the Schematic diagram of Ref. [46] (Figure 3 in present).

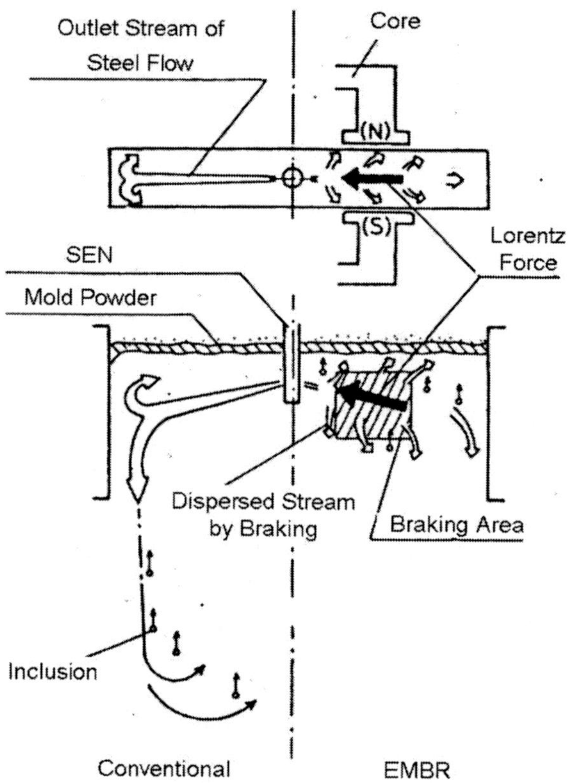

Figure 3. Schematic diagram of mechanism of local EMBr.

One static magnetic field, whose direction is perpendicular to the flow velocity, is imposed near the region of outlet from SEN. According to the Ohm's law, an induced current will be produced within the liquid metal. The interaction of the induced current and the added magnetic field will produce the Lorentz force, which always try to stop (brake) the fluid flow. The outlet streams can be decelerated and dispersed remarkably.

Up to now, the development of EMBr technique has experienced three stages. According to the type or style of the imposed magnetic fields, they are ①the localized magnetic fields, ②the level magnetic fields, and ③the coupled two level magnetic fields or FC (flow control) mold. Figure 4 shows the principles of this three stages[50].

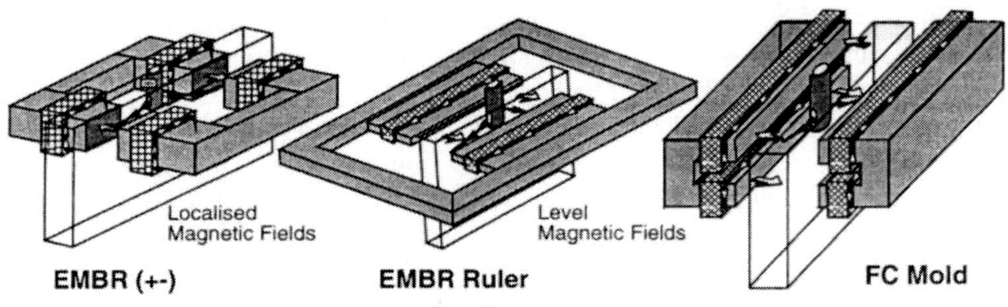

Figure 4. Structures of three types of EMBr devices.

The functions or advantages of EMBr may be reduced to the following[11, 12, 46-52]:

(1) The velocity of liquid flow ejected from bifurcated SEN is reduced and the heavy impingement on solidification shell can be avoided, further the breaking of the shell; the depth where the following down stream reaches becomes shallower then a uniform plug flow is achieved at outlet of the mold and more opportunity for inclusions and bubbles to float up to the top mold flux is given, hence the inner quality of slab can be improved.

(2) The striking of upper recirculating flow to free surface can be suppressed and then the oscillation of free surface can be reduced, consequently the possibility of entrapment of slag into molten metal is reduced and further to improve the surface quality of slab.

(3) Due to the shallower depth of down stream, the heat transfer within the upper part of the mold is enhanced, and temperature of liquid metal near the meniscus region is increased. Thus the realization of lower superheating casting is possible, and the fraction of coaxial crystal structure can be promoted.

(4) The mixture of different metals can be restrained during bimetallic slab continuous casting process by EMBr.

Takatani *et al.*[53] made the early researches on fluid flow, heat transfer and MHD with EMBr in continuous casting mold using 3D numerical simulations, Idogawa *et al.*[54] numerically and experimentally investigated the molten steel control by FC EMBr, and Bal *et al.*[55] further numerically studied the FC mold in which the solid skin effect was considered. In 1996, Moon *et al.*[56] adopted two different types of magnet core to check the EMBr effects on meniscus profile and surface velocity, and inversely to optimize the EMBr equipment in plant at Kwangyang Works, POSCO. In fact, the oscillation of meniscus is related with surface instability under electromagnetic field. Thus Cha and Yoon[57] analytically investigated the effect of EMBr on the Kelvin—Helmholtz instability, which is correlated to the instability of interface between two fluids, say, liquid flux and molten steel, with relative motion. The effects of EMBr on flow pattern, fluid control, or some parameter optimizations can refer to Refs. [58, 59].

EMBr can be used to control the liquid metal fluid flow within the mold, however the fluid phenomena will further influence the initial solidification in the mold. Ha *et al.*[60] made a 3D numerical simulation on fluid flow, heat transfer, and macro—solidification for continuous casting mold with EMBr, the similar work was finished by Shamsi *et al.*[61] to predict the solidification shell thickness with the help of commercial software FLUENT, Reza *et al.*[62] numerically did the similar work, and further made a comparison of flow modifications obtained by EMBr and by inserting a ceramic filter. Using the combination of EMBr and direct current to refine the solidification structure was performed by Usui *et al.*[63] experimentally by low molten Sn-10%Pb alloy.

The idea of combination of EMBr and direct current to control the fluid flow appeared before the work of Usui *et al.*[63], and it was used to uniform the pouring over the whole roll width and control the flow distribution in the twin-roll strip casting by Lei *et al.*[64]. Later, Yu and Lei *et al.*[65] also made the experimental research of EMBr effect on fluid flow and meniscus behavior using mercury.

Relying on the idea of electromagnetic valve, the flow redirection in the mold can be achieved, and this was verified by Pavlicevic[66] using a new concept EMBr experimentally. The EMBr effect on fluid flow in the mold fed through a slide gate system was predicted numerically by Morales *et al.*[67]. The steel flow in the mold is difficult to monitor, whereas, one can measure the temperature of copper plate and indirectly predict the molten flow pattern under ElectroMagnetic Level Stabilizer (EMLS) system[68].

Up to now, it is impossible to measure the velocity field of molten steel with or without electromagnetic fields. Nevertheless, the low molten metal or alloy can be used as model metal in the EMCC researches. Eckert *et al.*[69] made the investigation on EMBr effect on the filling process of aluminum investment casting, the mold was made of plexiglass, the eutectic melt GaInSn alloy was used as the molten metal, thereafter, the velocity field could be measured by Ultrasound Doppler Velocimeter (UDV), and more researches on bubble or inclusion entrapment[70], or modification of bubble-driven liquid metal flows[71] with EMBr were carried out.

As mentioned before, the EMBr has the function of suppressing mixture of two different metals during the bimetallic slab continuous casting process. Li *et al.*[72] numerically investigated the EMBr effect on bimetal, Sn—5%Pb and Sn, slab continuous casting. They found that, when the magnetic flux density is 0.2T, the flow of the molten metals can be strongly suppressed, and the depth of shoot region decreases remarkably with the increasing of magnetic flux density, when the magnetic flux density is up to 0.3T, the shoot region nearly disappeared. After this[72], more researches[73, 74] were carried out on this topic experimentally or numerically by the same group.

Beginning from the end of last century, scholars from Northeastern University performed lots of researches on EMBr for continuous casting process. Li *et al.*[75] first numerically studied the coupled effect of argon gas injection and EMBr on fluid flow in the mold, second, they[76] checked the influence of this combination on the biased flow phenomenon, third, Li and Tsukihashi[77] modeled the EMBr influence on vortex flows in thin slab continuous casting mold. In 2003, a revised k-ε turbulence model, in which the magnetic field application was considered, special for EMCC of molten metal was proposed by Huo and Li[78]. Taking the low molten Bi—Pb—Sn alloy as molten metal, and the silicon oil as mold powder, experiments were done to check the control effect of different magnetic fields on the molten metal flow in FC mold[79]. The inclusion collision growth in continuous caster with EMBr was specially regarded and numerically simulated by Lei *et al.*[80]. In 2007, the author and Tian[81-83] proposed a new type EMBr equipment for thin slab continuous casting, in which the surfaces of new magnets take the same shape of NURBS surface of funnel type thin slab continuous casting mold. First, the influences of casting speed and SEN depth on fluid flow were studied[81], then the EMBr efficiency of our new type was verified numerically[49, 82,

83]. It is worth to note that, compared with the conventional EMBr equipment, almost 40% electrical power may be saved by our new type for the same EMBr effects. Finally, further numerical simulations of coupled fluid flow, heat transfer and macro-solidification in this thin slab funnel type mold with new EMBr were executed by Tian *et al.*[49, 82, 83]. In 2009, another new type EMBr was proposed by Wang *et al.*[84], the so called V—EMBr can provide longitudinal vertical magnetic fields near the narrow face of mould from the meniscus to the impinging point of jet flow, and then efficiently control the molten steel flow within this region, but without considering the shape and angle of SEN and the depth of SEN depth.

Electromagnetic Stirring (EMS)

Let us go back to Figure 1. After the release of liquid metal from the SEN into the mold, the fraction of solidified metal will increase gradually with the continuous casting process going, the liquid part, whether in the mold or in the central part of the strand, will trend to be motionless. However, the steady— or quasi—steady—state of liquid metal is really detrimental to the round structure of solidification, further to the quality of final casting product. Fortunately, due to MHD theory, if a traveling magnetic field is imposed to the liquid metal, the eddy current will be generated. Then the imposed magnetic field and the induced current will give rise to the electromagnetic force, i.e., the Lorentz force, which can move the liquid metal—stirring so called.

If the imposed traveling magnetic field is parallel to the casting direction, the linear electromagnetic stirring (L—EMS) is called. On the other hand, if the magnetic field is rotating around the strand or the mold, the rotating electromagnetic stirring (R—EMS) is called. The final style of EMS is the combination of both, in which an L—EMS with relatively higher frequency in vertical direction produces downward force, and a R—EMS with relatively lower frequency horizontally produces rotating force, are co—axial installed. The helicoidal electromagnetic stirring (H—EMS) is called for this type. As the name implied the helicoidal flow can be generated for the molten metal. In some references, the double axial coils[12] or co—axial coils are called for H—EMS equipment.

One major function of EMS is due to its driving liquid metal to form the forced convection near the solidification front, hence to change the solidification process significantly. Similar to Ref. [85], the Figure 5 schematically indicates the mechanism of EMS. The solidification front juts may be broken or curved by the forced convection flow, the solute—rich melt near the front or columnar crystals may be entangled into the bulk region. The broken dendrite fragments can act as the new nucleus or will be re—melted into the liquid. This will enhance the transformation of dendrite to equi—axed crystals and further refine the grains, also dilute and disperse the solute in the molten metal. The liquid flow resulted from EMS in ladle, tundish, or mold, can impose Saffman force on nonmetallic inclusions[11], thus improve the floatation of them. Finally, the forced convection can enhance heat and mass transfer at the solidus—liquidus interface[86], hence to suppress superheat and uniform temperature and even more velocity of metal.

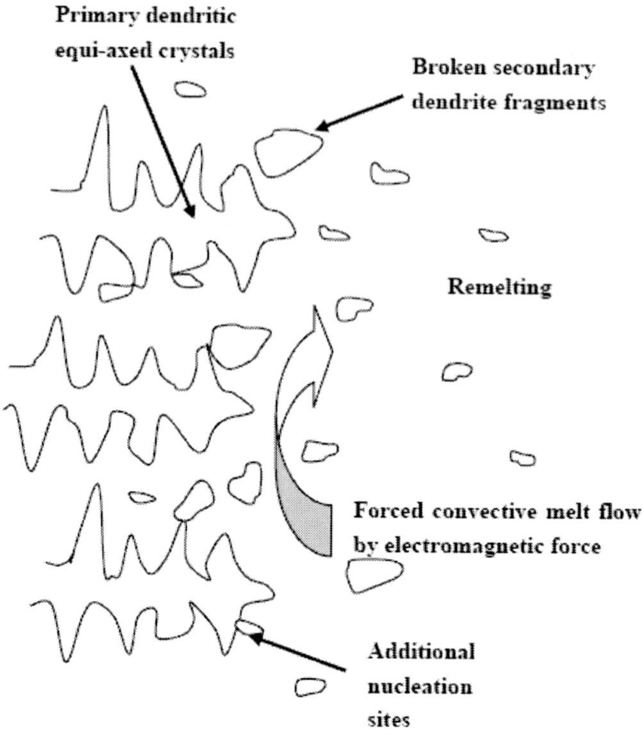

Figure 5. Schematic diagram of mechanism of EMS.

Certainly, the detail working principles of each EMS styles can be found individually in many references, say, the L—EMS in Refs. [85, 87], the R—EMS in Ref. [88], and the H—EMS in Ref. [12].

The functions or advantages of EMS may be summarized as following[1, 11, 12, 85-89]:

(1) Improving the transformation of solidification structure from dendrite crystal to equi—axed crystal, further refining the grains;

(2) Diluting and dispersing the solutes in the metal, hence reducing the segregation effect;

(3) Promoting the floatation of nonmetallic inclusions, like carbon, oxide bubble, slag particle, etc.;

(4) Suppressing turbulence of meniscus;

(5) Homogenizing temperature, solute distribution, and even velocity.

Based on above functions of EMS, the surface and inner qualities of the continuous casting products can be improved significantly, for example, reducing or eliminating pinholes and inclusions at the slab surface layer, suppressing oscillation marks and longitudinal cracks on the slab and cleanliness improvement of slab sub—surface, removing solute segregations within slab, refining and purifying microstructure thus improving the hardness and weariness and fatigue.

As shown in Figure 1 of Ref. [6], the EMS technique can be used for three stages during the continuous casting process, and the corresponding names are called like, the M—EMS for EMS effect in the mold, the S—EMS in the strand, and F—EMS at the end of the strand where the liquid metal still exits in the center. Besides these, the EMS can also be used in the ladle to separate the suspended particles and reduce the quantity of inclusions[90]; in the tundish to promote inclusions floatation[5].

The most widely researches of EMCC remain with EMS whether in ferrous or nonferrous metal industries. The reasons exit in at least two aspects: one is that the EMS technique can be used at every stage of the continuous casting process; the other is that the quality of final product is significantly related to the phase change, *i.e.*, solidification process under EMS effects.

Almost the earliest researches on tracer dispersion and deoxidation measurements under EMS effect were carried out by Nakanishi *et al.*[91]; the inclusions removal from steel under EMS was experimentally and numerically studied by Cremer and Driole[92] in 1982; a little bit later in 1984, Bridge and Royers[93] studied the effects of location of transverse electromagnetic fields on the structure and band segregation formation of strand cast steel; more widely, Spitzer *et al.*[88] and Dubke *et al.*[94] checked the influences of stirrer position, length and electromagnetic parameters on R—EMS and L—EMS separately. Another systematically investigation on electromagnetic separation of inclusions from liquid metal was achieved by Natarajan and El-Kadda[95, 96] numerically; and also investigation about R—EMS effects on macrosegregation of cast iron and separation of anomalous eutectics by Cao *et al.*[97, 98] respectively. One special and fundamental study on nonmetallic inclusions particles separation but under direct current (DC) electromagnetic field was made by Ogasawara[99]. Electromagnetic separation of nonmetallic inclusions from Al-Si melt is studied by Li *et al.*[100]; Using R—EMS, Willers *et al.*[101, 102] studied the influence of electromagnetic—driven convection on temperature field and solute distribution during directional solidification.

The most widely and fundamentally researches for EMS come from the effects of electromagnetic field on the solidification structure and grain refining during continuous casting. For example, Hu and Zheng[103] studied the EMS effects on dendrite structure and mechanical properties of low sulphur steel, Blazek[104] investigated the continuous rheocaster for ferrous and high melting pint alloys using a dual chamber casting machine utilizing separate EMS around each chamber, Vivès did the similar work[105], and further used the combination of electromagnetic vibrations and stationary magnetic fields to get more smaller grain size for aluminum alloys[106]. Griffiths and McCartney studied the solidification structure changes of columer—equiaxed transition (CET) under the R—EMS effect for alloys[107, 108]; Guo *et al.*[109] numerically simulated the EMS contribution to the CET through increasing the number of fractured dendrites; Zhang *et al.*[110] obtained the lamellar changes to rod—like eutectic structure of Al—$CuAl_2$ by R—EMS, and simultaneously they observed the structure of Al—Cu eutectic alloy solidified in complex field of centrifugal and EMS[111]; Zhang *et al.*[112] experimentally investigated the EMS effect on refining microstructures of pipe—line steel submerged arc welding (SAW) deposited metal. Wang *et al.*[113] made the experiments to check the EMS effect on steel solidification process, and they found that the equiaxed grain zone increased from 47% to 76% under EMS; Denisov *et al.*[114] observed the EMS effects on cylindrical moulds of continuous casting machines fabricated aluminum alloy, and they found that the finer grain structure, hence with high quality surface, product could be obtained; Nafisi *et al.*[115] studied the microstructural characteristics of Al—Si—Fe alloy under the hybrid conditions of EMS and superheat. The effects of electromagnetic parameters of M—EMS like, the frequency and the current intensity, on the quality of continuous casting round billets were studied by Kuchaev *et al.*[116]; while, the numerical simulation works were carried out by Barna *et al.*[117] to predict the effects of M—EMS for round bloom strands, and their results indicate that with the increasing of flow velocity, the columnar to equiaxed transition solidification was improved, the inclusions, segregations and bubbles were reduced. Using the combination of EMSC and EMS to aluminum or magnesium alloys casting, the microstructure, grain fining etc., and the related qualities were investigated in Refs. [118-120]. Adopting a electromagnetic vibration technique, in other words, by combining a static and a traveling magnetic fields, the solidification behavior of AZ31 magnesium alloys was studied by Li *et al.*[121]; a numerical study, in which the macrosolidification fraction distribution was included, on L—EMS for aluminum alloy was finished by Barman *et al.*[122]. Some other operational analysis and parameter optimal researches about EMS may be referred to Refs. [88, 123-126]. Under R—EMS, directional solidifications of different alloys were performed by Zimmermann *et al.*[127, 128], and unidirectional solidification of binary and ternary aluminum alloys by Kovacs[129].

The combinations of EMS itself and with others can control the fluid more efficiently and improve stirring significantly. Taniguchi *et al.*[130] and Shimasaki *et al.*[131] gave the investigations on the combination of rotational and vertical traveling magnetic fields, i.e., the mentioned H—EMS above; the similar researches can be found in Ref. [132] for the dual coil EMS, and in Ref. [133, 134] in which the two—frequency magnetic field was adopted; and the multi—mode EMS so—called in Ref. [135]. Ando *et al.*[136] presented a new equipment to produce a helical magnetic field which can stir and pump the liquid metal simultaneously; Wang *et al.*[137] invented another equipment using one pair of rotating permanent magnets to induce both rotating and traveling magnetic fields. The numerical simulation on flow control in the meniscus of continuous casting mold with opposing alternating current magnetic fields was conducted by Chang *et al.*[138]; while, the numerical simulation of fluid flow under the effects of both M—EMS and S—EMS was made by Harada *et al.*[139]. The research of combination of EMS and gas stirring was finished by Chungy *et al.*[140] in ASEA—SKF ladle in 1992, and later by Toh *et al.*[141] for steel continuous casting numerically and experimentally, and recently by Wang and Li[142] for the behavior of two—phase flow also numerically and experimentally. Similar to the works of combination of EMSC and EMS in Refs. [118-120], the study on the combination of EMBr and EMS was finished by Hirayama and Fujisaki[143]. Under the conditions of terrestrial and parabolic flight, if the EMS was applied to the solidification process, what would happen? This interesting topic can be found in Ref. [144].

The effects of EMS on the solidification process happen at the key stage of the phase change from liquid to solid. Therefore, lots of researches were conducted on the semi—solid alloys under the condition of EMS. The group of Mao *et al.*[145-147], Liu *et al.*[148] and Zhao *et al.*[149, 150] systematically investigated the effects of different EMS alone or EMS together with low superheat pouring condition on the solidification structure for varied alloys. Also the effect of EMS together with superheat but for Al-Si-Fe alloy was observed in Ref. [115]. The similar studies were finished by Bae *et al.*[151] for Al-7%Si alloy and Kim and Kang[152] for Al6061 aluminum alloys. Some further relationship between macro— and microstructure of the rheocast alloy A356 obtained by EMS was considered by Zoqui *et al.*[153].

Not only the primary elements can be refined by EMS during the solidification of alloys, say, the refinement of primary Si in hypereutectic Al-Si alloy by EMS reported in Ref. [154], but although the combination of EMS and additional refining particles can even more improve the grain refinement. Early in 1984, Patchett and Abbaschian[155] investigated the combination effect of EMS and addition of iron on grain refinement of copper; late in 2001, Jung *et al.*[156] checked the effect of EMS and Sr modification in A356 alloy; Liu *et al.*[157] studied the refinement role of EMS and Sr in AZ91 magnesium alloy; Metan *et al.*[158] observed the grain refinement of Al-Si alloy under the effects of both EMS and AlTi5B1 particle refiner; the other related works may be referred to Ref. [159-161]. A comparison between particle refiner (i.e. P and Sr) and electromagnetic stirring and vibration (ESV) was made to improve microstructure refinement of 390Al-Si hypereutectic alloys by Robles Hernández and Sokolowski[162], and they concluded that: the use of melt treatments assisted by chemical modifiers for Al-Si hypereutectic alloys are limited to reduce in size the primary Si particles by the multiplication of active nuclei sites (Si agglomerates), this limits the coarsening effect of the primary Si particles in the liquid state, the ESV melt treatments are effective in the liquid and semi-solid states.

The electromagnetic parameters should be appropriately determined especially for M—EMS due to its influences on free surface or meniscus. Thus some researches were conducted on this topic. For example, in Ref. [133], the experimental and theoretical studies of the motion generated by a two—frequency magnetic field at the free surface of a gallium pool were made; and in Ref. [138], a numerical simulation considering the meniscus with EMS was finished; the fluid flow, meniscus or drift flow, and the EMS were coupled simulated in Refs. [163, 164], respectively. Using Sn-32%Pb-52%Bi alloy as the model liquid, Deng *et al.*[165] experimentally studied the effects of a coupled AD and DC magnetic fields on the meniscus.

If a pulsed magnetic field is imposed during the continuous casting or solidification[166], the heat and mass transfer at the solidus—liquidus interface can be promoted[86], further to change crystal growth and grain refinement and hence to improve the quality[167, 168].

Usually, the numerical simulations of solidification are limited to macro—scale due to the huge computer cost, especially when the fluid flow or electromagnetic fields are coupled. While, the cellular automaton (CA) method, which is special for micro—solidification, was adopted to simulate the detail solidification structure formation under EMS by Zhang *et al.*[169] and Yamazaki *et al.*[170], separately.

In the group of Ren *et al.*[171-176], first, the combination of EMS and EMSC was applied for oxygen—free copper (OFC) processing[171]; and at the same time the new technique of special amplitude—modulated magnetic field (AMMF) was proposed for electromagnetic continuous casting[172]; later, they studied the influences of high static magnetic field, which is up to 10T, on unidirectional solidification of Al-4.5%Cu alloy[173]; recently, they studied the solidification of Zn-30wt%Bi monotectic alloy under hybrid of high static magnetic field and AC current[174]; the similar work was done by them but for horizontally continuous casting of super—thin slab[175]; finally, they gave a experiment on microstructure and properties of Cu-Cr-Zr alloy aged with DC electric current[176].

As to the group of Li *et al.*[177-182], first, the effects of combination of EMS and EMSC on meniscus behavior and the quality of continuously cast metal were studied[177]; then the R—EMS for copper tube billets continuous casting was investigated[178]; a numerical simulation of solidification process for Sn-3.5%Pb hollow billet with EMS was made[179, 180]; and later the research on semisolid processing AZ91 magnesium alloy by EMS after near—liquidus isothermal heat treatment was made[181]; recently, the effects of combination of EMS and power ultrasonic field on the horizontal continuous casting of Al-1%Si alloy was observed[182].

In another group of Cui *et al.*[183-188], first, the effects of EMS, the low—frequency electromagnetic casting (LFEC) is called there, on as—cast microstructure and mechanical properties of superhigh strength aluminum alloy was considered[183]; and then the same work for Al-Zn-Mg-Zr alloy[184]; and similar works for HDC2024[185] and 6063 aluminum alloy[186]; at the same time, a numerical simulation for four—strand casting aluminum alloy under EMS was finished and validated by experiments[187]; finally, the effects of combination of M—EMS and F—EMS with proper location on inner quality of high carbon steel bloom were studied[188].

All above works of three groups whether for steel continuous casting or for alloys solidification, the EMS effects were almost positive for grain refinement and macrosegregation reduction, and further help to improve the inner quality.

Some other researches or applications of EMS can be found. Inomoto *et al.*[189] used the EMS technique to promote desulphurization in ladle through slag emulsification; Thess *et al.*[190] applied the EMS on glass melts; similar to the work of Thess *et al.*, Cepite and Jakovics[191] also used the EMS on glass melts and further analyzed the influence of radiation heat transfer; Kolesnichenko[192] studied high frequency (50-60 Hz) EMS for continuous billet/bloom casting middle/high carbon steels; Yao *et al.*[193] proposed the new experimental system for the study of mold heat transfer under EMS; Cheng *et al.*[194] used the EMS to improve the microstructure and wear behavior of iron—based composite coating; etc.

With the review coming here, the author would like to say, the tremendous potential in using electromagnetic fields is not limited to the continuous casting or even metallurgy, but spreads to materials, just as the electromagnetic processing of materials (EPM) implied. Some researches or applications can be beyond the above three categories, i.e., EMBr, EMSC, and EMS. Some newly researches or applications of electromagnetic fields can be briefly reviewed in the following.

Using high frequency up 200kHz alternating magnetic field and high static magnetic field up to 10T, Yasuda *et al.*[195] studied the effect of the melt flow on the solidification structure of middle carbon steel; using both electrostatic levitator (ESL) and electromagnetic levitator (EML), Li *et al.*[196] investigated the microtexture formation of Ni99B1 alloys solidification; Mechighel *et al.*[197] numerically checked the damping effect of an external magnetic field on to alloy solidification; Su *et al.*[198] used a rotating magnetic field to produce swirling flow in immersion nozzle in continuous casting of steel; Timmel *et al.*[199] invented a new equipment named CONCAST special for steel continuous casting researches; researches on solidification of metals or alloys under high static magnetic fields, which can be more than 10T, were presented by Refs. [200-202]; etc.

Numerical Simulations

With the development of computer techniques and numerical skills, the numerical simulations play more and more important role in the EMCC researches due to their economics and reliability. On the other hand, the difficult or impossible measurements of physical fields or variables, say, velocity, electromagnetic fields, temperature, in EMCC, make the numerical simulation a necessary tool, or at least one important complement of experiments and theoretical analyzes, to understand the complex mechanisms and phenomena. From physical[203] and mathematical view, the EMCC includes thermodynamics, in which the Navier—Stokes equations together with energy equation for flow field and temperature, solute/bubble transport equations for solute/bubble trace, the electromagnetic process, in which the Maxwell equations together with Ohm's law for electromagnetic fields, balance euqation of hydrostatic and magnetic pressure for meniscus height, phase change Darcy source term for macro—solidification within mushy solidification region, should be solved coupledly.

The difficulties of numerical simulations come from many aspects, i.e., ①complexities of system, ②transient turbulent flow and accompanying phenomena like heat & mass transfer, deformation of solidification shell and even the mold, ③solidification itself, ④electromagnetic prosess, ⑤non-matellic slag (solute, inclusions) and/or gas bubbles movement, ⑥free surface (meniscus) oscillation, etc. In fact, the biggest trouble may be the interaction or coupling of two or more physical phenomena. These facts were also partly mentioned in Ref. [48].

Up to now, the Reynolds averaged Navier Stokes (RANS) approach is most wildly used to solve the time—avaraged Navier—Stokes equations. Together with the continuity equation for mass conservation, the hydrodynamics can be known. As to the normal situation the fluid flow is turbulent, then the well known $k-\varepsilon$ model is adopted in which another two equations for the turbulent kinematic energy and its dissipation should be solved to modify the turbulent effective viscosity. In order to capture the moderate scale information of turbulence, the LES (large eddy simulation) model is used gradually. While, the most modern and advanced DNS (direct numerical simulation) model is not found in EMCC due to the heavy computer cost allowing very samll grid and time step.

For static magnetic field, used in EMBr for example, the calculation is an easy work, and usually the electric potential is used in Ohm's low and charge free like

$$\vec{J}=\sigma\left(\vec{E}+\vec{v}\times\vec{B}\right) \quad , \quad \nabla\cdot\vec{J}=0 \tag{1}$$

with the definition of electric potential, $\vec{E}=\nabla\phi$, one can obtain

$$\nabla(\sigma\nabla\phi)=\nabla\cdot\sigma\left(\vec{v}\times\vec{B}\right) \tag{2}$$

In principle, the magnetic field $\vec{B}$ should include the induced one by induced current in the liquid metal, i.e., $\vec{B}=\vec{B}_0+\vec{b}$. However, if the induced magnetic field $\vec{b}$ is much smaller than the applied $\vec{B}_0$ as the ordinary cases of liquid metal, in another words, the magnetic Reynolds number $\mathrm{Re}_m=vL\mu\sigma\ll 1$, then it can be neglected[204].

If the unsteady magnetic field is considered as the usual cases in EMSC or EMS, and the induced magnetic field can not be neglected, the work of Tallbäck *et al.* for example[205], then the magnetic field can be obtained by two ways[206]. First, the Maxwell equations are transformed to magnetic induction equations which need the magnetic flux density boundary conditions; second, the vector scalar $\vec{A}$ for magnetic field and potential ϕ for electric field are introduced to give the similar form of transport equations of Navier—Stokes equations and with easy boundary condition treatments. Besides these two ways, the FDTD method can be adopted to solve the Maxwell equations directly if the geometric system is rectangular[33, 207]. After the solution of electromagnetic fields, and further the induced current density, the time averaged Lorentz force will act as source terms in momentum equations.

The oscillation of free surface (meniscus) can be determained by VOF (volume of fluid) method iteratively companying with fluid flow solution[208]. Another approximation of meniscus shape is the pressure balance[19], i.e., the sum of magnetic pressure, hydraulic pressure, and the pressure due to surface tension, give a constant at each node.

The particle (solute, inclusion, even gas bubble) movement can be traced by Lagrangian methods[48]. While another more rigorous approach name mass—population—conservation was used by Lei *et al.*[80], in which the collision and coalescence of inclusions can be considered, and the inclusion collision growth governing equations take the same form of general transport equation.

The numerical simulations of solidification in the EMCC can be classified into two categories according to the intensity required or computer cost, macro— and micro—solidification. For the detailed review on this topic one can refer Ref. [209].

Most of wildly used numerical methods in CFD (computational fluid dynamics), e.g., finite element method (FEM), finite difference method (FDM), boundary element method (BEM), were adopted in the solution of PDEs appeared in above described mathematical models for EMCC process. Yoon[47] reviewed the numerical methods used in continuous casting technology. Here the author will give some complementary description.

The very early numerical simulation of combined fluid flow and electromagnetic field in EPM may be tracked to the work of Evans *et al.*[210]. In fact, the electromagnetic fields used in EMS could be approximated analytically if the result was not accurately required[94]. After the introdution of vector scalar potential, the magnetic variables were subsitituted by Fourier polynomial expression, and then the fast Fourier transform together with eigenvalue decomposition was adopted to solve magnetic field by Lavers[211]. Kageyama and Evans[32] used a "modified hybrid technique" to solve Maxwell equations and Ohm's law; due to the easy adaptable to the complex nonlinear differential equations with variable coefficients governing the moving boundary nonlinear problem, a pseudospectral collocation method, in combination with a coordinate tranformation continuously adapting to the interface, was used for modeling the dynamics of magnetic semilevitation melting by Bojarevics *et al.*[21]; Bermúdez *et al.*[16] used an asymptotic approximation in the electromagnetic casting. Bai *et al.*[23] used the FEM/FDM—joint method for the transport phenomena in the electromagnetic casting.

Similarly, with the development of computer techniques and numerical skills, the comerical softwares are getting more and more reliable and robust. The applications of softwares in EMCC simulation are no longer expensive. From 1990s, the FLUENT for CFD and ANSYS for electromagnetic fields have been wildly used in EMCC althrough they were combined to one unit finally. Especially with the MHD mode and UDF (user difined function), the FLUENT becomes more powerful in EMCC or EPM now. Some other softwares can also be found special for CFD or electromagnetic processes. In 1989, the PHOENICS wase applied to solve the fluid flow by Ilegbusi and Szekely[212] for R—EMS; the OPERA—3D for electromagnetic fields was used by Kim *et al.*[85] in their EMS simulation; the ELEKTRA and ANSOFT were used for electromagnetic processes by Chang *et al.*[138] and Lavers *et al.*[132], separately; the OpenFOAM was used by Maiwald *et al.*[213].

From now on, the works concered with the author will be reviewed briefly.

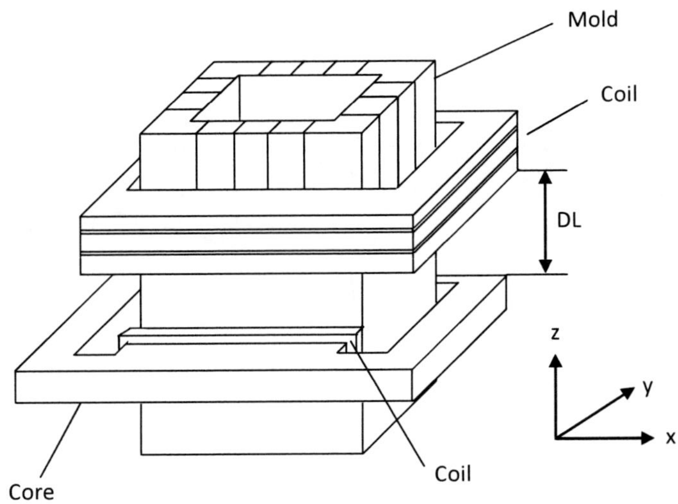

Figure 6. Schematic diagram of the hybrid electromagnetic mold.

It is well known that the FDTD method[214] is wildly used for time—varied Maxwell equations in electrodynamics. However, it was originally proposed for electromagnetic fields with high frenguency up to 10^7 Hz[214, 215]. In EMCC, the frequency of interest is usually within the range of $10 \sim 10^4$ Hz, and this fact limited the application of FDTD method due to two challenges. First, the temperature of metal is more than 760°C, the magnetic permeability μ and the permittivity ε of the melt metal take the values of vacuum respectively, i.e., the unit. To ensure stability of the solution, the time step should be $\delta / 2c$. Assuming δ has the value of 5mm, the iterations will be great up to 10^8 times. Then the computer cost is very heavy and the accuracy of the computational results will be viewed with quite skepticism because of the accumulation of iterative errors of computation. Second, the metal concerned in the numerical simulations[216, 217] is taken to be perfectly conducting scatter when the frequency is up to 10^7 Hz. But it's not the fact in present numerical simulation and great errors may be conducted. Refered the work of Gandhi and Chen[218], Qian *et al.*[33] modified the scaled-frequency FDTD method to simulate the electromagnetic fields in metal at low frequency. A simplified physical model with analytical solutions was used to validate the modified method. For another problem of metal and dielectric interface treatment at low frequency existing in FDTD method, a new improvement has been proposed[207] and experiments have been made to verify the method. After the solution of electromagnetic fields, further the induced current in liguid metal, the combination with fluid flow and meniscus shape were numerically simulated for a rectangular mold, on which the hybrid EMSC and EMBr were added. In figure 6, the upper coil is for EMSC, the lower one is for EMBr, and the symbol "DL" means distance between two coils. Finally, the influences of operational parameters on the fluid flow were analysised. For details one can refer Ref. [219, 220]. Furthermore, Qian and Wu has made the LES on this topic[221] later.

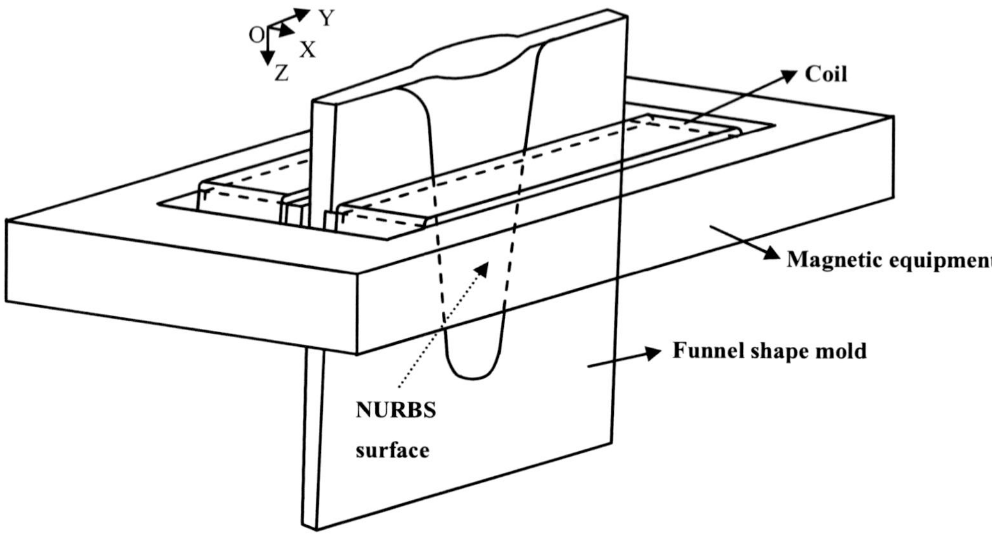

Figure 7. Schematic of funnel shape thin slab mold with level EMBr.

Thin slab casting can provide near—net—shape products and short rolling process, hence to save energy, thus it is more economical than conventional slab casting. Compared with conventional slab continuous casting, the speed of thin slab casting is higher even more than 6m/min, and its mold has a large ratio of wide to narrow sides, thus the fluid flow in the mold is much more difficult to control to prevent non uniform of initial very thin solidification shell, severe oscillation of meniscus, and some other defects in the casting process. Therefore, EMBr technology is more attractive to control and redistribute the melt steel flows in a mold region using a transverse static magnetic field. First, the author and colleagues[81] made numerical analysis for fluid flow in a thin slab funnel shape mold, some structural and operational parameters were optimized. Afterward, Tian *et al.*[82] made detail numerical analysis for EMBr effects on fluid flow using a newly designed magnet. This time, as shown in figure 7, the mold is irregular with NURBS surfaces. Therefore, the software FLUENT together with MHD mode was used to simulate fluid flow and EMBr. While, the steady magnetic fields were computed by the software ANSYS first, and then the datum was transport into FLUENT after interpolation. For the funnel shape mold, the EMBr due to the new magnet, which takes the same NURBS surface to fit the mold as shown in figure 8, can provide more uniform downstream velocity profile than due to conventional magnet, and almost 40% reduction in ampere turns (electric power unit), this fact implies lots of electrical energy could be saved. Finally, more extensively and deeply simulations considering heat transfer, macro—solidification were carried out by Tian *et al.*[49, 83].

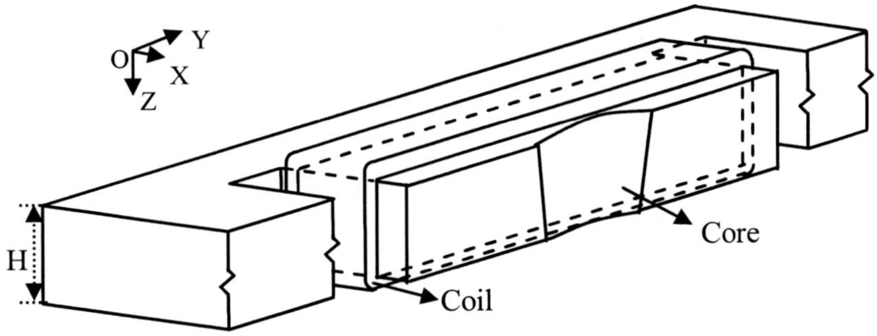

(a)

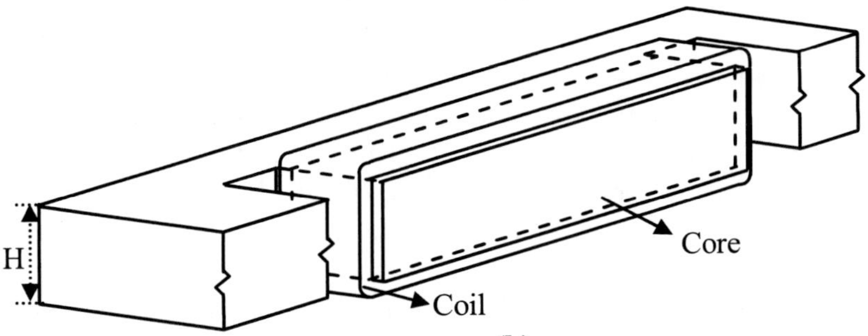

(b)

Figure 8. Half of the magnet with coils: (a) new type magnet, (b) conventional magnet.

Prospects and Remarks

The applications of MHD in metallurgy and material industries improved theoritical analysis and research of EMCC and EPM. To promote technologies and skills, only the consideration of Lorentz force is not enough. With more strict limitation of energy consumption and friendly enviroment requirement, the MHD can play more important role in many areas. For example, the magnetization force can be used to remove impurities of material, or abstract noble metal from wastes. Besides this, more extensive researches of MHD also should be focused, for example, the MHD instability analysis, time evolution of MHD fluid, the complex phenomena of combination of MHD with heat transfer, solidification, etc.

Acknowledgment

Our researches concered with MHD were financially sponsored by the National Nature Science Foundation of China under Contract No. 10772044. The author acknowledges Dr. Xi-Yan Tian for her edition work.

References

[1] Asai S. Electromagnetic processing of materials, past, present and future. In *Proc. 5th Int. Conf. on Electromagnetic Processing of Materials*, 2006, Sendai, Japan, ISIJ. pp. 3-8.

[2] *Proceedings of a Symposium of the Int. Union of Theoretical and Applied Mechanics held at Cambridge*, UK, Sep. 1982.

[3] Aoyagi R.; Asanuma M. *Proceedings of the Third Meeting*, 22-42, Int. Symp. New Magneto-science' 99, JST, 1999, Omiya, Japan.

[4] Yasuda H. Application of high magnetic fields on solidification processing. In: *Proc. 5th Int. Conf. on Electromagnetic Processing of Materials*, 2006, Sendai, Japan, ISIJ, pp. 345-350.

[5] Miki Y. Applications of MHD to continuous casting of steel. In: *Proc. 5th Int. Conf. on Electromagnetic Processing of Materials*, 2006, Sendai, Japan, ISIJ, pp. 26-30.

[6] Barglik J.; Smagor A. Some aspects of electromagnetic stirring during process of continual steel casting. In: *Proc. 6th Int. Conf. on Electromagnetic Processing of Materials*, 2009, Dresden, Germany, Forschungszentrum Dresden-Rossendorf. pp. 684-687.

[7] Furui M.; Kojima Y.; Matsuo M. Fabrication of small aluminum ingot by electromagnetic casting. *ISIJ Int.*, 1999, 33, 400-404.

[8] Kageyama R.; Evans J.W. Development of a three dimensional mathematical model of the electromagnetic casting of steel. *ISIJ Int.*, 2002, 42, 163-170.

[9] Park J.; Kim H.; Jeong H.; Kim G.; Cho M.J.; Chung J.S.; Yoon M.; Kim K.R.; Choi J. Continuous casting of steel billet with high frequency electromagnetic field. *ISIJ Int.*, 2003, 43, 813-819.

[10] Takeuchi E. EPM in continuous casting and its extensive prospect. In: *Proc. 4th Int. Conf. on Electromagnetic Processing of Materials*, 2003, Lyon, France, pp. 137-144.

[11] Matsumiya T. Recent topics of research and development in continuous casting. *ISIJ Int.*, 2006, 46, 1800-1804.

[12] Yasuda H.; Toh T.; Iwai K.; Morita K. Recent progress of EPM in steelmaking, casting, and solidification processing. *ISIJ Int.*, 2007, 47, 619-626.

[13] Laber J.D. Computational methods for the analysis of molten metal electromagnetic confinement systems. *ISIJ Int.*, 1989, 29, 993-l005.

[14] Viviès C.; Ricou R. Fluid flow phenomena in a single phase coreless induction furnace. *Metall. Trans. B*, 1985, 16, 227-235.

[15] Li B.Q.; Evans J.W.; Cook D.P. An improved mathematical model for electromagnetic casters and testing by a physical model. *Metall. Trans. B*, 1991, 22, 121-134.

[16] Bermúdez A.; Muňiz M.C.; Salgado P. Asymptotic approximation and numerical simulation of electromagnetic casting. *Metall. Trans. B*, 2003, 34, 83-91.

[17] Park J.; Jeong H.; Kim G.; Kim H. Electromagnetic continuous casting of aluminum billet using slit mold. In: *Proc. 4th Int. Conf. on Electromagnetic Processing of Materials*, 2003, Lyon, France, pp. 200-205.

[18] Makino H.; Kuwabra M.; Asai S. Process analysis of non-contact continuous casting of materials using cold crucible. *ISIJ Int.*, 1996, 36, 380-387.

[19] Cha P.R.; Hwang Y.S.; Nam H.S.; Chung S.H.; Yoon J.K. 3D numerical analysis on electromagnetic and fluid Dynamic pheonomena in a soft contact electromagnetic slab caster. *ISIJ Int.*, 1998, 38, 403-410.

[20] Toh T.; Yamamura H.; Kondo H.; Wakoh M.; Shimasaki S.I.; Tanichi S. Evaluation of inclusions removal during levitation melting of steel in cold crucible. *ISIJ Int.*, 2007, 47, 1625-1632.

[21] Bojarevics V.; Pericleous K.; Cross M. Modeling the dynamics of magnetic semilevitation melting. *Metall. Trans. B*, 2000, 31, 179-189.

[22] Cao Z.Q.; Jia F.; Zhang X.Q.; Hao H.; Jin J.Z. Microstructures and mechanical characteristics of electromagnetic casting and direct-chill casting 2024 aluminum alloys. *Mater. Sci. Eng. A*, 2002, 327, 133-137.

[23] Bai Y.F.; Xu D.M.; Mao L.H.; Guo J.J.; Fu H.Z. FEM/FDM-joint simulation for transport phenomena in directionally solidifying shaped TiAl casting under electromagnetic field. *ISIJ Int.*, 2004, 44, 1173-1179.

[24] Wang P.; Shi L.F.; Lu G.M.; Cui J.Z. Microstructure and properties of ZL201 alloy obtained by near-liquidus electromagnetic casting. *Acta Metall. Sin. (Engl. Lett.)*, 2005, 18, 653-658.

[25] Zhu Y.F.; Zhang W.Q.; Yang Y.S.; Liu Q.M.; Hu Z.Q. Microstructure and macrosegregation of Al-7%Si alloy solidified under complex effects of electromagnetic and centrifugal forces. *Acta Metall. Sin. (Engl. Lett.)*, 1997, 10, 105-114.

[26] Nakata H.; Etay J. Meniscus shape of molten steel under alternating magnetic field. *ISIJ Int.*, 1992, 32, 521-528.

[27] Deng A.Y.; Wang E.G.; He J.C. Meniscus behavior in electromagnetic soft-contact continuous casting round billet mold. *J. Iron & Steel Res. Int.*, 2006, 13, 13-16.

[28] Zhang L.T.; Wang E.G.; Deng A.Y.; He J.C. Distribution of magnetic flux density in soft-contact EMCC rectangular mold. *J. Iron & Steel Res. Int.*, 2006, 13, 14-17.

[29] He J.C.; Wang E.G.; Deng A.Y.; Yu G.W.; Wang Q.; Zhang Y.J.; Chen Z.P.; Zhou Y.M.; Feng C.B. Experiment of soft-contact electromagnetic continuous casting of steel. In: *Proc. 4th Int. Conf. on Electromagnetic Processing of Materials*, 2003, Lyon, France, pp. 195-199.

[30] Deng A.Y.; Wang E.G.; He J.C. Effect of structure parameters on power and magnetic field in electromagnetic soft-contact continuous casting system. *J. Iron & Steel Res. Int.*, 2008, 15, 19-24.

[31] Deng K.; Ren Z.M.; Zhong Y.B.; Jiang G.C.; Hu S.M. The computation and experiment analysis of electromagnetic continuous casting with soft-contacted mould. *Acta Metall. Sin. (Engl. Lett.)*, 2000, 13, 708-714.

[32] Kageyama R.; Evans J.W. A mathematical model for the dynamic behavior of melts subjected to electromagnetic forces: Part I. model development and comparison of predictions with published experimental results. *Metall. Trans. B*, 1998, 29, 919-928.

[33] Qian Z.D.; Li B.W.; Jia G.L.; He J.C. Numerical computation of electromagnetic fields in metals using a modified finite-difference time domain method. *ISIJ Int.*, 2001, 41, 683-688.

[34] Deng A.Y.; Jia G.L.; He J.C. Three-dimensional characteristics and homogenization of electromagnetic field in soft-contact continuous casting mold. *Acta Metall. Sin. (Engl. Lett.)*, 2001, 14, 137-142.

[35] Zhang Y.J.; Chen X.Y.; Yang Y.S.; He J.C. Numerical analysis of the influence of geometric parameters on distributions of 3D electromagnetic fields in soft-contact electromagnetic continuous molds of round billets. In: *Proc. 4th Int. Conf. on Electromagnetic Processing of Materials*, 2003, Lyon, France, added.

[36] Deng A.Y.; He J.C. Coupled numerical simulation of steel flow and solidification in soft-contact billet mold. *Acta Metall. Sin. (Engl. Lett.)*, 2002, 15, 471-478.

[37] Na X.Z.; Xue M.; Zhang X.Z.; Gan Y. Numerical simulation of heat transfer and deformation of initial shell in soft contact continuous casting mold under high frequency electromagnetic field. *J. Iron & Steel Res. Int.*, 2007, 14, 14-21.

[38] Kim H.; Park J.P.; Jeong H.; Kim J.k. Continuous casting of billet with high frequency electromagnetic field. *ISIJ Int.*, 2002, 42, 171-177.

[39] Park J.; Jeong H.; Kim H.; Kim J. Laboratory scale continuous casting of steel billet with high frequency magnetic field. *ISIJ Int.*, 2002, 42, 385-391.

[40] Kim G.; Kim H.; OH K.; Park J.; Jeong H.; Lee E.W. Level meter for the electromagnetic continuous casting of steel billet. *ISIJ Int.*, 2003, 43, 224-229.

[41] Jacobson N.; Lehman A.; Svensson E. Improved mould for electromagnetic casting of steel cuts energy consumption. In: *Proc. 4th Int. Conf. on Electromagnetic Processing of Materials*, 2003, Lyon, France, pp. 206-211.

[42] Li T.J. Effect of intermittent high frequency magnetic field on initial solidification in continuous casting. *Acta Metall. Sin. (Engl. Lett.)*, 1997, 10, 151-156.

[43] Nakata H.; Inoue T.; Mori H.; Ayata K.; Murakami T.; Kominami T. Improvement of billet surface quality by ultra-high-frequency electromagnetic casting. *ISIJ Int.*, 2002, 42, 264-272.

[44] Lei Z.S.; Ren Z.M.; Deng K.; Li W.X.; Zhong Y.B. Amplitude-modulated magnetic field coupled with mold oscillation in electromagnetic continuous casting. *ISIJ Int.*, 2006, 46, 680-686.

[45] Huang S.Y.; Meng Z.H.; Li Y.C.; Zhang B.; Ouyang W. Investigation on the density uniformity of pressed parts by low-voltage electromagnetic compaction. *Acta Metall. Sin. (Engl. Lett.)*, 2004, 17, 805-811.

[46] 永井潤; 児玉正範; 宮崎容治; 日和佐章一; 小島信司; 村田賢二. 連続鋳造における鋳型内電磁グレキの適用. 鉄と鋼, 1982, 68, s270.

[47] Yoon J.K. Applications of numerical simulation to continuous casting technology. *ISIJ Int.*, 2008, 48, 879-884.

[48] Thomas B.G.; Chaudhary R. State of the art in electromagnetic flow control in continuous casting of steel slabs: modeling and plant validation. In: *Proc. 6th Int. Conf. on Electromagnetic Processing of Materials*, 2009, Dresden, Germany, Forschungszentrum Dresden-Rossendorf, pp. 9-14.

[49] Tian X.Y. Numerical analysis of the electromagnetic continuous casting process in the thin slab funnel shape mold. *Doctorial dissertation*, Northeastern University, 2009, Shenyang, China.

[50] 鈴木健一郎; 村田賢治; 中西恭二; 新良正典; 児玉正範; 小島信司; 宮崎容治. 電磁グレキによる連鋳鋳型内の溶鋼流動の制御. 鉄と鋼, 1983, 68, s912.

[51] Kollberg S.G.; Hackl H.R.; Hanley P.J. Improving quality of flat rolled products using electromagnetic brake (EMBr) in continuous casting. *Iron and Steel Engineering*, 1996, 73, 24-28.

[52] Zeze M.; Harada H.; Takeuchi E. Rayleigh-Taylor instability in the process of clad steel slab casting with level DC magnetic field. *ISIJ Int.*, 1999, 39, 563-569.

[53] Takatani K.; Nakai K.; Kasai N.; Watanabe T.; Nakajima H. Analysis of heat transfer and fluid flow in the continuous casting mold with electromagnetic brake. *ISIT Int.*, 1989, 29, l063-l068.

[54] Idogawa A.; Sugizawa M.; Takeuchi S.; Sorimachi K.; Fujii T. Control of molten steel flow in continuous casting mold by two static magnetic fields imposed on whole width. *Mater. Sci. Eng. A*, 1993, 173, 293-297.

[55] Bal D.; Brockhoff J.; Damen W.; Peeters T. Numerical modelling of an FC-II mould in a vertical bending continuous caster. In: *Proc. 5th Int. Conf. on Electromagnetic Processing of Materials*, 2006, Sendai, Japan, ISIJ, pp. 108-113.

[56] Moon K.H.; Shin H.K.; Kim B.J.; Chung J.Y.; Hwang Y.S.; Yoon K.J. Flow control of molten steel by electromagnetic brake in the continuous casting mold. *ISIJ Int.*, 1996, Supplement, 36, S201-S203.

[57] Cha P.R.; Yoon J.K. The effect of a uniform direct current magnetic field on the stability of a stratified liquid flux/molten steel system. *Metall. Trans. B*, 2000, 31, 317-326.

[58] Cukierski K.; Thomas B.G. Flow control with local electromagnetic braking in continuous casting of steel slabs. *Metall. Trans. B*, 2008, 39, 94-107.

[59] Yu H.Q.; Wang B.F.; Li H.Q.; Li J.C. Influence of electromagnetic brake on flow field of liquid steel in the slab continuous casting mold. *J. Mater. Process. Technol.*, 2008, 202, 179-187.

[60] Ha M.Y.; Lee H.G.; Seong S.H. Numerical simulation of three-dimensional flow, heat transfer, and solidification of steel in continuous casting mold with electromagnetic brake. *J. Mater. Process. Technol.*, 2003, 133, 322-339.

[61] Shamsi M. R. R. I.; Ajmani S. K. Three dimensional turbulent fluid flow and heat transfer mathematical model for the analysis of a continuous slab caster. *ISIJ Int.*, 2007, 47, 433-442.

[62] Aboutalebi M. Reza; Guthrie R.I.L.; Seyedein S.H. Mathematical modeling of coupled turbulent flow and solidification in a single belt caster with electromagnetic brake. *Applied Mathematical Modelling*, 2007, 31, 1671-1689.

[63] Usui M.; Asai S.; Iwai K. Refinement of solidified structure using DC electromagnetic field. *ISIJ Int.*, 2008, 48, 330-333.

[64] Lei Z.S.; Deng K.; Li Q.S.; Ren Z.M.; Li W.X.; Zhong Y.B. Physical simulation on nozzle electromagnetic brake in twin-roll strip casting. In: *Proc. 5th Int. Conf. on Electromagnetic Processing of Materials*, 2006, Sendai, Japan, ISIJ, pp. 89-93.

[65] Yu Z.; Lei Z.S.; Zhang Z.Q.; Zhang H.B.; Ren Z.M.; Deng K. Experimental investigation on influence of static electromagnetic field on the fluid flow in slab continuous casting mold. In: *Proc. 6th Int. Conf. on Electromagnetic Processing of Materials*, 2009, Dresden, Germany, Forschungszentrum Dresden-Rossendorf, pp. 227-230.

[66] Pavlicevic M.; Codutti A.; Kapaj N.; Poloni A. Experimental verification of liquid metal flow redirection in mould for slab continuous casting process by a new innovative concept of electromagnetic Brake. In: *Proc. 4th Int. Conf. on Electromagnetic Processing of Materials*, 2003, Lyon, France, pp. 160-165.

[67] Morales R.D.; Garcia-Hermandez S.; Hackt G.; Nitzl G.; Barreto J.de. Flow control in a curved slab mold with sliding gate using electromagnetic fields. In: *Proc. 6th Int. Conf.*

on Electromagnetic Processing of Materials, 2009, Dresden, Germany, Forschungszentrum Dresden-Rossendorf, pp. 567-570.

[68] Kubo N.; Kubota J.; Suzuki M. Molten steel flow monitoring and control by the EMLS in a continuous casting mold. In: *Proc. 5th Int. Conf. on Electromagnetic Processing of Materials*, 2006, Sendai, Japan, ISIJ, pp. 45-50.

[69] Eckert S.; Galindo V.; Witke W.; Gerbeth G.; Gerke-Cantow R.; Nicolai H.; Steinrücken U. Application of DC magnetic field to the filling process of aluminum investment casting. In: *Proc. 4th Int. Conf. on Electromagnetic Processing of Materials*, 2003, Lyon, France, pp. 373-378.

[70] Gerbeth G.; Eckert S.; Galindo V.; Willers B.; Hewelt U.; Katz H.W.; Ziemann M. Use of magnetic fields in aluminum investment casting. In: *Proc. 5th Int. Conf. on Electromagnetic Processing of Materials*, 2006, Sendai, Japan, ISIJ, pp. 323-328.

[71] Zhang C.J.; Eckert S.; Gerbeth G. Modification of bubble-driven liquid metal flows under the influence of a DC magnetic field. *ISIJ Int.*, 2007, 47, 795-801.

[72] Li Y.T.; Wen B.; Li T.J.; Yu Y.X.; Zhao Y.H.; Jin J.Z. The study of continuous casting of bimetal slab under electromagnetic field. In: *Proc. 4th Int. Conf. on Electromagnetic Processing of Materials*, 2003, Lyon, France, pp. 185-190.

[73] Gao X.P.; Li T.J.; Wu C.H.; Wu Y.P.; Jin J.Z. Process for bimetallic slab of aluminum alloy under the level DC electromagnetic field. *ISIJ Int.*, 2006, 46, 1033-1039.

[74] Sun J.B.; Li J.; Yan Z.M.; Cao Z.Q.; Li T.J. Study on preparation of clad aluminum alloy slab by electromagnetic continuous casting. In: *Proc. 6th Int. Conf. on Electromagnetic Processing of Materials*, 2009, Dresden, Germany, Forschungszentrum Dresden-Rossendorf, pp. 595-598.

[75] Li B.K.; Okane T.; Umeda T. Modeling of molten metal flow in a continuous casting process considering the effects of argon gas injection and static magnetic-field application. *Metall. Trans. B*, 2000, 31, 1491-1503.

[76] Li B.K.; Okane T.; Umeda T. Modeling of biased flow phenomena associated with the effects of static magnetic-field application and argon gas injection in slab continuous casting of steel. *Metall. Trans. B*, 2001, 32, 1053-1066.

[77] Li B.K.; Tsukihashi F. Effects of electromagnetic brake on vortex flows in thin slab continuous casting mold. In: *Proc. 5th Int. Conf. on Electromagnetic Processing of Materials*, 2006, Sendai, Japan, ISIJ, pp. 102-107.

[78] Huo H.F.; Li B.K. Revised k-ε turbulence model in electromagnetic continuous casting of melt. *Acta Metall. Sin. (Engl. Lett.)*, 2003, 16, 395-400.

[79] Xu G.J.; He J.C. Distribution of the magnetic field and its control effect on the molten metal flow in flow control mold. *Acta Metall. Sin. (Engl. Lett.)*, 2000, 13, 1093-1100.

[80] Lei H.; Xu G.J.; He J.C. Magnetic field, flow field and inclusion collision growth in a continuous caster with EMBR. *Chem. Eng. Technol.*, 2007, 30, 1650-1658.

[81] Li B.W.; Tian X.Y.; Wang E.G.; He J.C. Influences of casting speed and SEN depth on fluid flow in the funnel type mold of a thin slab caster. *Acta Metall. Sin. (Engl. Lett.)*, 2007, 20, 15-26.

[82] Tian X.Y.; Li B.W.; He J.C. Electromagnetic brake effects on the funnel shape mold of a thin slab caster based on a new type magnet. *Metall. Trans. B*, 2009, 40, 596-604.

[83] Tian X.Y.; Zou F.; Li B.W.; He J.C. Numerical analysis of coupled fluid flow, heat transfer and macroscopic solidification in the thin slab funnel shape mold with a new type EMBr. *Metall. Trans. B*, 2010, 41, 112-120.

[84] Wang E.G.; Kang L.; Li F.; He J.C. Numerical simulation of magnetic field and fluid flow in continuous casting mold with a new pattern electromagnetic brake. In: *Proc. 6th Int. Conf. on Electromagnetic Processing of Materials*, 2009, Dresden, Germany, Forschungszentrum Dresden-Rossendorf, pp. 583-586.

[85] Kim H.; Jeong H.; Kim G.; Cho M.; Lee S. Metal flow analysis of strand electromagnetic stirrer for a vertical bloom caster. In: *Proc. 4th Int. Conf. on Electromagnetic Processing of Materials*, 2003, Lyon, France, pp. 166-171.

[86] Kolesnichenko A.; Kolesnichenko A.; Buryak V. Magneto-pulse mold stirring and centerline defects by continuous steel casting. In: *Proc. 5th Int. Conf. on Electromagnetic Processing of Materials*, 2006, Sendai, Japan, ISIJ, pp. 57-62.

[87] Genma N.; Soejima T.; Saito T.; Kimura M.; Kaihara Y.; Fukumoto H.; Ayata K. The linear-motor type in-mold electromagnetic stirring technique for the slab continuous Caster. *ISIJ Int.*, 1989, 29, 1056-l062.

[88] Spitzer K.H.; Dubke M.; Schwerdtreger K. Rotational electromagnetic stirring in continuous casting of round strands. *Metall. Trans. B*, 1986, 17, 119-131.

[89] Kolesnichenko A.F.; Kolesnichenko A.A.; Buryak V.V. Improvement of in-mold electromagnetic stirrer by feeding of magnetic system with polyharmonic current. IS*IJ Int.*, 2004, 44, 342-349.

[90] Meyer J.L.; Durand F.; Ricou R.; Viver C. Steady flow of liquid aluminum in a rectangular-vertical ingot mold, thermally or electromagnetically activated. *Metall. Trans. B*, 1984, 15, 471-478.

[91] Nakanishi K.; Szekely J.; Fujii T.; Mihara Y.; Iwaoka S. Stirring and its effect on aluminum deoxidation of steel in the ASEA-SKF furnace: part I plant scale measurements and preliminary analysis. *Metall. Trans. B*, 1975, 6, 111-118.

[92] Cremer P.; Driole J. Effects of the electromagnetic stirring on the removal of inclusions of oxide from liquid steel. *Metall. Trans. B*, 1982, 13, 45-52.

[93] Bridge M.R.; Royers G.D. Structural effects and band segregation formation during the electromagnetic stirring of strand-cast steel. *Metall. Trans. B*, 1984, 15, 581-589.

[94] Dubke M.; Tacke K.H.; Spitzer K.H.; Schwerdtreger K. Flow fields in electromagnetic stirring of rectangular strands with linear inductors: theory and experiments with cold models Part I. *Metall. Trans. B*, 1988, 19, 581-593.

[95] Natarajan T.T.; El-Kadda N. Finite element analysis of electromagnetically driven flow in sub-mold stirring of steel billets and slabs. *ISIJ Int.*, 1998, 38, 680-689.

[96] Natarajan T.T.; El-Kaddah N. A new method for three-dimensional numerical simulation of electromagnetic and fluid-flow phenomena in electromagnetic separation of inclusions from liquid metal. *Metall. Trans. B*, 2002, 33, 775-785.

[97] Cao Z.Q.; Zhang L.P.; Ren Z.M.; Jin J.Z. Influence of electromagnetic stirring on macrosegregation of cast iron. *Acta Metall. Sin. (Engl. Lett.)*, 1992, 5, 311-314.

[98] Cao Z.Q.; Piao L.Y.; Zhang X.G.; Li Z.; Jin J.Z. Surface composite gear and pipe made by separated eutectic technology. *Acta Metall. Sin. (Engl. Lett.)*, 2000, 13, 883-887.

[99] Ogasawara T.; Yoshikawa N.; Tanigushi S.; Asai T. Flow of conducting liquid around two nonconducting particles in DC electromagnetic field and the electromagnetic migration force. *Metall. Trans. B*, 2004, 35, 847-855.

[100] Li T.X.; Xu Z.M.; Sun B.D.; Shu D.; Zhou Y.H. Remove inclusions from aluminum melt in electromagnetic field. *Acta Metall. Sin. (Engl. Lett.)*, 2000, 13, 1068-1074.

[101] Willers B.; Eckert S.; Michel U.; Zouhar G.; Nikrityuk P.; Eckert K. Experimental and numerical study about the solidification of a PbSn alloy exposed to a rotating magnetic field. In: *Proc. 4th Int. Conf. on Electromagnetic Processing of Materials*, 2003, Lyon, France, pp. 435-440.

[102] Willers B.; Nikrityuk P. A.; Eckert S.; Eckert K. The influence of electromagnetically-driven convection on temperature field and solute distribution during directional solidification. In: *Proc. 5th Int. Conf. on Electromagnetic Processing of Materials*, 2006, Sendai, Japan, ISIJ, pp. 407-412.

[103] Hu H.Q.; Zheng R.Q. Influence of electromagnetic stirring on dendrite structure and mechanical properties of low sulphur steel. *Acta Metall. Sin. (Engl. Lett.)*, 1991, 4, 75-78.

[104] Blazek K. E.; Kelly J.E.; Pottore N.S. The development of a continuous rheocaster for ferrous and high melting point alloys. *ISIJ Int.*, 1995, 35, 813-818.

[105] Vivès C. Elaboration of semisolid alloys by means of new electromagnetic rheocasting processes. *Metall. Trans. B*, 1992, 23, 189-206.

[106] Vivès C. Effect of forced electromagnetic vibrations during the solidification of aluminum alloys: part II solidification in the presence of colinear variable and stationary magnetic fields. *Metall. Trans. B*, 1996, 27, 457-464.

[107] Griffiths W.D.; McCartney D.G. The effect of electromagnetic stirring during solidification on the structure of A1-Si alloys. *Mater. Sci. Eng. A*, 1996, 216, 47-60.

[108] Griffiths W.D.; McCartney D.G. The effect of electromagnetic stirring on macrostructure and macrosegregation in the aluminium alloy 7150. *Mater. Sci. Eng. A*, 1997, 222, 140-148.

[109] Guo D.Y.; Yang Y.S.; Tong W.H. Simulation of Growth, Deformation and Fracture of Dendrites with Electromagnetic-Force-Driven Melt Flow. *5th Int. Conf. on Electromagnetic Processing of Materials*, 2006, Sendai, Japan, ISIJ, pp. 341-344.

[110] Zhang W.Q.; Liu Q.M.; Zhu Y.F.; Yang Y.S.; Hu Z.Q. Study on structures of Al-$CuAl_2$ eutectic solidified electromagnetic stirring. *Acta Metall. Sin. (Engl. Lett.)*, 1997, 10, 461-464.

[111] Zhang W.Q.; Zhu Y.F.; Yang Y.S.; Hu Z.Q. Structure of Al-Cu eutectic alloy solidified in complex field of centrifugal and electromagnetic forces. *Acta Metall. Sin. (Engl. Lett.)*, 1997, 10, 120-126.

[112] Zhang Y.; Qian B.N.; Guo X.M. Study of electromagnetic stirring refining microstructures of pipe-line steel saw deposits. *Acta Metall. Sin. (Engl. Lett.)*, 2002, 15, 396-400.

[113] Wang E.G.; Deng A.Y.; He J.C. Experiment of steel solidification process under electromagnetic stirring. In: *Proc. 4th Int. Conf. on Electromagnetic Processing of Materials*, 2003, Lyon, France, pp. 191-194.

[114] Denisov S.; Khripchenko S.; Kolesnichenko I.; Yudakov A. MHD-stirrer for cylindrical moulds of continuous casting machines fabricated aluminum alloy ingots. In: *Proc. 4th Int. Conf. on Electromagnetic Processing of Materials*, 2003, Lyon, France, pp. 178-183.

[115] Nafisi S.; Emadi D.; Shehata M.T.; Ghomashchi R. Effects of electromagnetic stirring and superheat on the microstructural characteristics of Al-Si-Fe alloy. *Mater. Sci. Eng. A*, 2006, 432, 71-83.

[116] Kuchaev A.; Kasyan G.; Jakobshe R. Using of electromagnetic stirring in continuous casting of round billets. In: *Proc. 6th Int. Conf. on Electromagnetic Processing of Materials*, 2009, Dresden, Germany, Forschungszentrum Dresden-Rossendorf, pp. 243-246.

[117] Barna M.; Javurek M.; Gittler Ph.; Reiter J.; Lechner M. Numerical simulations of mould-electromagnetic stirring for round bloom strands with full coupling methods. In: *Proc. 6th Int. Conf. on Electromagnetic Processing of Materials*, 2009, Dresden, Germany, Forschungszentrum Dresden-Rossendorf, pp. 571-574.

[118] Park J.; Lee G.C.; Kim M. Application of electromagnetic force to melting and casting of aluminum and magnesium alloys. In: *Proc. 5th Int. Conf. on Electromagnetic Processing of Materials*, 2006, Sendai, Japan, ISIJ, pp. 391-394.

[119] Kim W.J.; Lee K.E.; Park J.P.; Kim M.G.; Wang J.Y.; Yoon U.S. Microstructure and superplasticity of Mg-Al-Ca electromagnetic casting alloys after hot extrusion. *Mater. Sci. Eng. A*, 2008, 494, 391-396.

[120] Kim M.G.; Kim J.H.; Park J.P.; Lee G.C.; Cho I.S. Designing of continuous casting machine for near net shape aluminum billets. In: *Proc. 6th Int. Conf. on Electromagnetic Processing of Materials*, 2009, Dresden, Germany, Forschungszentrum Dresden-Rossendorf, pp. 323-325.

[121] Li M.J.; Tamura T.; Miwa K. Solidification behavior of AZ31 magnesium alloys using an electromagnetic vibration technique. *ISIJ Int.*, 2008, 48, 320-329.

[122] Barman N.; Kumar P.; Dutta P. Studies on transport phenomena during solidification of an aluminum alloy in the presence of linear electromagnetic stirring. *J. Mater. Process. Technol.*, 2009, 209, pp. 5912-5923.

[123] Liu C.T.; Chen Y.M.; Chen J.H.; Lu M.J. Operational characteristics analyses of an in-mold electro-magnetic stirrer. In: *Proc. 5th Int. Conf. on Electromagnetic Processing of Materials*, 2006, Sendai, Japan, ISIJ, pp. 126-130.

[124] Kang C.G.; Bae J.W.; Kim B.M. The grain size control of A356 aluminum alloy by horizontal electromagnetic stirring for rheology forging. *J. Mater. Process. Technol.*, 2007, 187-188, 344-348.

[125] Iwai K.; Wan J.; Kohama T. Solidified structure control of metallic alloy using electromagnetic field. In: *Proc. 6th Int. Conf. on Electromagnetic Processing of Materials*, 2009, Dresden, Germany, Forschungszentrum Dresden-Rossendorf, pp. 285-287.

[126] Mapelli C.; Gruttadauria A.; Peroni M. Application of electromagnetic stirring for the homogenization of aluminium billet cast in a semi-continuous machine. *J. Mater. Process. Technol.*, 2010, 210, 306-314.

[127] Zimmermann G.; Sturz L.; Dagner J. Directional solidification of AlSiMg0.3-alloys in a rotating magnetic field. In: *Proc. 5th Int. Conf. on Electromagnetic Processing of Materials*, 2006, Sendai, Japan, ISIJ, pp. 375-380.

[128] Zimmermann G.; Sturz L. Grain orientation during directional solidification of AlSi7-alloy in a rotating magnetic field. In: *Proc. 6th Int. Conf. on Electromagnetic Processing of Materials*, 2009, Dresden, Germany, Forschungszentrum Dresden-Rossendorf. pp. 288-291.

[129] Kovacs J.; Ronaöldi A.; Roosz A. Unidirectional solidification of binary and ternary aluminum alloys in a rotating magnetic field. In: *Proc. 6th Int. Conf. on*

Electromagnetic Processing of Materials, 2009, Dresden, Germany, Forschungszentrum Dresden-Rossendorf, pp. 664-667.

[130] Taniguchi S.; Maitake K.; Okubo M.; Ando T.; Ueno K. Rotary stirring of liquid metal without free surface deformation by combination of rotational and vertical traveling magnetic fields– development of hybrid stirrer-. In: *Proc. 4th Int. Conf. on Electromagnetic Processing of Materials*, 2003, Lyon, France, pp. 339-343.

[131] Shimasaki S.-I.; Maitake K.; Hori S.; Ueno K. Application of double-axis-electromagnetic stirring to solidification of Al-Bi alloy and agglomeration of SiC particle. In: *Proc. 5th Int. Conf. on Electromagnetic Processing of Materials*, 2006, Sendai, Japan, ISIJ, pp. 419-423.

[132] Lavers J.D.; Tallbäck G.R.; Lavers E.D.; Beitelman L.S.; Curran C.P. Flow control in continuous casting mold with dual coil EMS: computational simulation study. In: *Proc. 5th Int. Conf. on Electromagnetic Processing of Materials*, 2006, Sendai, Japan, ISIJ, pp.33-38.

[133] Perrier D.; Fautrelle Y.; Etay J. Experimental and theoretical studies of the motion generated by a two-frequency magnetic field at the free surface of a gallium pool. *Metall. Trans. B*, 2003, 34, 669-678.

[134] Fu H.Z.; Shen J.; Liu L.; Hao Q.T.; Li S.M.; Li J.S. Electromagnetic shaping and solidification control of Ni-base superalloys under vacuum. *J. Mater. Process. Technol.*, 2004, 148, 25-29.

[135] Kunstreich S.; Dauby P.H.; NovéM.C. Multi-mode EMS design of an automatic on-line process control system. In: *Proc. 4th Int. Conf. on Electromagnetic Processing of Materials*, 2003, Lyon, France, pp. 172-177.

[136] Ando T.; Ueno K.; Taniguchi S.; Takagi T. A stirring and axial drive MHD pump using helical coil stator for molten steel—molten gallium experiment. In: *Proc. 4th Int. Conf. on Electromagnetic Processing of Materials*, 2003, Lyon, France, pp. 321-326.

[137] Wang X.D.; Li T.J.; Fautrelle Y.; Dupouy M.D.; Jin J.Z. Two kinds of magnetic fields induced by one pair of rotating permanent magnets and their application in stirring and controlling molten metal flows. *J. Crystal Growth*, 2005, 275, e1473-e1479.

[138] Chang F.C.; Hull J.R.; Bettelman L. Simulation of flow control in the meniscus of a continuous casting mold with opposing alternating current magnetic fields. *Metall. Trans. B*, 2004, 35, 1129-1137.

[139] Harada H.; Isono T.; Wakagi A.; Kon-no T.; Sawai T.; Toh T. Control of molten steel flow with in-mold and strand electromagnetic stirring in bending type continuous caster. In: *Proc. 5th Int. Conf. on Electromagnetic Processing of Materials*, 2006, Sendai, Japan, ISIJ, pp. 39-44.

[140] Chungy S.L.; Shin Y.H.; Yoon J.K. Flow characteristics by induction and gas stirring in ASEA-SKF ladle. *ISIJ Int.*, 1992, 32, 1287-1296.

[141] Toh T.; Hasegawa H.; Harada H. Evaluation of multiphase phenomena in mold pool under in-mold electromagnetic stirring in steel continuous casting. *ISIJ Int.*, 2001, 41, 1245-1251.

[142] Wang F.; Li B.K. Behavior of two-phase flow of argon/steel in metallurgical reactor with the rotating magnetic field. In: *Proc. 6th Int. Conf. on Electromagnetic Processing of Materials*, 2009, Dresden, Germany, Forschungszentrum Dresden-Rossendorf, pp. 247-250.

[143] Hirayama R.; Fujisaki K. Combined system of AC and DC electromagnetic field for stabilized flow in continuous casting. In: *Proc. 5th Int. Conf. on Electromagnetic Processing of Materials*, 2006, Sendai, Japan, ISIJ, pp. 84-88.

[144] Kolbel M.; Gao J.; Lierfeld T.; Schneider S.; Eggeler G.; Herlach D. M. The influence of electromagnetic processing under terrestrial and parabolic flight conditions on solidifying metallic melts. In: *Proc. 5th Int. Conf. on Electromagnetic Processing of Materials*, 2006, Sendai, Japan, ISIJ, pp. 357-362.

[145] Mao W.M.; Zhao A.M.; Yun D.; Zhang L.P.; Zhong X.Y. Preparation study of SEMI-solid 60Si2Mn spring steel slurry. *Acta Metall. Sin. (Engl. Lett.)*, 2003, 16, 483-488.

[146] Mao W.M.; Zhao A.M.; Zhong X.Y. Spherical microstructure formation of the semi-solid high chromium cost iron Gr20Mo2. *Acta Metall. Sin. (Engl. Lett.)*, 2004, 17, 77-82.

[147] Mao W.M.; Zhen Z.S.; Chen H.T.; Zhong X.Y. Microstructural formation mechanism of SEMI-Solid AZ91D alloy. *Acta Metall. Sin. (Engl. Lett.)*, 2005, 18, 539-546.

[148] Liu Z.; Mao W.M.; Zhao Z.D. Effect of grain refining on primary α phase in semi-solid A356 alloy prepared by low superheat pouring and slight electromagnetic stirring. *Acta Metall. Sin. (Engl. Lett.)*, 2008, 21, 57-64.

[149] Zhao Z.D.; Mao W.M. Preparation of SEMI-Solid AlSi7Mg alloy slurry. *Acta Metall. Sin. (Engl. Lett.)*, 2008, 21, 139-145.

[150] Zhao Z.D.; Mao W.M.; Li S.; Zhong R.M. Preparation of semisolid AlSi7Mg alloy slurry through weak traveling-wave electromagnetic stirring. *Int. J. Minerals, Metallurgy and Materials*, 2009, 16, 554-558.

[151] Bae J.W.; Kim T.W.; Kang C.G. Experimental investigation for rheology forming process of Al–7% Si aluminum alloy with electromagnetic system. *J. Mater. Process. Technol.*, 2007, 191, 165-169.

[152] Kim H.H.; Kang C.G. Study of the microstructure and mechanical properties of Al6061 aluminum alloys rheo-forged by vacuum pump. *Mater. Sci. Eng. A*, 2008, 497, 309-321.

[153] Zoqui E.J.; Paes M.; Es-Sadiqi E. Macro- and microstructure analysis of SSM A356 produced by electromagnetic stirring. *J. Mater. Process. Technol.*, 2002, 120, 365-373.

[154] Lu D.H.; Jiang Y.H.; Guan G.S.; Zhou R.F.; Li Z.H.; Zhou R. Refinement of primary Si in hypereutectic Al–Si alloy by electromagnetic stirring. *J. Mater. Process. Technol.*, 2007, 189, 13-18.

[155] Patchett J.A.; Abbaschian G.J. Grain refinement of copper by the addition of iron and by electromagnetic stirring. *Metall. Trans. B*, 1985, 16, 505-511.

[156] Jung B.I.; Jung C.H.; Han T.K.; Kim Y.H. Electromagnetic stirring and Sr modification in A356 alloy. *J. Mater. Process. Technol.*, 2001, 111, 69-73.

[157] Liu S.F.; Liu L.Y.; Kang L.G. Refinement role of electromagnetic stirring and strontium in AZ91 magnesium alloy. *J. Alloys and Compounds*, 2008, 450, 546-550.

[158] Metan V.; Eigenfeld K.; Räbiger D.; Leonhardt M.; Eckert S. Grain size control in Al-Si alloys by grain refinement and electromagnetic stirring. *J. Alloys and Compounds*, 2009, 487, 163-172.

[159] Zoqui E.J.; Gracciolli J.I.; Lourençao L.A. Thxo-formability of the AA6063 alloy: Conventional production processes versus electromagnetic stirring. J. *Mater. Process. Technol.*, 2008, 198, 155-161.

[160] Robles Hernández F.C.; Sokolowski J.H. Effects and on-line prediction of electromagnetic stirring on microstructure refinement of the 319 Al–Si hypoeutectic alloy. *J. Alloys and Compounds*, 2009, 480, 416-421.

[161] Hao H.; Gu S.W.; Dong H.W.; Yao L.; Zhang X.G. Improvement of grain refining effect of magnesium alloys by imposing external physical fields. In: *Proc. 6th Int. Conf. on Electromagnetic Processing of Materials*, 2009, Dresden, Germany, Forschungszentrum Dresden-Rossendorf, pp. 300-303.

[162] Robles Hernández F.C.; Sokolowski J.H. Comparison among chemical and electromagnetic stirring and vibration melt treatments for Al–Si hypereutectic alloys. *J. Alloys and Compounds*, 2006, 426, 205-212.

[163] Fujisaki K.; Tomizawa Y. Downward flow diffusion and meniscus flow by electromagnetic stirring. In: *Proc. 5th Int. Conf. on Electromagnetic Processing of Materials*, 2006, Sendai, Japan, ISIJ, pp. 98-101.

[164] Sato K.; Hirayama R.; Fujisaki K. Analysis of drift flow phenomena and effect of the in-mold electromagnetic stirrer in continuous casting. In: *Proc. 5th Int. Conf. on Electromagnetic Processing of Materials*, 2006, Sendai, Japan, ISIJ, pp. 51-56.

[165] Deng A.Y.; Wang E.G.; Xu Y.Y.; Zhang X.W.; He J.C. Experimental study on meniscus behavior affected by alternating and static magnetic field. In: *Proc. 6th Int. Conf. on Electromagnetic Processing of Materials*, 2009, Dresden, Germany, Forschungszentrum Dresden-Rossendorf, pp. 235-238.

[166] Tani M.; Tanaka K.; Fukunaga S.; Toh T.; Tsunenari K.; Umetsu K.; Hayashi K.; Zeze M.; Isogami K.; Okumura H. The pulsative electromagnetic casting technique for slab casting. In: *Proc. 5th Int. Conf. on Electromagnetic Processing of Materials*, 2006, Sendai, Japan, ISIJ, pp. 63-68.

[167] Yang Y.S.; Wang B.; Ma X.P.; Tong W.H.; Li Y.J.; Luo T.J.; Feng X.H. Crystal growth and grain refinement by pulsed magnetic field. In: *Proc. 6th Int. Conf. on Electromagnetic Processing of Materials*, 2009, Dresden, Germany, Forschungszentrum Dresden-Rossendorf, pp. 281-284.

[168] Hua J.S.; Zhang Y.J.; Hu Z.G.; Wang E.G.; Yang Y.S. Theoretical analysis and model experiment on grain refinement under pulsed magnetic-vibrating field. In: *Proc. 6th Int. Conf. on Electromagnetic Processing of Materials*, 2009, Dresden, Germany, Forschungszentrum Dresden-Rossendorf, pp. 342-345.

[169] Zhang Z.F.; Kim J.M.; Hong C.P. Numerical simulation of grain structure evolution in solidification of an Al-5.0wt%Cu alloy under electromagnetic stirring and its experimental verification. *ISIJ Int.*, 2005, 45, 183-191.

[170] Yamazaki M.; Natsume Y.; Harada H.; Ohsasa K. Numerical simulation of solidification structure formation during continuous casting in Fe–0.7mass%C alloy using cellular automaton method. *ISIJ Int.*, 2006, 46, 903-908.

[171] Deng K.; Ren Z.M.; Lei Z.S.; Li W.X. A new technology of EMS-EMC for OFC processing and its industrial experiment. In: *Proc. 5th Int. Conf. on Electromagnetic Processing of Materials*, 2006, Sendai, Japan, ISIJ, pp. 401-406.

[172] Lei Z.S.; Ren Z.M.; Deng K.; Li W.X.; Zhong Y.B. Application of special amplitude-modulated magnetic field in electromagnetic continuous casting. In: *Proc. 5th Int. Conf. on Electromagnetic Processing of Materials*, 2006, Sendai, Japan, ISIJ, pp. 74-79.

[173] Ren Z.M.; Li X.; Ren W.L.; Deng K.; Xu K.D. The influence of a high magnetic field on unidirectional solidification of Al-4.5%Cu alloy. In: *Proc. 5th Int. Conf. on Electromagnetic Processing of Materials*, 2006, Sendai, Japan, ISIJ, pp. 369-374.

[174] Wang J.; Zhong Y.B.; Wang Z.Q.; Ren W.L.; Lei Z.S.; Ren Z.M. Solidification of Zn-30wt%Bi monotectic alloy by superimposing high static magnetic field and AC current simultaneously. In: *Proc. 6th Int. Conf. on Electromagnetic Processing of Materials*, 2009, Dresden, Germany, Forschungszentrum Dresden-Rossendorf, pp. 326-329.

[175] Huang J.; Deng K.; Ren Z. Solidification of horizontally continuous casting of super-thin slab with coupling of magnetic field and alternating current. In: *Proc. 6th Int. Conf. on Electromagnetic Processing of Materials*, 2009, Dresden, Germany, Forschungszentrum Dresden-Rossendorf, pp. 320-322.

[176] Wang Z.Q.; Zhong Y.B.; Ren Z.M. Microstructure and properties of Cu-Cr-Zr alloy aged with dc electric current. In: *Proc. 6th Int. Conf. on Electromagnetic Processing of Materials*, 2009, Dresden, Germany, Forschungszentrum Dresden-Rossendorf, pp. 316-319.

[177] Li T.J.; Li X.T.; Zhang Z.F.; Jin J.Z. Effect of multielectromagnetic field on meniscus shape and quality of continuously cast metals. *Ironmaking and Steelmaking*, 2006, 33, 57-60.

[178] Li X.T.; Guo Z.X.; Zhao X.W.; Wei B.; Chen F.B.; Li T.J. Continuous casting of copper tube billets under rotating electromagnetic field. *Mater. Sci. Eng. A*, 2007, 460-461, 648-651.

[179] Thang Q.; Wang T.M.; Li T.J.; Jin J.Z. Numerical simulation of solidification process of Sn-3.5%Pb hollow billet with stirring magnetic field. *Trans. Nonferrous Met. Sci.* China, 2007, 17, 940-946.

[180] Zhang Q.; Li T.J.; Wang T.M.; Jin J.Z. Effects of AC magnetic field on solidification structure of continuously cast hollow billet. *Ironmaking and Steelmaking*, 2009, 36, 2115-2119.

[181] Zhang X.L.; Li T.J.; Teng H.T.; Xie S.S.; Jin J.Z. Semisolid processing AZ91 magnesium alloy by electromagnetic stirring after near-liquidus isothermal heat treatment. *Mater. Sci. Eng. A*, 2008, 475, 194-201

[182] Zhang Z.; Li J.; Shi T.; Yue H.; Li T. Effect of electromagnetic field and power ultrasonic field on the horizontal continuous casting of Al-1%Si alloy. In: *Proc. 6th Int. Conf. on Electromagnetic Processing of Materials*, 2009, Dresden, Germany, Forschungszentrum Dresden-Rossendorf, pp. 312-315.

[183] Zuo Y.B.; Cui J.Z.; Dong J.; Yu F.X. Effects of low frequency electromagnetic field on the as-cast microstructures and mechanical properties of superhigh strength aluminum alloy. *Mater. Sci. Eng. A*, 2005, 408, 176-181.

[184] Dong J.; Cui J.Z.; Yu F.X.; Zhao Z.H.; Zhuo Y.B. A new way to cast high-alloyed Al–Zn–Mg–Cu–Zr for super-high strength and toughness. *J. Mater. Process. Technol.*, 2006, 171, 399-404.

[185] Zhu Q.F.; Zhao Z.H.; Cui J.Z. Effect of low-frequency electromagnetic field on the AS-casting microstructures and mechanical properties of HDC 2024. *Acta Metall. Sin. (Engl. Lett.)*, 2008, 21, 205-210.

[186] Wang X.J.; Zhang H.T.; Zuo Y.B.; Zhao Z.H.; Zhu Q.F.; Cui J.Z. Experimental investigation of heat transport and solidification during low frequency electromagnetic hot-top casting of 6063 aluminum alloy. *Mater. Sci. Eng. A*, 2008, 497, 416-420.

[187] Li J.C.; Wang B.F.; Cui J.Z. Modeling and experimental research of four-strand low-frequency electromagnetic casting aluminum alloy. *Acta Metall. Sin. (Engl. Lett.)*, 2007, 20, 157-166.

[188] Li J.C.; Wang B.F; Ma Y.L.; Cui J.Z. Effect of complex electromagnetic stirring on inner quality of high carbon steel bloom. *Mater. Sci. Eng. A*, 2006, 425, 201-204.

[189] Inomoto T.; Ogawa Y.; Toh T. Promotion of desulphurization in ladle through slag emulsification by stirring with stationary AC electromagnetic field. *ISIJ Int.*, 2003, 43, 828-835.

[190] Thess A.; Conrad G.; Halbedel B.; Hülsenberg D.; Kolesnikov Y.; Lüdtke U. Electromagnetic Stirring of Glass Melts. In: *Proc. 4th Int. Conf. on Electromagnetic Processing of Materials*, 2003, Lyon, France, pp. 350-352.

[191] Cepite D.; Jakovics A. Radiation heat transfer influence on EM and natural convection of the semi-transparent glass melt. In: *Proc. 6th Int. Conf. on Electromagnetic Processing of Materials*, 2009, Dresden, Germany, Forschungszentrum Dresden-Rossendor, pp. 680-683.

[192] Kolesnichenko A.A.; Buryak V.V.; Kolesnichenko A.F. High frequency (50-60 Hz) electromagnetic stirrer for continuous billet/bloom casting of middle/high carbon Steels. In: *Proc. 4th Int. Conf. on Electromagnetic Processing of Materials*, 2003, Lyon, France, pp. 100-105.

[193] Yao M.; Yin H.B.; Wang J.C.; Fang D.C.; Liu X.; Yu Y.; Liu J.J. Development of an experimental system for the study of the effects of electromagnetic stirring on mold heat transfer. *Metall. Trans. B*, 2005, 36, 475-478.

[194] Cheng J.B.; Xu B.S.; Liang X.B.; Wu Y.X.; Liu Z.J. Effect of electromagnetic stirring on the microstructure and wear behavior of iron-based composite coatings. *Journal of University of Science and Technology Beijing*, 2008, 15, 451-456.

[195] Yasuda H.; Yoshimoto T.; Mizuguchi T.; Tamura Y.; Nagira T.; Yoshiya M. Effect of the melt flow on the solidified structure of middle carbon steel by means of the levitation method using alternating and static magnetic fields. *ISIJ Int.*, 2007, 47, 612-618.

[196] Li M.J.; Ishikawa T.; Nagashio K.; Kuribayashi K.; Yoda S. Microtexture formation of Ni99B1 alloys solidified on an ESL and an EML—a study based on the EBSP technique. *Mater. Sci. Eng. A*, 2007, 449-451, 684-688.

[197] Mechighel F.; Pateyron B.; Ganaoui M. El; Kadja M. Damping effect of an external magnetic field applied to alloy solidification in an enclosure. In: *Proc. 6th Int. Conf. on Electromagnetic Processing of Materials*, 2009, Dresden, Germany, Forschungszentrum Dresden-Rossendorf, pp. 330-333.

[198] Su Z.J.; Li D.W.; Sun L.W.; Marukawa K.; He J.C. Swirling flow induced by rotating magnetic field in immersion nozzle in continuous casting of steel. In: *Proc. 6th Int. Conf. on Electromagnetic Processing of Materials*, 2009, Dresden, Germany, Forschungszentrum Dresden-Rossendorf, pp. 587-590.

[199] Timmel K.; Eckert S.; Gerbeth G.; Claussner J.; Schlenk R.; Woigtlaender J. Concast: A liquid metal model for continuous steel casting. In: *Proc. 6th Int. Conf. on Electromagnetic Processing of Materials*, 2009, Dresden, Germany, Forschungszentrum Dresden-Rossendorf, pp. 251-256.

[200] Wang Q.; Wang E.G.; He J.C.; Hu K.; Takahashi K.; Watanabe K. Effects of strong magnetic fields on solidified structures of metals with different susceptibilities. In: *Proc. 4th Int. Conf. on Electromagnetic Processing of Materials*, 2003, Lyon, France, pp. 464-468.

[201] Zhang L.; Wang E.G.; Zuo X.W.; He J.C. Solidification structure of Cu-40wt%Pb monotectic alloy under high magnetic field. In: *Proc. 5th Int. Conf. on Electromagnetic Processing of Materials*, 2006, Sendai, Japan, ISIJ, pp. 351-356.

[202] Gao J.; Zhang Y.K.; Reutzel S.; Herlach D. M.; He J.C. Undercooling and solidification of peritectic Cu-Ge alloys in strong magnetic field. In: *Proc. 5th Int. Conf. on Electromagnetic Processing of Materials*, 2006, Sendai, Japan, ISIJ, pp. 363-368.

[203] Fautrelle Y. Electromagnetic processing of liquid metal: review of the research activity in France. In: *Proc. 5th Int. Conf. on Electromagnetic Processing of Materials*, 2006, Sendai, Japan, ISIJ. pp. 9-14.

[204] Li B.Q. A finite-element analysis of magnetically driven recirculating flow in electromagnetic near net shape casting. *J. Materials Proc. Tech.*, 1995, 55, 351-359.

[205] Tallbäck G.R.; Lavers J.D.; Beitelman L.S. Simulation and measurement of EMS induced fluid flow in billet/bloom casting systems. In: *Proc. 4th Int. Conf. on Electromagnetic Processing of Materials*, 2003, Lyon, France, pp. 154-159.

[206] Toh T.; Takeuchi E.; Matsumiya T. Recent advances in MHD applications to steelmaking processes. In: *Proc. 5th Int. Conf. on Electromagnetic Processing of Materials*, 2006, Sendai, Japan, ISIJ, pp. 21-25.

[207] Qian Z.D.; Li B.W.; Wang E.G.; Jia G.L.; He J.C. A new treatment for interface of metal and dielectric in FDTD method. *Acta Metall. Sin. (Engl. Lett.)*, 2001, 14, 42-46.

[208] Zhu X.R.; Harding R.A.; Campbell J. Calculation of the free surface shape in the electromagnetic processing of liguid metals. *Appl. Math. Modelling*, 1997, 21, 207-214.

[209] Nikrityuk P.A.; Eckert K. Recent progress in the modeling of solidification of binary metal alloys under the influence of electromagnetic fields: macro- and microscale approaches. In: *Proc. 6th Int. Conf. on Electromagnetic Processing of Materials*, 2009, Dresden, Germany. Forschungszentrum Dresden-Rossendorf. pp. 292-295.

[210] Evans J.W.; Zundelevich Y.; Sharma D. A mathematical model for prediction of currents, magnetic fields, melt velocities, melt topography and current efficiency in hall-héroult cells. *Metall. Trans. B*, 1981, 12, 353-360.

[211] Lavers J.D. Computational methods for the analysis of molten metal electromagnetic confiniement systems. *ISIJ Int.*, 1989, 29, 993-1005.

[212] Ilegbusi O.J.; Szekely J. Three-dimensional velocity fields for Newtonian and non-Newtonian melts produced by a rotating magnetic field. *ISIJ Int.*, 1989, 29, 462-468.

[213] Maiwald A.; Timmel K.; Schwarze R.; Gerbeth G. Comparison of numerical and experimental flow data from a liquid metal model of a continuous casting mould. In: *Proc. 6th Int. Conf. on Electromagnetic Processing of Materials*, 2009, Dresden, Germany. Forschungszentrum Dresden-Rossendorf. pp. 239-242.

[214] Yee K.S. Numerical solution of initial boundary value problems involving Maxwell's equations in isotropic media. *IEEE Transactions on Antennas and Propagation*, 1966, 3, 302-307.

[215] Taflove A.; Brodwin M.E. Numerical solution of steady—state electromagnetic scattering problems using the time—dependent Mawell's equations. *IEEE Transactions on Microwave Theory and Techniques*, 1975, 23, 623-630.

[216] Taflove A.; Umashankar K.R.; Beker B.; Harfoush F.; Yee K.S. Detailed FD-TD analysis of electromagnetic fields penetrating narrow slots and lapped joints in thick conducting screens. *IEEE Transaction on Antennas and Propagation*, 1988, 36, 247-257.

[217] Taflove A.; Umashankar K.R.; Jurgens T.G. Validation of FDTD modeling of the radar cross sections. *IEEE Transaction on Antennas and Propagation*, 1985, 33, 662-666.

[218] Gandhi O.P.; Chen J.Y. Numerical dosimetry at power—line frequencies using anatomically based models. *Bioelectromagnetics Supplement*, 1992, 1, 43-60.

[219] Qian Z.D.; Wu Y.L.; Li B.W.; He J.C. Numerical analysis of the influences of operational parameters on the fluid flow in mold with hybrid magnetic fields. *ISIJ Int.*, 2002, 42, 1259-1265.

[220] Qian Z.D. Simulation of electromagnetic field, flow field and movement of different phases in hybrid magnetic fields mold. *Doctoral dissertation of Northeastern University*, 2002, Shenyang, China.

[221] Qian Z.D.; Wu Y.L. Large eddy simulation of turbulent flow phenomena with the effects of DC magnetic field and vortex brake application in continuous casting. *ISIJ Int.*, 2004, 44, 100-107.

In: Structural Steel and Castings
Editor: Lena M. Becker, pp. 259-276
ISBN: 978-1-61728-104-4

Chapter 8

MACHINE-LEARNING-BASED DEFECT PREDICTION IN HIGH-PRECISION FOUNDRY PRODUCTION

***Igor Santos**, *Javier Nieves*†, *Pablo G. Bringas*‡ and *Yoseba K. Penya*§**
University of Deusto, Bilbao, Spain

1. Introduction

Foundry has evolved from the ancient magic-surrounded activity it used to be, to become a key industry that maintains the world as we know it. Thereby, it supplies major pieces to another industries such as, automotive, naval, aeronautic, weapon and so on. As one may think, many of these pieces play key roles in the final product as in the case of brakes, aeroplane components or wind energy castings and this reason imposes high-quality standards: the tiniest defect may become fatal.

Therefore, there are very strict quality standards to assure the exclusion of faulty pieces. Unfortunately, these controls are all performed in an ex-post manner, when the production effort is already done. In this sense, error prediction, firstly, allows avoiding the production of defective items to fulfil quality standards, and secondly, it also helps not to squander resources on that activity.

Nowadays, the most used techniques for the assurance of failure-free foundry processes are exhaustive production control and diverse simulation techniques [1] but they are extremely expensive and only achieve good results in an a posteriori fashion.

On one hand, microshrinkage, also called secondary contraction, consists of tiny porosities that appear when the casting is cooling down, and almost all process parameters interact on its apparition making it impossible to avoid so far [2, 3]. The biggest problem associated to microshrinkages is that pieces flawed with this defect must be discarded. Moreover,

*E-mail address: isantos@deusto.es
†E-mail address: javier.nieves@deusto.es
‡E-mail address: pablo.garcia.bringas@deusto.es
§E-mail address: yoseba.penya@deusto.es

caused either by an increment on the amount of disposed castings in the routine quality inspections (with random-picked pieces) or after a client's reclamation, security measures stipulate that all castings of that production series must be X-ray-scanned in order to discover new possible faulty pieces. This procedure entails the subsequent cost increment which has to be added to the cost of the discarded castings themselves (transport, energy to remelt, new production process and still no guaranty that this time is going to work).

On the other hand, mechanical properties are the ability of the material of a piece to take several forces and tensions. In this chapter we focus on the so-called *ultimate tensile strength* that is the force that a casting can withstand until it breaks. Therefore, the foundry castings that are manufactured have to reach a certain value or threshold of ultimate tensile strength in order to pass the strict quality tests. More accurately, current standard procedures to determine that value are the employment of destructive inspections. Unfortunately, these procedures make the piece useless after the inspection and thus, as happens with microshrinkages, they also incur a cost increment.

Hence, providing effective ex-ante methods can help to increase the quality standards and to save resources in the process (i.e. saving money). To this extent, *machine-learning* classifiers have been applied in domains alike with outstanding results, for instance, neural networks [4], Bayesian networks [5] or the K-nearest neighbour algorithm [6]. In this way, successful applications of artificial neural networks include for instance spam filtering [7], intrusion detection [8], or industrial fault diagnosis [9]. Moreover, successful applications of Bayesian networks include for instance email classification for spam detection [10], failure detection in industrial production lines [11] [12], weather forecasting [13] [14], intrusion detection over IP networks [15] [16] or reconstruction of traffic accidents [17] [18]. Similarly, K-nearest neighbour algorithm, despite its simplicity, has been applied for instance to visual category recognition [19], weather forecasting [20], malware detection [21] or image retrieval [22].

Against this background, this chapter advances the state of the art in two main ways. First, we propose a methodology to adapt machine learning classifiers to the prediction of microshrinkages or ultimate tensile strength and we describe the method for training them. Second, we evaluate the classifiers with a historical dataset from a real foundry process in order to compare the accuracy and suitability of each method.

The remainder of this chapter is organised as follows. Section 2. discusses related work. Section 3. details the foundry process of iron castings. Section 5. introduces and describes the *machine-learning algorithms* we tailor to iron foundries. Section 6. describes the experiments performed and examines the obtained results and explains feasible enhancements. Finally, section 7. concludes and outlines the avenues of future work.

Keywords: Defect forecasting, machine-learning, data-mining.

2. Related Work

There has been a hectic activity around the applications of neural networks to several other problems of foundry process, for instance on the prediction of the ferrite number in stainless steel arc welds [23]. Similarly, successful experiments involving K-nearest neighbour algorithm include fault detection of semiconductor manufacturing processes [24].

In a verge closer to our view, neural networks have been used in several works, for instance, classifying foundry pieces [25], optimising casting parameters [26], detection of causes of casting defects [27] and in other problems [28]. Nevertheless, Bayesian networks are used as previous methods in Bayesian neural networks methodology (i.e. predicting the ferrite number in stainless steel [29]). In addition, K-nearest neighbour algorithm and artificial neural networks have been applied for enhance quality of steel [30] that achieves an overall root mean square error of 0.38 (a comparison between their results and ours can be further analysed in section 6.).

The good results obtained for these works have encouraged us to tailor these approaches into our concrete problem domain.

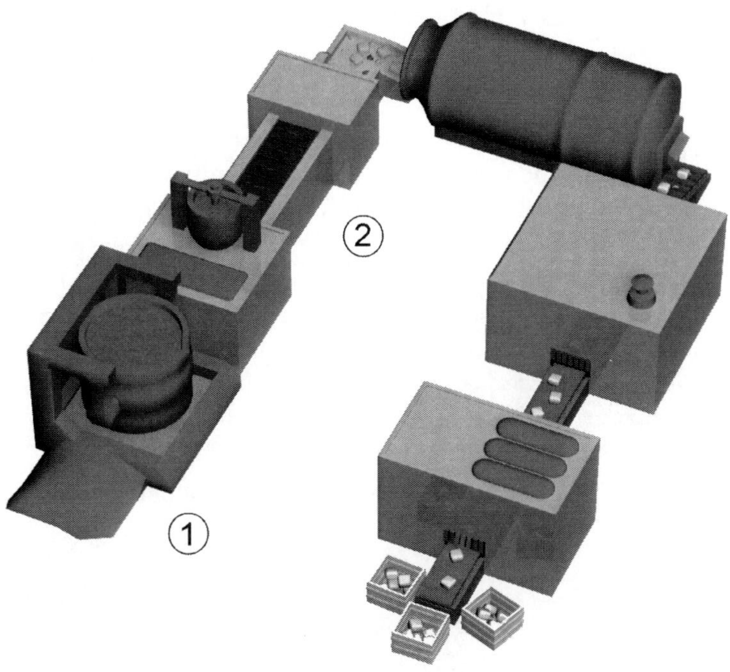

Figure 1. Moulding and cooling in the casting production.

3. Foundry Process

There are several factors that make foundry production a very complex process, such as, the extreme conditions in which it is performed. In this way, starting from the raw material to the manufactured item, this procedure involves numerous phases, some of which may be performed in parallel. More accurately, when it refers to iron ductile castings, this process presents the following phases:

- **Melting and pouring:** The raw metals are melt, mixed and poured onto the sand shapes.

- **Moulding:** The moulding machine forms and prepares the sand moulds.
- **Cooling:** The solidification of the castings is controlled in the cooling lines until this process is finished.

Figure 1 shows the moulding and cooling phases. Once the raw material is melt, it is poured onto the moulds (made out of sand mixed in the sand-mill) and shaped in (1). The cooling lines (2) accelerate the natural cooling process of the castings. When they are properly solidified, the sand moulds are detached from them and return to the sand-mill, so the sand can be reused to mould further castings.

4. Defects on Foundry Processes

4.1. Microshinkages

Microshrinkages appear during the cooling process of the metal but are only visible once the production is accomplished. More accurately, this flaw consists of minuscule internal porosities or cavities. Since metals are less dense as a liquid than as a solid, the density of the metal increases while it solidifies and the volume decreases in parallel. In this process, diminutive, microscopically undetectable interdendritic voids may appear leading to a reduction of the casting's hardness and, in the cases faced here (where the casting is a part of a very sensitive piece), rendering the piece useless [31, 3]. Unfortunately, the only way to examine finished parts is the employment of non-destructive inspections. In this way, the most usual techniques are X-ray and ultrasound emissions but both require suitable devices, specialised staff and quite a long time to analyse all the produced parts. Therefore, postproduction inspection is not an economical alternative to the pre-production detection of microshrinkages.

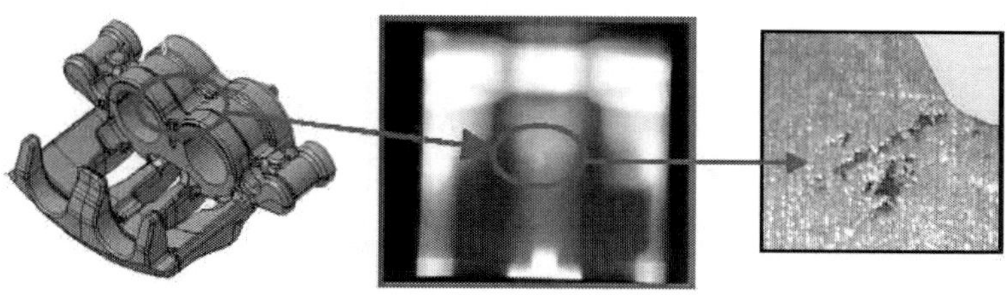

Figure 2. Computer simulation of microshrinkage formation in a double-cylinder break calliper.

Figure 2 shows one of the castings examined here, the calliper of a front double-cylinder brake that, as exposed in section IV, tends to present microshrinkages. Figure 3 illustrates the gradual formation of a microshrinkage as the piece is cooling down. This computer simulation shows a transversal cut of the casting and the risk that the defect appears (the closer to red, the higher the risk). The first two images illustrate the solidification process and the last one shows the final presence of the microshrinkage.

As aforementioned, the complexity of detecting secondary contractions arises principally from the high number of variables that participate in production process and, there-

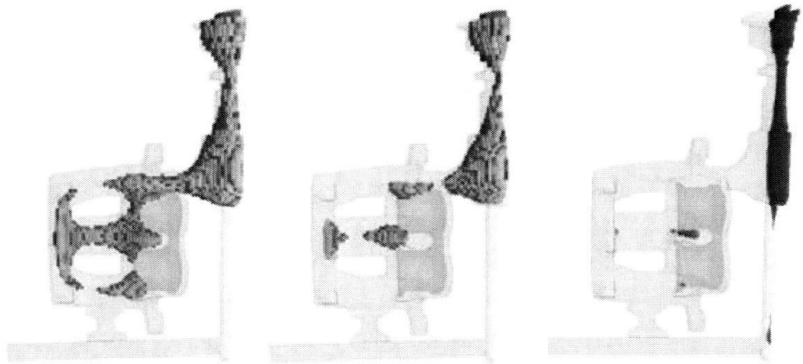

Figure 3. Formation of the microshinkage in the double-cylinder break calliper.

fore, may have influence on the final design of a casting. In this way, the main variables to control in order to predict the apparition of microshrinkages are:

- **Metal-related:**
  - *Composition:* Type of treatment, inoculation and charges.
  - *Nucleation potential and melt quality:* Obtained by means of a thermal analysis program [32].
  - *Pouring:* Duration of the pouring process and temperature.
- **Mould-related:**
  - *Sand:* Type of additives used, sand-specific features and carrying out of previous test or not.
  - *Moulding:* Machine used and moulding parameters

Commonly, the dimension and geometry of the casting play a very important role in this practice and, thus, we also include several variables to control this two features. In the same way, the system should take into account parameters related to the configuration of each machine that works in the manufacturing process [33].

Furthermore, there are some variables that may influence the apparition of second contraction during the foundry process, such as the composition [34], the size of the casting, the cooling speed and thermal treatment [35] [36]. The system must take into account all of them in order to issue a prediction on those mechanical properties. In this way, the machine-learning classifiers used in our experiments are composed of about 24 variables.

4.2. Mechanical Properties

Once the phases shown in section 3. have been completed, the resultant casting is subject to forces (loads). Engineers have to calculate the value of these forces and how the material deforms or breaks as a function of applied load, time or other conditions. Hence, it is a very important theme knowing how mechanical properties affect to iron castings [35], since they directly affect the quality of the final piece. More accurately, the most important mechanical properties of foundry materials are the following ones [37]:

- **Strength:** it is the property that enables a metal to resist deformation under load. There are many kinds of strength such as ultimate strength and ultimate tensile strength.
- **Hardness:** it is the property to resist permanent indentation.
- **Toughness:** it is the property that enables a material to withstand shock and to be deformed without rupturing. This property is considered as a combination as strength and plasticity.
- **Resilience:** it is the property of a material to absorb energy when it is deformed elastically.
- **Elasticity:** it is the ability of a material to return to its original shape after the load is removed.
- **Plasticity:** it is the ability of a material to deform permanently without breaking or rupturing. This property is the opposite of strength.
- **Brittleness:** it is the opposite of plasticity. A brittle metal is one that breaks or shatters before it deforms. Generally, brittle metals have a high value in compressive strength but a low value in tensile strength.
- **Ductility:** it is the property that enables a material to stretch, bend or twist without cracking or breaking.
- **Malleability:** in comparison with ductility, it is the property that enables a material to deform by compressive forces without developing defects. A malleable material can be stamped, hammered, forged, pressed or rolled into thin sheets.

Furthermore, there are common or standard procedures (i.e. ASTM standards [38, 39]) for testing the value of mechanical properties of the materials in a laboratory. Unfortunately, they need to use destructive inspections and they are *ex-post* (i.e. performed after production). Moreover, the process demand appropriate instruments, specialised staff and time resources in order to inspect the materials.

Regarding the ultimate tensile strength, which we focus here on, its testing method is conducted as follows. First, a scientist prepares a testing specimen from the original casting (see (1) in Figure 4). Second, the specimen is placed on the tensile testing machine (2). And, finally, this machine pulls the sample from both ends and measures the force required to break the specimen apart and how much the sample stretches before breaking.

Moreover, there are some variables that may influence the mechanical properties of the metal during the foundry process, such as the composition [34], the size of the casting, the cooling speed and thermal treatment [35, 36]. The system must take into account all of them in order to issue a prediction on those mechanical properties. In this way, the machine-learning models used in our experiments are composed of about 25 variables.

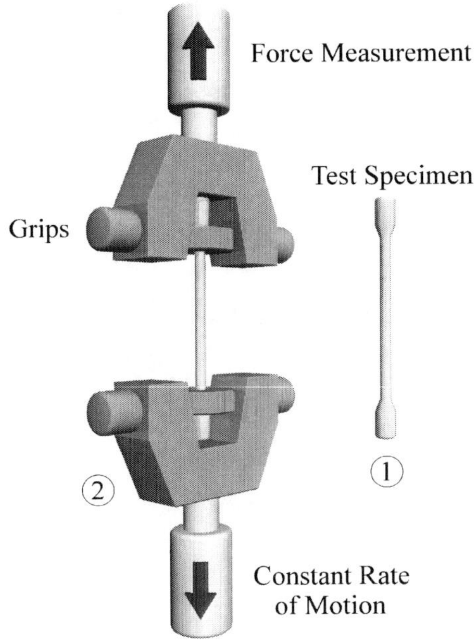

Figure 4. Ultimate Tensile Strength Test.

5. Machine-learning Classifiers

5.1. Bayesian Networks

The research on cause-consequence relationships was pioneered by Reverend Thomas Bayes [40], and his main work is known as the "Bayes theorem" in his honour. According to its classical formulation, given two events A and B, the conditional probability $P(A|B)$ that A occurs if B occurs can be obtained if we know the probability that A occurs, $P(A)$, the probability that B occurs, $P(B)$, and the conditional probability of B given A, $P(B|A)$ (as shown in equation 1):

$$P(A|B) = \frac{P(B|A) \cdot P(A)}{P(B)} \tag{1}$$

Extending this model, Bayesian networks are probabilistic models for multivariate analysis. We can represent a Bayesian network as an acyclic directed graph and the probability distribution function associated to that graph [41]. On the one hand, the graphical model represents the set of probabilistic relationships among the collection of variables modelling a particular problem. On the other hand, the probability function illustrates the strength of these relationships or edges in the graph.

We use this kind of model for several activities, for example, machine learning based on historical data, pattern matching over ambiguous or incomplete data , data mining for relationship discovery and inference of non-observable variables given the rest of the set [42]. In particular, this inference capability fits to our experiments. These capability represents a semantical super-set of those expert systems based on rule chaining, both for forward and backward style (in fact, Bayesian models allow a third further kind of inference, that is known as explanation or justification [41]). Moreover, a Bayesian network can grow extending its knowledge base with new evidences without reducing its performance level [41] whilst adapts to the problem and maintain an updated procedure.

To our needs, the most important ability of Bayesian networks is their capability of inferring the probability that a certain hypothesis becomes true (i.e. the probability of microshrinkage apparition).

Besides, the sensitive module [43] (SM), that we used for both microshrinakages and mechanical properties [44], provided a decision support system for the operators in the foundry rely on Bayesian theory. Specifically, SM studies the different values that each variable adopts in order to trace the influence of such values in the apparition of a range of the ultimate tensile strength. Note that a variable may represent, for example, the use of one or another product in a certain phase of the process, applying one certain methodology or not, and so on. This is, the SM evaluates the results obtained by the Bayesian network and calculates the causal relationship between each amount of magnesium and the probability that a range of ultimate tensile strength appears.

5.2. K-Nearest Neighbours

The *K- nearest neighbour* [6] algorithm is one of the simplest supervised machine learning algorithms for classifying instances. This algorithm is based on the class of the most nearest instances of an unknown instance.

Specifically, the training phase of this algorithm comprises representing a set of training data instances $S = s_1, s_2, .., s_m$ in a $n-dimensional$ space where n is the amount of variables for each instance (i.e. in our case the variables of the casting production).

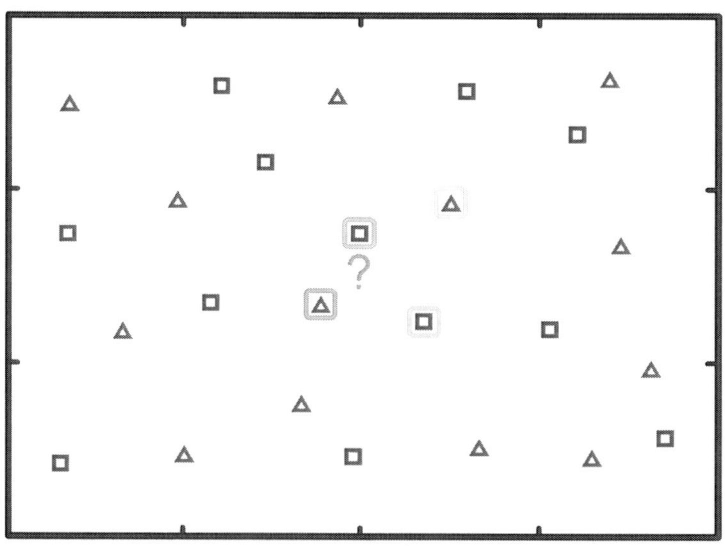

Figure 5. Example of a knn classifier in a bidimensional space.

Furthermore, the classification phase of an unknown instance (whose class is not known) is performed by measuring the distance between the training instances and the unknown instance. In this way, establishing the distance between two points X and Y in a $n-dimensional$ space, can be achieved by using any distance measure, in our case we used Euclidean distance (shown in equation 2).

$$\sqrt{\sum_{i=0}^{n}(X_i - Y_i)^2} \tag{2}$$

Finally, there are several ways to choose the class of the unknown instance, the most used technique is to classify the unknown instance as the most common class amongst the *K-nearest neighbours*.

5.3. Artificial Neural Networks

Artificial Neural Networks (ANN) is a machine learning model that simulates the behaviour of neurons in the human brain [4]. Formally, a neuronal network consists on interconected neurons. In this way, the activation of a neuron depends on its set of inputs, where y_i is the activation of the current neuron, f_i is the activation function, $W_{j,i}$ is the weight of the neuron and a_j is the activation of the input neuron (shown in equation 3).

$$y_i = f_i\left(\sum_{j=0}^{n} W_{j,i} \cdot a_j\right) \tag{3}$$

More accurately, *Multi-Layer Perceptron* (MLP) is a kind of artificial neural network model of simple neurons called perceptrons that are structured in layers. The layers can be classified as *input layers*, *hidden layers* and *output layer*. We perform the training of the model using *backpropagation algorithm* [45] that calculates the weights W_j of the activation function for each neuron.

6. Experiments and Results

6.1. General Methodology

We have collected data from a foundry specialised in safety and precision components for the automotive industry, principally in disk-brake support with a production over 45000 tons a year. These experiments are focused on prediction of microshrinkages and ultimate tensile strength. Note that, as aforementioned, microshrinkages are detected by the employment of X-Ray inspections, and, in a similar vein, the only way to examine the mechanical properties is the employment of destructive inspections. Therefore, the evaluation must be performed after the production is done

Moreover, the acceptance/rejection criterion of the studied models resembles the one applied by the final requirements of the customer (i.e, in the examined cases, the automotive industry). According to the very restrictive quality standards imposed by these clients, pieces flawed with microshrinkages or an invalid ultimate tensile strength must be rejected.

Specifically, we conducted the next methodology in order to evaluate properly the machine-learning classifiers:

- **Cross validation:** We have performed a *k-fold cross validation* [46] with $k = 10$. In this way, our dataset is 10 times split into 10 different sets of learning (66% of the total dataset) and testing (34% of the total data).
- **Learning the model:** For each fold, we have made the learning phase of each algorithm with each training dataset, applying different parameters or learning algorithms depending on the model. More accurately, we have use this three different models:
  - *Bayesian networks:* In order to train Bayesian networks we have used different structural learning algorithms; K2 [47], Hill Climber [48] and Tree Augmented Naive (TAN) [49]. Moreover, we have also performed experiments with a Naïve Bayes Classifier.
  - *K-nearest Neighbour:* For *knn* we have performed experiments with $k = 1$, $k = 2$, $k = 3$, $k = 4$, and $k = 5$.
  - *Artificial neural networks:* We have used a Multilayer Perceptron (MLP) learned with *backpropagation* algorithm.
- **Testing the model:** For each fold, we evaluated the error rate between the predicted value set X and the real value set Y (both with size of the testing dataset m) with mean absolute error (MAE) (shown in equation 4).

$$MAE(X,Y) = \sum_{i=1}^{m} \frac{|X_i - Y_i|}{m} \tag{4}$$

Similarly, we have used root mean square error (RMSE) (shown in equation 5)

$$RMSE(X,Y) = \frac{1}{m} \cdot \sqrt{\sum_{i=1}^{m} (X_i - Y_i)^2} \tag{5}$$

6.2. Microshrinkage Experiments

Figure 6 shows the obtained results in terms of prediction accuracy and Figure 7 shows the error rate of the three classifiers (mean absolute error and root mean square error). In this way, nearly every algorithm achieves good results, however both artificial neural networks and Bayesian networks trained with Tree Augmented Naïve seem more suitable if we focus on the results. Still, Naïve Bayes classifier behaves worse than other classifiers. Please note that Naïve Bayes is a type of Bayesian network where the input variables are assumed to be linear independent. In this way, it skips the causal dependency that may be within the variables, therefore it cannot achieve as good results as the other classifiers.

Moreover, K-nearest neighbour algorithm, witch is a non-linear classifier, achieves better results than one may think in beforehand. In this way, K-nearest neighbour has no training phase itself (only a little data preprocessing), it only focuses on the resemblance between the instances. Therefore, it behaves worse than more robust methods such as ANN and Bayesian networks.

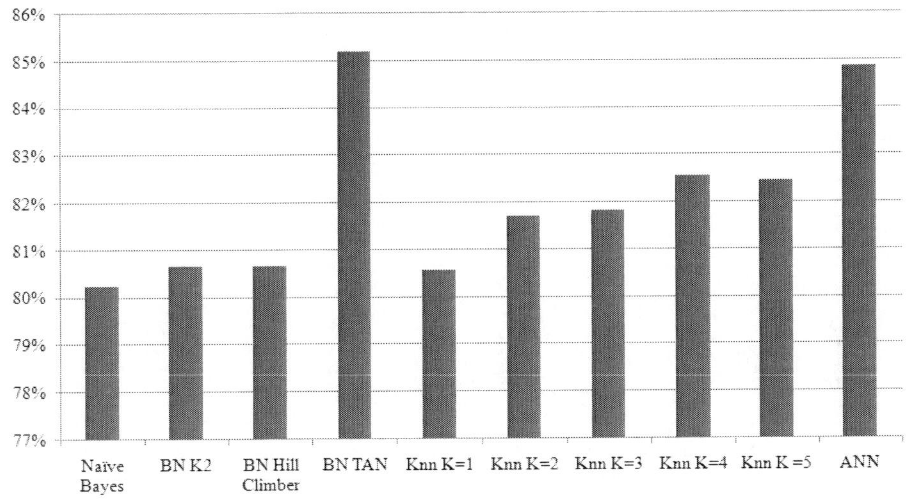

Figure 6. Accuracy of Evaluated Classifiers.

Actually, even the classifiers have not reached a 100% accuracy level, they have interesting results for being used in a high-precision foundry. Remarkably, the good results achieved by the ANN show that it can be used in a similar way as we have used the Bayesian networks in a previous work [50]. Thereby, combining the better classifiers and using them for the defects that suit better, we can reduce in a significative manner the cost and the duration of the actual testing methods, apart from providing an effective *ex-ante* method.

6.3. Ultimate Tensile Strength Experiments

As we mentioned before, we evaluated the classifiers in terms of prediction accuracy and error (i.e. MAE and RSME). In this experiments we have also performed an analysis of the optimal training dataset size. Therefore we have performed the experiments for the following training sizes n: $n = 100$, $n = 200$, $n = 300$, $n = 400$, $n = 500$, $n = 600$, $n = 700$, $n = 800$ and $n = 889$.

In this way, Figure 8 shows the obtained results in terms of prediction accuracy. For a size of the training dataset of 100 the overall prediction of the *machine-learning classifiers* is low, however, Bayesian networks trained with *Hill Climbing* outperformed the rest of classifiers with a prediction accuracy of 81.82%. Despite these results, this approach for Bayesian networks did not perform well when increasing the size of the dataset. On the other hand, the rest of classifiers, obtained their best results for a training size between 700 and 800 instances and not for the whole training dataset ($n = 889$) where there is an interesting accuracy reduction. This fact may be result of the data acquiring phase that is performed manually and, thus, it is subject to numerous errors that can include noise in the dataset. Therefore, techniques for reducing the noise of the dataset such as Principal Component Analysis [51] will be studied in further work.

More accurately, the algorithms showed quite different performances. In this way, *k-nearest neighbour* did it surprisingly well for being a lazy algorithm. Specifically, for a value of $k = 1$, *k-nearest neighbour* achieved the best results and the second best classifier

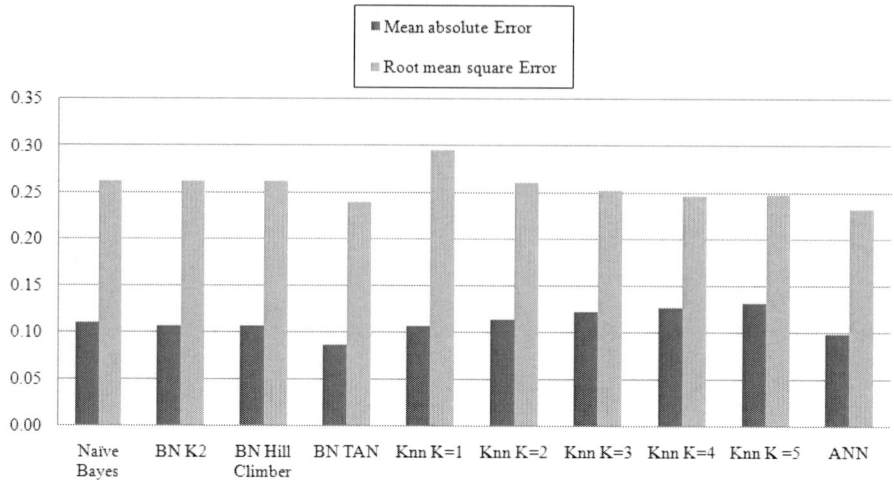

Figure 7. Error Rate of Evaluated Classifiers.

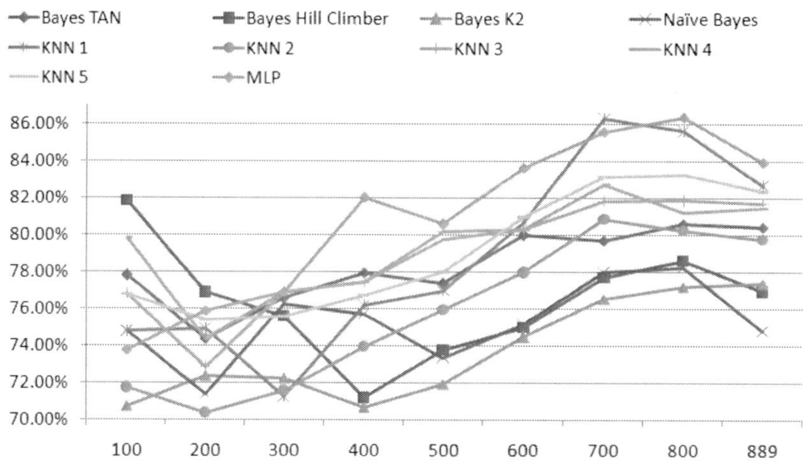

Figure 8. Accuracy of the evaluated classifiers.

tested. Second, Bayesian networks do achieve overall good results, specifically, Bayesian networks trained with TAN perform as expected and the results were similar to the ones obtained in [44]. Finally, *MLP* outperformed the rest of the classifiers, showing that can be a interesting classifier for predicting the values of the mechanical properties.

Furthermore, Figure 9 shows the Mean Absolute Error and Figure 10 shows the Root Mean Square Error. Thereby, the results obtained are similar to the ones of prediction accuracy and MLP also outperformed the rest of algorithms in terms of error.

In addition, the classifiers have interesting results and can be used in a high-precision foundry. Remarkably, the outstanding results achieved by Multilayer Perceptron show that it can be used in a similar way as we have used the Bayesian networks in a previous work [44].

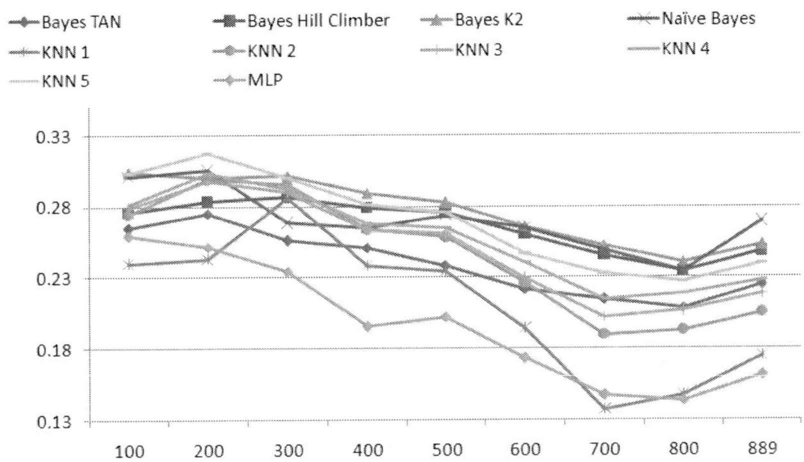

Figure 9. Mean absolute error of the evaluated classifiers.

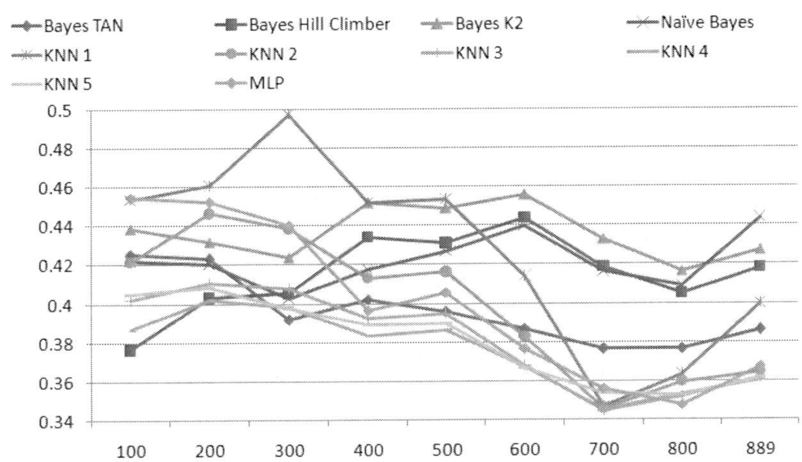

Figure 10. Root mean square error of the evaluated classifiers.

7. Conclusions

On one hand, microshrinkages are tiny porosities that appear when the casting is cooling down. On the other hand, ultimate tensile strength is the capacity of a metal to resist deformation when subject to a certain load. Predicting the apparition of microshrinkages and the value of ultimate tensile strength render as the hardest issues in foundry production, due to many different circumstances and variables that are involved in the casting process and determine it.

Our work in [50] pioneers the application of Artificial Intelligence to the prediction of microshrinkages. Likewise, regarding ultimate tensile strength we presented a model to the prediction of mechanical properties [44].

To this extent, we focus here on comparing *machine-learning* classifiers used for the prediction of ultimate tensile strength and microshrinkage apparition. Specifically, we have included and adapted to our particular problem domain three classifiers that have a widely use in similar issues. All of them behave well, but Bayesian networks performed the best in the microshrinkage experiment whilst artificial neural networks outperformed the rest of the classifiers regarding ultimate tensile strength. Still, the ability of Bayesian theory and specifically, the sensitivity module cannot be ignored, since it is an effective method that provides a decision support system for the operators in the foundry plant. Therefore, with the combination of the sensitivity module and the best classifiers we can reduce in a significative manner the cost and the duration of the actual testing methods, apart from providing an effective *ex-ante* method.

The future development of this predictive tool is oriented in three main directions. First, we plan to extend our analysis to the prediction of other defects in order to develop a global system of incident analysis. Second, we will compare more supervised and semi-supervised machine learning algorithms in order to prove their effectiveness to predict foundry defects. Finally, we plan to integrate the best classifiers in meta-classifier combining the partial results.

References

[1] J. Sertucha, A. Loizaga, and R. Suárez. Improvement opportunities for simulation tools. In *Proceedings of the 16th European Conference and Exhibition on Digital Simulation for Virtual Engineering*, 2006. invited talk.

[2] A. Zabala, R. Suárez, and J Izaga. Iron castings, advanced prediction tools, foundry process control and knowledge management. In *Proceedings of the 68'h WFC - World Foundry Congress*, page 355360, 2008.

[3] R. Siclari, T. Margaria, E. Berthelet, and J. Fourmann. Microshrinkage in ductile iron mechanism and solution. In *Proceedings of the Keith Millis Symposium on Ductile Cast Iron*, pages 355–360, 2003.

[4] Christopher M. Bishop. *Neural Networks for Pattern Recognition*. Oxford University Press, 1995.

[5] J. Pearl. Reverend bayes on inference engines: a distributed hierarchical approach. In *Proceedings of the National Conference on Artificial Intelligence*, pages 133–136, 1982.

[6] E. Fix and J. L. Hodges. Discriminatory analysis: Nonparametric discrimination: Small sample performance. *Technical Report Project 21-49-004, Report Number 11*, 1952.

[7] S.A. Elfayoumy, Y. Yang, and S.P.: Ahuja. Anti-spam filtering using neural networks. In *Proceedings of the International Conference on Artificial Intelligence*, volume 4, pages 984–989. IC-AI.

[8] C. Zhang, J. Jiang, and M. Kamel. Intrusion detection using hierarchical neural networks. *Pattern Recognition Letters*, 26(6):779–791, 2005.

[9] Silvio Simani and Cesare Fantuzzi. Neural networks for fault diagnosis and identification of industrial processes. In *Proceedings of 10th Eurorean Symposium on Artificial Neural Networks (ESANN)*, pages 489–494, 2002.

[10] Zhen Yang, Xiangfei Nie, Weiran Xu, and Jun Guo. An approach to spam detection by naive Bayes ensemble based on decision induction. In *Proceedings of the Sixth International Conference on Intelligent Systems Design and Applications (ISDA'06)* , pages 861–866. IEEE Computer Society, 2006.

[11] N. A. Masruroh and K. L. Poh. A Bayesian network approach to job-shop rescheduling. In *Proceedings of the IEEE International Conference on Industrial Engineering and Engineering Management*, pages 1098–1102, 2007.

[12] Y. Liu and Shi-Qi Li. Decision support for maintenance management using Bayesian networks. In *Proceedings of the International Conference on Wireless Communications, Networking and Mobile Computing (WiCom 2007)*, pages 5713–5716, 2007.

[13] Bruce Abramson, John Brown, Ward Edwards, Allan Murphy, and Robert L. Winkler. Hailfinder: A Bayesian system for forecasting severe weather. *International Journal of Forecasting*, 12(1):57–71, 1996.

[14] A. S. Cofiño, R. Cano, C. Sordo, and J. M. Gutiérrez. Bayesian networks for probabilistic weather prediction. In *Proceedings of the European Conference on Artificial Intelligence (ECAI)*, pages 695–699, 2002.

[15] Christopher Krügel, Darren Mutz, William K. Robertson, and Fredrik Valeur. Bayesian event classification for intrusion detection. In *Proceedings of the 19th Annual Computer Security Applications Conference*, pages 14–23, 2003.

[16] A. Faour, P. Leray, and B Eter. A SOM and Bayesian network architecture for alert filtering in network intrusion detection systems. In *Proceedings of the International Conference on Information and Communication Technologies* , pages 3175–3180, 2006.

[17] G.A. Davis and J. Pei. Bayesian networks and traffic accident reconstruction. In *Proceedings of the 9th international conference on Artificial intelligence and law (ICAIL)* , pages 171–176, 2003.

[18] G.A. Davis. Bayesian networks, falsification, and belief revision in accident reconstruction. In *Proceedings of the Transportation Research Board 85th Annual Meeting*, 2006.

[19] H. Zhang, A. Berg, M. Maire, and J. Malik. Svm-knn: Discriminative nearest neighbor classification for visual category recognition. In *Proceedings of IEEE Computer Society Conference on Computer Vision and Pattern Recognition (CVPR)* , volume 2, pages 2126–2136, 2006.

[20] M. Sharif and D.H. Burn. Improved k-nearest neighbor weather generating model. *Journal of Hydrologic Engineering*, 12:42, 2007.

[21] Igor Santos, Yoseba K. Penya, Jaime Devesa, and Pablo G. Bringas. N-grams-based file signatures for malware detection. In *Proceedings of the 11th International Conference on Enterprise Information Systems (ICEIS), Volume AIDSS*, pages 317–320, 2009.

[22] Z.M. Lu and H. Burkhardt. Fast Image Retrieval Based on Equal-average Equal-variance K-Nearest Neighbour Search. In *Proceedings of 18th International Conference on Pattern Recognition*, volume 2.

[23] J.M. Vitek, S.A. David, and C.R. Hinman. Improved ferrite number prediction model that accounts for cooling rate effects part 1 model development. *Welding Journal*, 82(10).

[24] H.Q. Peter and J. Wang. Fault detection using the k-nearest neighbor rule for semiconductor manufacturing processes. *IEEE Transactions on Semiconductor Manufacturing*, 20(4).

[25] A. Lazaro, I. Serrano, J.P. Oria, and C. de Miguel. Ultrasonic sensing classification of foundry pieces applying neuralnetworks. In *$5^t h$ International Workshop on Advanced Motion Control*, pages 653–658, 1998.

[26] Pei Zhang, Zhiqiang Xu, and Fengshan Du. Optimizing casting parameters of ingot based on neural network and genetic algorithm. In *ICNC '08: Proceedings of the 2008 Fourth International Conference on Natural Computation* , pages 545–548, Washington, DC, USA, 2008. IEEE Computer Society.

[27] M. Perzyk and A. Kochanski. Detection of causes of casting defects assisted by artificial neural networks. In *Proceedings of the Institution of Mechanical Engineers Part B Journal of Engineering Manufacture*, volume 217, pages 1279–1284, 2003.

[28] H. Bhadeshia. Neural networks in materials science. *ISIJ international*, 39(10):966–979, 1999.

[29] M. Vasudevan, M. Muruganath, and A. K. Bhaduri. Application of bayesian neural network for modelling and prediction of ferrite number in austenitic stainless steel welds. *ser. Mathematical Modelling of Weld Phenomena - VI. London: Institute of Materials*, pages 1079–1100, 2002.

[30] K.D. Schnelle and R.S.H. Mah. Product quality management using a real-time expert system. *ISIJ International*, 34(10):815–821, 1994.

[31] Inoculation alloy against microshrinkage cracking for treating cast iron castings. Patent US 2005/0180876 A 1.

[32] P. Larrañaga, J. Sertucha, and R. Suárez. Análisis del proceso de solidificación en fundiciones grafíticas esferoidales. *Revista de Metalurgia*, 42(2):244–255, 2006.

[33] J. Sertucha, R. Suárez, J. Legazpi, and P. Gacetabeitia. Influence of moulding conditions and mould characteristics on the contraction defects appearance in ductile iron castings. *Revista de Metalurgia*, 43(3):188–195, 2007.

[34] J.F. Carrasquilla and R. Ríos. A fracture mechanics study of nodular iron. *Revista de Metalurgia*, 35(5):279–291, March–April 1999.

[35] R. Gonzaga-Cinco and J. Fernández-Carrasquilla. Mechanical properties dependency on chemical composition of spheroidal graphite cast iron. *Revista de Metalurgia*, 42:91–102, March–April 2006.

[36] M. Hecht and F. Condet. Shape of graphite and usual tensile properties of sg cast iron: Part 1. *Fonderie, Fondeur d'aujourd'hui*, 212:14–28, 2002.

[37] C. W. Lung and Norman H. March. *Mechanical Properties of Metals: Atomistic and Fractal Continuum Approaches*. World Scientific Pub Co Inc, 1992.

[38] American Society for Testing and Materials. ASTM D1062 - Standard Test Method for Cleavage Strength of Metal-to-Metal Adhesive Bonds, 2008.

[39] American Society for Testing and Materials. ASTM B489 - e1 Standard Practice for Bend Test for Ductility of Electrodeposited and Autocatalytically Deposited Metal Coatings on Metals, 2008.

[40] T. Bayes. An essay towards solving a problem in the doctrine of chances. *Philosophical Transactions of the Royal Society*, 53:370–418, 1763.

[41] Enrique Castillo, José M. Gutiérrez, and Ali S. Hadi. *Expert Systems and Probabilistic Network Models*. Springer, New York, NY, USA, erste edition, December 1996.

[42] J. Pearl and S. Russell. Bayesian networks. Technical Report Tech. Rep. R-216, Computer Science Department, University of California, Los Angeles, 2000.

[43] Chan K. Saltelli, A. and and Scott E. M. *Sensitivity Analysis: Gauging the Worth of Scientific Models*. Springer, New York, NY, USA, wiley edition, August 2000.

[44] Javier Nieves, Igor Santos, Yoseba K. Penya, Sendoa Rojas, Mikel Salazar, and Pablo G. Bringas. Mechanical properties prediction in high-precision foundry production.

[45] Donald Michie, D. J. Spiegelhalter, C. C. Taylor, and John Campbell, editors. *Machine learning, neural and statistical classification* . Ellis Horwood, Upper Saddle River, NJ, USA, 1994.

[46] R. Kohavi. A study of cross-validation and bootstrap for accuracy estimation and model selection. In *International Joint Conference on Artificial Intelligence*, volume 14, pages 1137–1145, 1995.

[47] Gregory F. Cooper and Edward Herskovits. A bayesian method for constructing bayesian belief networks from databases. In *Proceedings of the 7th conference on Uncertainty in artificial intelligence*, 1991.

[48] S. J. Russell and Norvig. *Artificial Intelligence: A Modern Approach (Second Edition)*. Prentice Hall, 2003.

[49] Dan Geiger, Moises Goldszmidt, G. Provan, P. Langley, and P. Smyth. Bayesian network classifiers. In *Machine Learning*, pages 131–163, 1997.

[50] Y.K. Penya, P. García Bringas, and A. , Zabala. Advanced fault prediction in high-precision foundry production. In *Proceedings of the* 6th *IEEE International Conference on Industrial Informatics*, pages 1673–1677, 2008.

[51] C.G. Thomas, R.A. Harshman, and R.S. Menon. Noise reduction in BOLD-based fMRI using component analysis. *Neuroimage*, 17(3):1521–1537, 2002.

In: Structural Steel and Castings
Editor: Lena M. Becker, pp. 277-308
ISBN: 978-1-61728-104-4

Chapter 9

CASTING TECHNOLOGY AT NANOSCALE

***Yong X. Gan*[1,*], *Dwight K. Inkrott*[1], *Lu Yang*[1], *Bo J. Gan*[2] and *Joseph G. Lawrence*[3]**

[1]Department of Mechanical, Industrial and Manufacturing Engineering, College of Engineering, University of Toledo, Toledo, OH 43606, USA
[2]Ottawa Hills High School, 2532 Evergreen Road, Toledo, OH 43606, USA
[3]Department of Bioengineering, College of Engineering, University of Toledo, Toledo, OH 43606, USA

Abstract

Nanocasting refers to the special fabrication technology that can be used to produce nanoscale features. This technology has the capability of generating well-ordered, very fine entities such as nanofibers, coaxial nanotubes, and porous structures. In this chapter, an overview of the process of nanocasting technology will be provided followed by a description of the mechanism of generating controlled nanostructures. Different types of molds used for nanocasting will also be introduced. Some details on how to use nanocasting to make nanoscale hollow shells, nanofibers with core structures and nanoscale networks will be presented. Finally, the applications of nanocasting for preparing high surface area, electrochemically and photoelectrochemically active nanomaterials and structures will be discussed.

Keywords: Nanocasting, nanomaterials and ordered structures, high surface area, electrochemically active materials, catalysis, fuel cell, photoelectrochemical behavior.

1. Introduction

In general, casting is defined as the solidification process whereby a molten metal is transferred into the cavity of a mold [1]. The cavity of the mold has the desired shape and dimension. The molten metal solidifies within the cavity and takes the shape of the mold

*E-mail address: yong.gan@utoledo.edu, Tel: +1-419-530-6007, Fax: +1-419-530-8206. (Corresponding Author)

cavity. It must be noted that the metal may undergo some shrinkage due to the contraction caused by cooling. What are the typical advantages of casting as compared with other materials fabrication and manufacturing processes? First, since casting is a high temperature process, some brittle materials which can not be shaped by cutting and cold working (rolling) may be easily processed through die casting. Second, large and complicated structures can be generated via casting as long as the rheological property of the raw materials permit. Third, casting could be used to generate net-shape parts or components with minimum machining. To that end, casting could be a very cost-effective technique for processing and manufacturing of various materials [2].

Traditional casting techniques can be classified into three categories as discussed below. Sand casting is the most commonly used one, especially suitable for casting iron based alloys. In this method, sand is used as the mold material. One of the challenges in casting is gas trapping during the solidification process, which is the main reason for void defect formation in finished parts. Nevertheless, the defects generated by sand casting can be minimized by the careful control of degassing.

Metal die casting is another frequently used casting technique. The salient feature of metal die casting is that high heat transfer rate can be obtained due to the high conductivity of the die. The formation of harmful dendrite structure can be suppressed and the segregation of alloying elements can be prevented effectively. Other advantages include the good surface finish of such metal die cast parts. Metal mold has low gas permeability. Porosity and shrinkage could become problems in such casting.

The third type of casting is investment casting. Low melting temperature materials such as wax and plastics are used to form various shapes and patterns. After that, a liquid or semisolid slurry is poured onto the surface of the wax/plastics patterns. Upon heating or burning, the wax and/or low melting point plastics are removed and the slurry form the mold required for the casting. Investment casting is very useful in getting parts with high dimensional accuracy. It may also be used to generate very fine features with very low surface roughness.

Casting can be combined with rolling, which results in a highly automated and more efficient process. This technique is called continuous casting or semisolid casting. The molten metal is cast into a large ingot. The ingot is subjected to a hot-rolling process. In some cases, the molten metal is cast into the space of water cooled dies so that the metal is in a mixed liquid and solid state. The semisolid slurry is under the subsequent extrusion or rolling. Since the solidification time is very short, segregation is inhibited and the chemical composition and mechanical properties are very uniform [3].

Inspired by the casting technique, researchers have considered making nanoscale components or features using the similar process which is used for making large components. This is the origin of nanocasting technology. Nanocasting deals mainly with the processing of nonmetallic materials. Significant progress has been achieved in this field for the last two decades. Göltner et al. [4-8] investigated casting of nanoporous materials using oligomeric surfactants and block copolymers. The template morphology can be copied from soft into hard matter on the nanometer length scale. For example, porous silica materials were obtained via nanocasting on a pseudopolyrotaxanes template synthesized from α-cyclodextrin and polyamines [9]. The obtained structure was wormhole-like. Short tube-like pores were also found on transmission electron microscopic (TEM) micrographs. Such features were considered originating from the linear pseudopolyrotaxane chain structure. This indicates the pseudopolyrotaxane structure was maintained during the nanocasting and copied into the hard silica materials.

The objective of this chapter is to provide an overview on nanocasting technology. First, different types of mold used for nanocasting will be reviewed. Then, details on how to use nanocasting to make nanoscale hollow shells, nanofibers with coaxial cable structures and nanoscale networks will be introduced. After that, applications of nanocasting for preparing high surface area, electrochemically and photoelectrochemically active nanomaterials and structures will be presented. Finally, a case study on preparation of nanostructured cobalt sulphide using different templates as the casting molds will be shown.

2. Nanocasting Procedures and Casting Molds Used

2.1. Nanocasting through Self-assembled Colloidal Crystal Templates

Nanocasting of two-dimensionally (2D) ordered porous arrays using monolayer colloidal crystal templates has the advantage of generating hierarchical structures at micro and nanometer length scales. For example, in the work performed by Xia et al. [10], polystyrene (PS) beads with the diameter of 1 micron were used to form 2D ordered arrays. The ordered arrays were served as the nanocasting molds to make Co_3O_4 hierarchical structures. Electrodeposition was performed to deposit Co within the interstitial sites of the PS beads using a nitrate electrolyte. The as-prepared Co_3O_4 array materials have a hierarchical porous structure. The skeleton is composed of ordered monodispersed submicrometer porous bowls. The wall of these bowls is nanoporous. The interstices between the bowls are filled with Co_3O_4 nanoflakes.

Obviously, such a nanocasting technique can also be used to duplicate the feature of various self-assembled polymer beads through hydrothermal processes. As shown in Fig. 1(a), polymer beads can be aligned on the substrate by self assembly. The solution containing a metal salt can be cast into the interstitial space of the template as shown in Fig. 1(b). Upon drying or calcination, the metal salt solution is converted into a solid compound. The polymer beads may be removed by an organic solution to form the ordered features on the substrate as illustrated in Fig. 1(c). The cast feature may look like hollow bowls or spheres as shown in Fig. 1(d), if there is sufficient amount of the filled solution material.

Gels and polymer microspheres may be co-cast to prepare nanocrystalline materials with a porous structure. The polymer microspheres are taken as the sacrificial templates. For example, Cruz et al. [11] made macroporous nanocrystalline 4 mol% yttria-stabilised zirconia ceramics by such a casting technique. The ZrO_2-Y_2O_3 nano-powders with modified surface and polymethylmethacrylate (PMMA) sacrificial templates were the starting materials. The diameter of the PMMA spheres was around 1 μm. Porous monolithic samples of nanocrystalline stabilised ZrO_2 were successfully fabricated at laboratory scale. In order to scale-up this method for the production of samples with larger dimensions, it is necessary to use a slower drying process and a more delicate burn-out step.

Polymer colloidal crystal beads can be self-assembled into three dimensional (3D) templates as nanocasting molds. Chai et al. [12] used PS microspheres with the diameter of about 660 nm to make ordered opal structure for nanocasting ordered $Cs_{2.5}H_{0.5}PW_{12}O_{40}$.

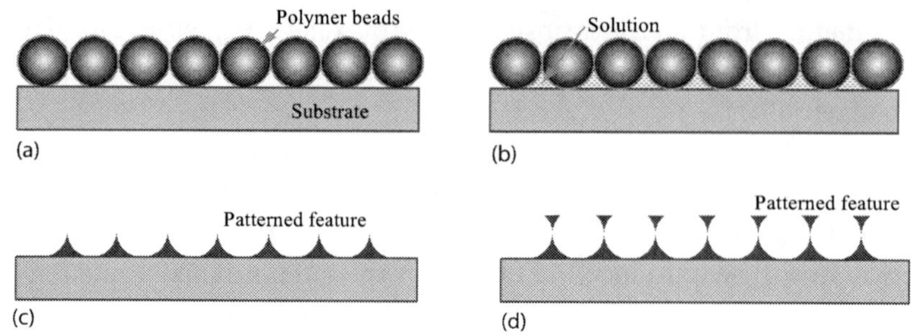

Figure 1. Nanocasting through self-assembled polymer beads: (a) polymer beads on substrate, (b) casting solution into the space of the beads, (c) patterned feature after heating, (d) patterned hollow sphere feature.

The procedures are given in Fig. 2. The PS spheres can be synthesized by emulsion polymerization. The 3D template structures consisting of microspheres can be assembled on glass, quartz, and silicon substrates by the vertical deposition technique as discussed in literature [13]. PMMA colloidal crystal monoliths were also used as the starting template for nanocasting [14, 15]. After infiltration with a silica source in the PMMA template, and adding a mesostructure-directing agent, condensation of silica and calcination to remove PMMA resulted in the three dimensional ordered SiO_2 as illustrated in Fig. 3. The mesoporous silica monolith was further used as a hard template for making porous carbon or carbon/silica nanocomposites. The scanning electron microscope image of the PMMA mold is shown in Fig. 4(a). Fig. 4(b) is the morphology of the mesoporous silica monolith and Fig. 4(c) shows the deposited carbon/silica nanocomposites [15].

2.2. Nanocasting through Porous Polymer Membranes

Zhang and Wong [16] used a track etched polycarbonate (PC) polymer membrane as the template for growing one-dimensional (1D) nanostructure. Various discrete, individual motifs and arrays of crystalline and pure semiconducting transition metal sulfide including CuS, PbS, and CdS nanowires were made via template-directed technique. The synthesis process consists of nanocasting octadecyltetrachlorosilane self-assembled monolayers (OTS-SAMs) on the front and back surfaces of the polycarbonate template. It is noted that OTS-SAMs serve as passivation layers [17]. The nanocasting procedures are illustrated in Fig. 5. In Fig. 5(a), the nanoporous polymer membrane is shown. Fig. 5(b) illustrates OTS-SAMs on a polymer stamp. According to Zhang and Wong [16], a homemade PDMS (polydimethylsiloxane) stamp was initially inked with a low concentration (10 mM) hexane solution of the OTS. After drying with nitrogen, the OTS-SAMs was formed on the surface of the stamp. The stamp containing the OTS-SAMs was then placed into contact with the PC membranes (Fig. 5(c)) for 30 s and then peeled off carefully. The OTS-SAMs was transferred onto the surface of the PC mold as schematically shown in Fig. 5(d).

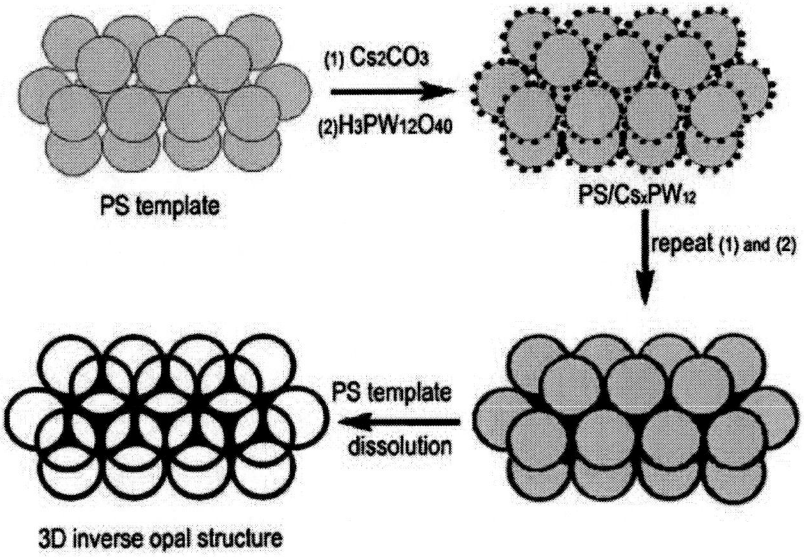

Figure 2. Schematic illustration of using a 3D PS colloidal crystal template mold to do nanocasting of $Cs_xH_{3-x}PW_{12}O_{40}$ (after Chai et al. [12]).

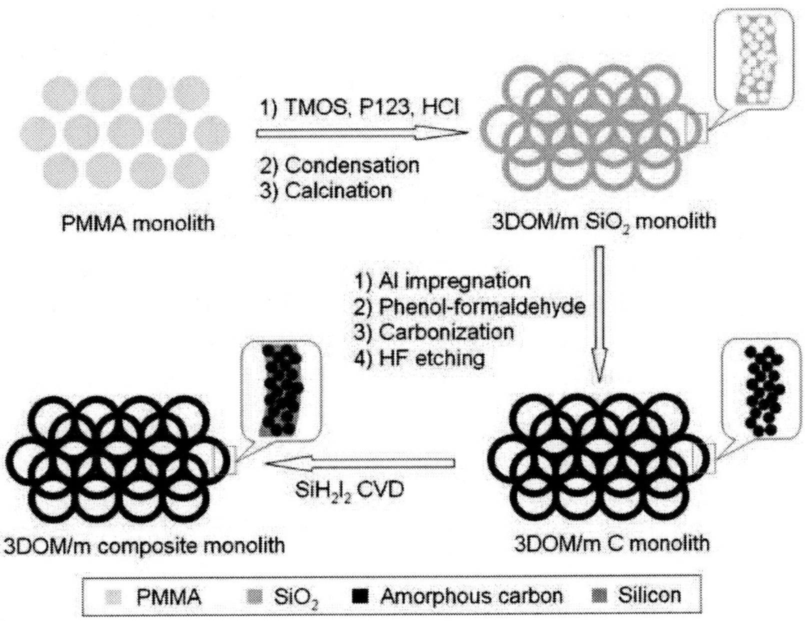

Figure 3. Schematic illustration of using a 3D PMMA colloidal crystal template mold to do nanocasting of SiO_2 and C/SiO_2 nanocomposites (after Wang et al. [15]).

2.3. Nanocasting through Anodic Aluminium Oxide (AAO) Templates

Preparation of anodic aluminium oxide (AAO) templates has been studied for a considerably long time [18-31]. In the work performed by Zhao et al. [24], a two-step anodic

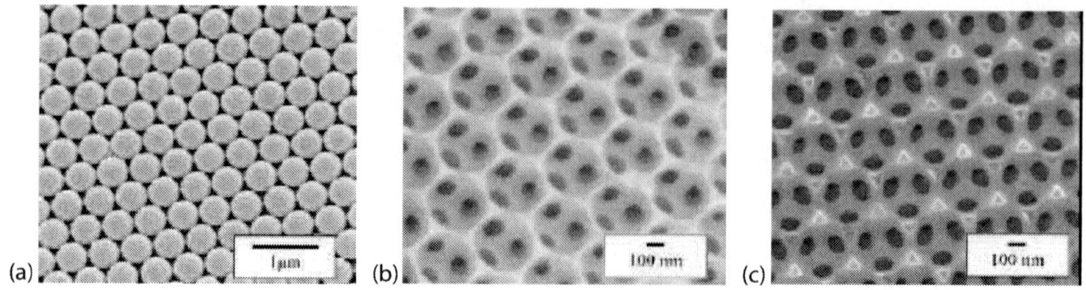

Figure 4. Scanning electron microscope images showing: (a) PMMA colloidal crystal template mold for nanocasting, (b) ordered porous SiO_2, (c) C/SiO_2 nanocomposites (after Wang et al. [15]).

oxidization method was used to obtain nanopores with uniform sizes and thin barrier layers. High-purity aluminum sheets were anodized on a single side using a regulated DC power supply in a 0.5 M oxalic acid solution. Lower concentration of oxalic acid (0.3 M) also works [19, 20]. In some papers, other types of acidic solution for oxidation were used. For example, H_3PO_4 was used as the electrolyte for oxidation [31]. Before anodization, the aluminum plates may be degreased in sodium hydroxide solution or trichloroethylene, followed by ultrasonic cleaning in acetone or ethanol. Then the samples were rinsed with methanol and distilled water, respectively. After that, the aluminum plates were etched in 5.0 M NaOH at 60°C for 20 minutes and subsequently rinsed with distilled water.

Electropolishing of the aluminum plates may be also conducted in a 30% HNO_3 methanol solution at −20°C so that the surface of the metal is fresh. In the work performed by Yu et al. [19], anodization was performed in 0.3 M $H_2C_2O_4$ at 0°C held at 20°. The cleaned rolled aluminum foils were used as the anode. Pt foil or graphite plate can be used as the cathode. In the process introduced in [19], the first anodization takes about 5 h. After the first anodization, the strip-off process was carried out in a solution mixture

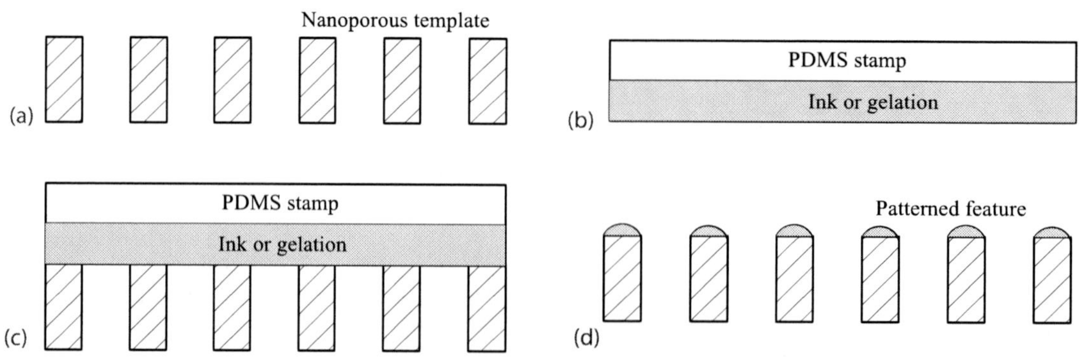

Figure 5. Nanocasting on a porous polymeric membrane: (a) nanoporous polymeric membrane, (b) ink carried by a PDMS stamp, (c) pressing on the polymer membrane, (d) nanosized feature on the membrane.

of H_3PO_4 and H_2CrO_4. The exposed and well-ordered concave patterns on the aluminum substrate acted as a self-assembled mask for the second anodization process. The second anodization takes about 12 h. After the second anodization, AAO templates with uniform nanopores were obtained for nanocasting. The depth of the nanopores can be controlled in the range of 750 nm to 1000 nm, and the diameter of the pores will be controlled in the range from 20 to 200 nm.

AAO templates can be used to cast nanofibers or nanotubes. The nanocasting procedures can be schematically shown in Fig. 6. Fig. 6(a) illustrates the AAO template. In Fig. 6(b), a reaction solution is poured into the nanopores of the template. The second reactant may be added to react with the first solution and form gelation as shown in Fig. 6(c). Upon drying or calcination, the product in the form of nanofibers is obtained as sketched in Fig. 6(d).

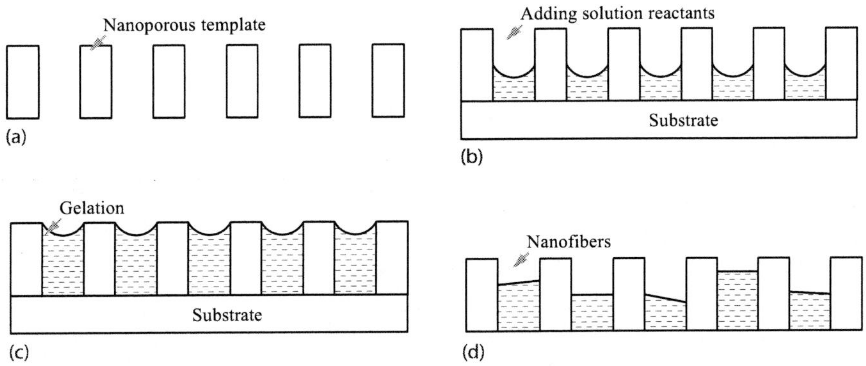

Figure 6. Nanocasting of ordered fibers via porous templates such as anodic aluminium oxide (AAO) membrane: (a) porous template, (b) solution in the pores, (c) gelation formation due to chemical reactions, (d) nanofiber formation due to drying or calcination.

2.4. Nanocasting through Plant Leaves

Plant leaves have a variety of surface features. The complex surface morphologies exhibit optimized functional properties, which stimulates the interests of some researchers to use plant leaves as the templates for nanocasting. In the work performed by Singh et al. [32], thin polymeric films were cast onto Lotus and Colocasia leaves to make replica of the surface structures. As shown in Fig. 7, the nanocasting was conducted on the poly(methyl methacrylate) (PMMA) film spin coated on a silicon wafer using a poly (dimethyl siloxane) (PDMS) mold. One of the interesting properties of such casting film is that the drastic reduction in frictional coefficient as compared with the control film without any plant leaf surface features. The possible explanation for such a microtribological behavior is due to the reduced real area of contact projected by the surfaces.

Similar work was carried out by Yuan et al. [33]. They prepared lotus-leaf-like superhydrophobic poly(vinyl chloride), PVC, film through nanocasting. The scanning electron microscope (SEM) images are shown in Fig. 8. Fig. 8(a) is the morphology of the starting nanocasting mold, the fresh lotus leaf. Fig. 8(b) shows the surface structure of the negative

PDMS template. In Fig. 8(c), the lotus-leaf-like super-hydrophobic poly(vinyl chloride), PVC, film is revealed. It is evident from the morphological analysis that compared with a common smooth PVC film, a lotus-leaf-like surface structure was clearly observed on the super-hydrophobic PVC film. From the water contact angle measurement results, it was found that the lotus-leaf-like PVC film is highly hydrophobic; the water contact angle of the as-prepared lotus-leaf-like PVC film is as high as 157±1.88°.

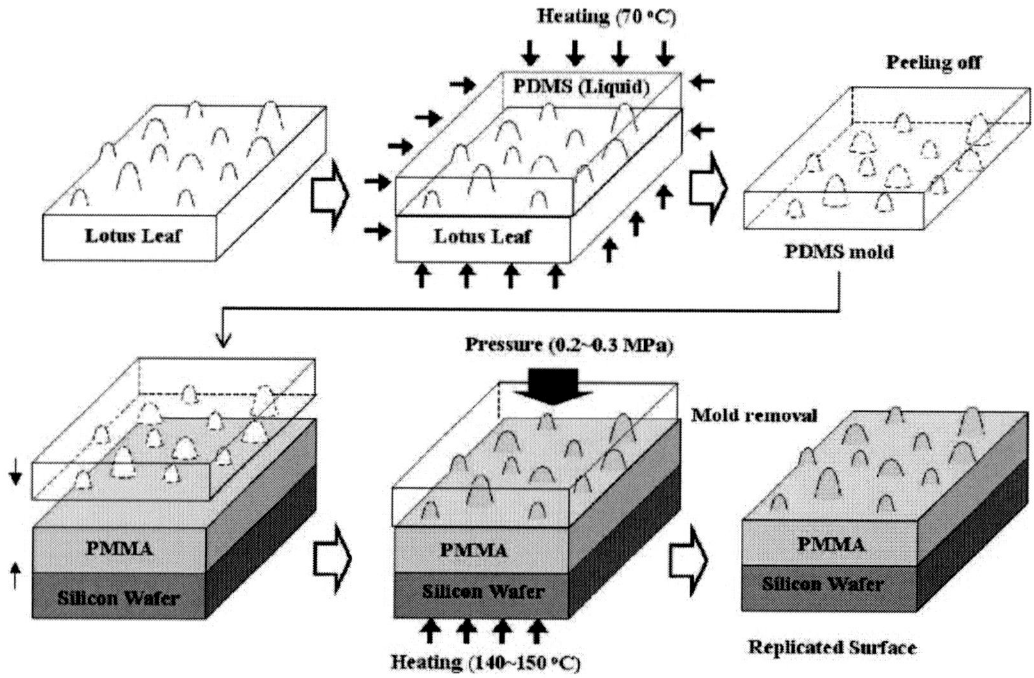

Figure 7. Nanocasting to get the surface features of plant leaves, In this carton, a lotus leaf was taken as the starting material. The PDMS polymer liquid was cast on to the surface of the lotus leaf to duplicate the surface morphology of the lotus leaf after the separation of the polymer and the leaf (after Singh et al. [32]).

2.5. Nanocasting through Mesoporous Silicas

Mesoporous silicas have ordered structures. Many types of mesoporous silicas, such as SBA-15, KIT-6, FDU-12 and SBA-16 were discovered and they have been extensively used for nanocasting [34-66]. Since various mesoporous silicas have different pore geometries, the expected morphologies of porous materials templated by them could be very different. Yue et al. [67] reviewed nanocasting transition metal nanoporous oxides. The predicted porous structures are shown in Fig. 9. The two interpenetrating pores in KIT-6 as shown in Fig. 9(b) are highlighted with black and white, respectively. Oxides such as Cr_2O_3, Co_3O_4, In_2O_3, NiO, CeO_2, WO_3, Fe_2O_3 and MnO_2, were synthesized using these mesoporous silicas. The pore size is less than 10 nm. Such porous transition metal oxide materials have the potential applications in catalysis, Li-ion rechargeable batteries and gas sensors [68].

2.6. Nanocasting through Porous Carbons

Nanoporous carbon materials have been widely used for adsorption, catalysis, and energy storage/conversion for a long time. Recently, porous carbon has been studied as the intermediate template for nanocasting synthesis. Many papers can be found in this field [69-87]. Typically, such nanocasting involves the preparation of a micro-mesoporous carbon mold using aggregated silica spheres assembled template and a carbon source. The diameter of these silica spheres is about 12 nm [69]. Lu et al. [70] prepared ordered mesoporous carbon containing highly dispersed coppersulphur compounds in the carbon framework via nanocasting. The mesoporous silica, SBA-15, was used as the starting template and copper(II) phthalocyanine-tetrasulfonic acid tetrasodium salt (PcS) was used as the single precursor. At the pyrolysis temperatures lower than 600°C, stable nanocast carbon was obtained. If the treatment temperature is higher than 600°C, highly dispersed copper-sulphur compounds were formed in the carbon framework. Porous carbon has also been used as the template to nanocast oxides. For example, Lai et al. [71] synthesized crystalline mesoporous indium oxide by using a mesoporous carbon (CMK-3) as the hard template.

Wu et al. [72] described the formation of unimodal and bimodal mesoporous carbon by nanocasting. Methylated β-cyclodextrin (CD) and tetramethyl orthosilicate (TMOS) were used as the starting materials. By controlling the ratio of CD to TMOS, both unimodal and bimodal porous carbons can be obtained as schematically shown in Fig. 10. The mechanism for both types of porous carbon may be explained as follows: With the proper CD/TMOS ratio, silica frame forms the mesoporous structure. By increasing the TMOS content, many silica gel frameworks in CD/silica wet gels could collapse due to lack of the support of CD during the drying process and then stack into many large silica nano-particles. Consequently, silica component in the dried CD/silica composites with low CD/TMOS ratios has two structures. One is the gel skeletal structure, and the other is the nano-particle structure. Based on this, Wu et al. [72] found a simple one-step nanocasting method to prepare bimodal mesoporous carbon.

Carbon templates can be made in fiber or tubular forms, as illustrated by Popp et al. [88]. Fig. 11 represents the schematic of the template technique to form a freestanding

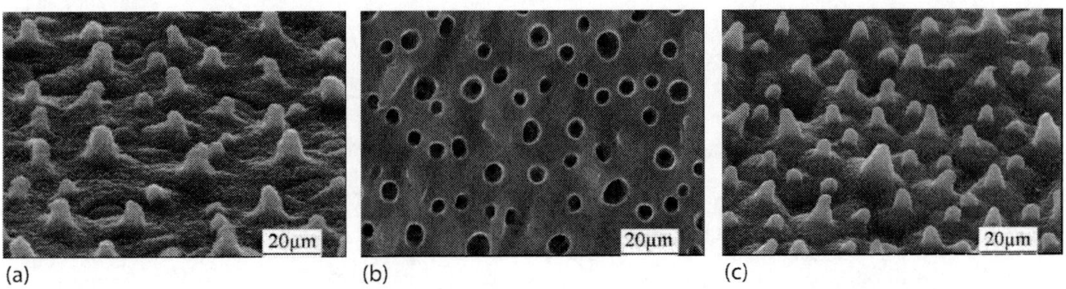

Figure 8. Scanning electron microscope (SEM) images showing the surface features of a fresh lotus leaf duplicated on PVC film (after Yuan et al. [33]): (a) the fresh natural lotus leaf, (b) the PDMS negative template, (c) the lotus-leaf-like super-hydrophobic PVC film.

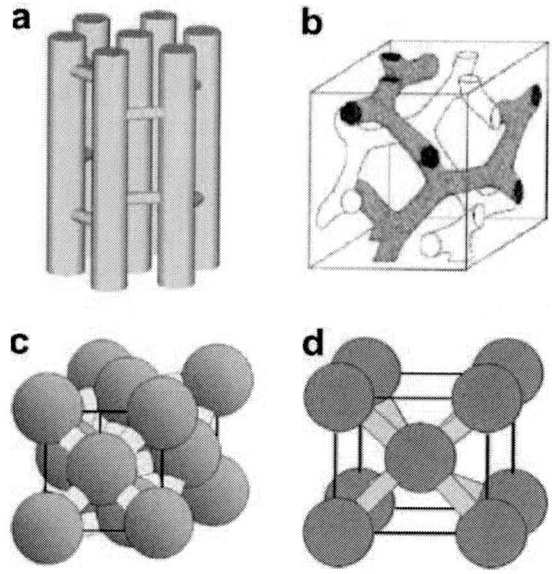

Figure 9. Predicted morphologies of porous crystals of oxides templated by different mesoporous silicas (after Yue et al. [67]): (a) SBA-15, (b) KIT-6, (c) FDU-12 and (d) SBA-16.

carbon monolith. Carbon with such a structure can then be reused for a second templating step. In Fig. 11(a), a porous alumina template with hexagonally ordered pores is shown. By nanocasting polymers into the pores, alumina/carbon nanotube (CNT) composites can be made as shown in Fig. 11(b). Even though the nanocasting conditions were carefully controlled, there is still free space within the internal pores. It must be noted that the top

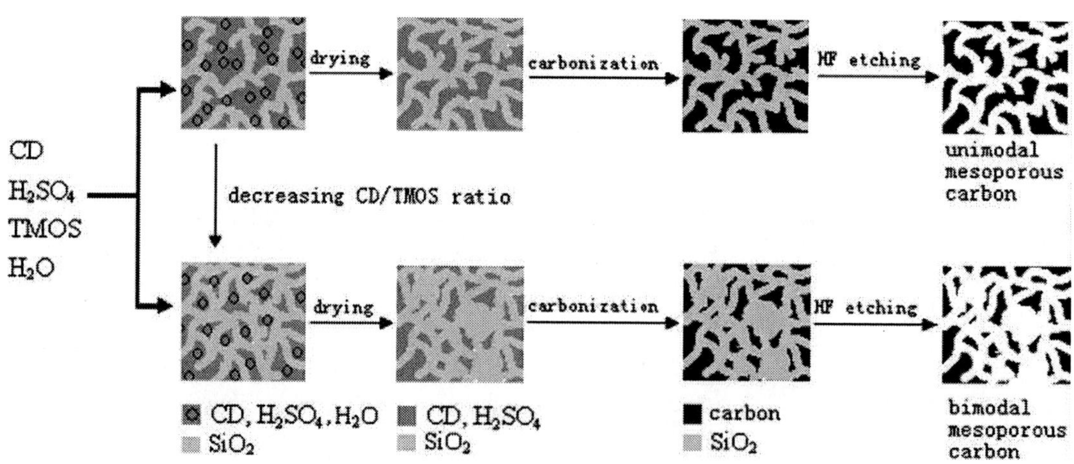

Figure 10. Schematic representation of the formation of unimodal and bimodal mesoporous carbons by changing the methylated β-cyclodextrin (CD) and tetramethyl orthosilicate (TMOS) (after Wu et al. [72]).

and bottom carbon layers are deposited onto the alumina frontal surfaces perpendicular to the tube axes, which forms the monolithic carbon structure that consists of parallel, aligned CNTs with top and bottom carbon layers as illustrated in Fig. 11(c). Popp et al. [88] suggested that prior to removal of the alumina using hydrofluoric acid, the inner CNT space should be protected by an acrylate polymer. After the second templating step and thermal removal of the acrylate layer, CNT reinforced material took the new composite structure as shown in Fig. 11(d). This structure mimics the original alumina template of Fig. 11(a).

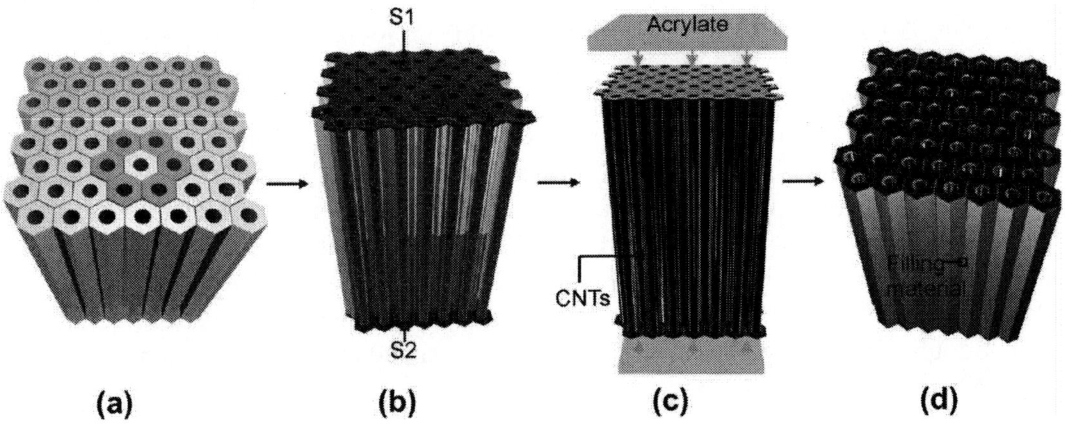

Figure 11. Schematic diagram of the formation of fiber or tubular carbon templates (after Popp et al. [88]).

Carbon micro-tubes with the diameter at micron level can be prepared by nanocasting as shown by Pérez-Cabero et al. [89]. In the review article [90], functionalization of mesoporous structured materials including porous carbon was introduced. The incorporation of NiO nanocrystallites with different content within the wall of ordered porous carbon (CMK-3) was carried out via the synthesis procedure which is considered combining the "nanocasting" strategy with the thermolysis method as described by Li et al. [91]. Recently hollow spherical carbon with mesoporous shell (HCMSC) was synthesized by replication through nanocasting of solid core/mesoporous shell (SCMS) silica [92]. The solid core/mesoporous shell (SCMS) silica is shown in Fig. 12(a), while Fig. 12(b) reveals the hollow spherical carbon with mesoporous shell. Carbon materials with such a structure are very useful in catalysis due to the high specific area.

2.7. Nanocasting through Zeolites

Zeolites are minerals with regularly aligned porous arrays. The composition of zeolites could change in a great extent and their structure could be very complicated. For example, one type of zeolites has the formula of $(Na_2,Ca)(Al_2Si_4)O_{12}\cdot 8H_2O$. Zeolites are mainly used in heterogeneous catalysis. Recently, they have been considered as nanocasting templates to make various functional materials and structures [93-96]. Srinivasu et al. [93] made the microporous carbon nitride (CN) material with very high surface area and pore volume by nanocasting technique using MCM-22 zeolite as the template through the polymerization reaction between ethylenediamine and carbon tetrachloride. The carbon nitride material with microporous structure may find many potential applications for catalysis, gas storage and lubrication. The material also possesses the size selective adsorption function. Small organic molecules such as amino acids and vitamins could be captured by the porous CN.

Parmentier et al. [94] investigated the effect of zeolite crystal size on the structure and properties of carbon replicas made by nanocasting. Zoelite particles with different sizes (from 100 nm to 1 μm) were used as the nanocasting molds for the synthesis of microporous carbon. The synthesis procedures include high temperature infiltration and calcination of a carbon source precursor and etching of the zeolite templates. Propylene was assumed to be the carbon source and the pyrolysis was carried out at 800°C for 2 h. The carbon replicas formed by dissolution of zeolite with 40% HF. Fig. 13(a) shows the morphology of the starting material, zeolite template. Fig. 13(b) is the SEM image of the carbon deposited template. After etching, the zeolite was removed and the carbon replica remained as shown in Fig. 13(c). In the work performed by Ducrot-Boisgontier et al. [95], different precursors were used to obtain carbon replica on zeolite (EMC-2) mold. Acetylene, furfuryl alcohol and acetonitrile were used as the carbon precursors. The nanocasting consisted of either a single or double infiltration procedure. The pore size ranged from 1.0 to 1.5 nm.

2.8. Nanocasting through Other Molds

Commercially available polystyrene (PS) beads were used as molds. For example, mesoporous $Ce_{1-x}Zr_xO_2$ beads were synthesized in high yields by a simple nanocasting approach using PS beads [97]. Shang et al. [98] used polyethylene glycol (PEG) as the scaffold for nanocasting oxide. A sonochemical approach was used to synthesize $BiVO_4$ photocatalyst with relative high photocatalytic activity. Fig. 14(a) shows the morphology of $BiVO_4$ obtained after ultrasonic irradiation for 30 min with PEG. It shows well sepa-

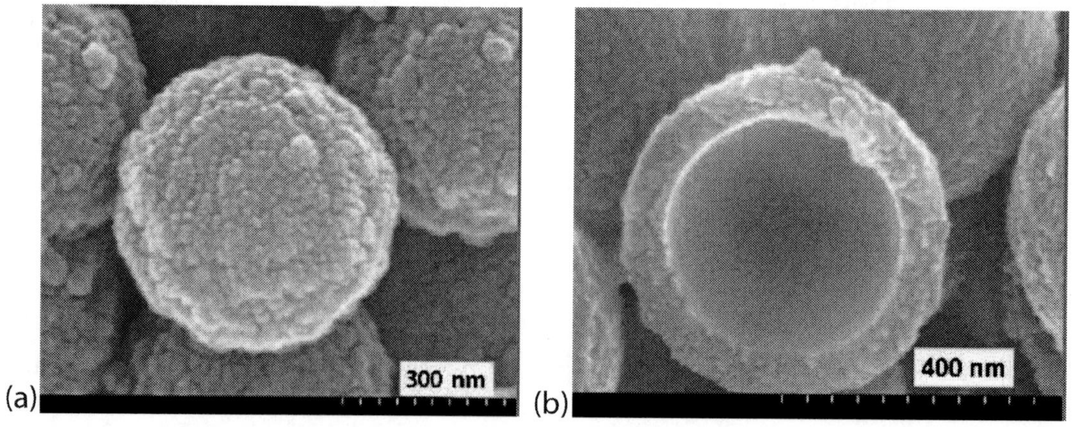

Figure 12. SEM images showing: (a) the solid core/mesoporous shell (SCMS) silica, (b) the nanocast hollow spherical carbon with mesoporous shell (HCMSC) (after Kim et al. [92]).

rated nanoparticles with an average size of about 60 nm. Without ultrasonic treatment, the sample is composed of rod-like blocks with some particles on them. The average size of the blocks is in the micrometer range as shown in Fig. 14(b). Fig. 14(c) shows the image of a control specimen without using PEG as the template or ultrasonic irradiation. It can been seen that the $BiVO_4$ particle size ranges from 100 to 300 nm for the control specimen, which is much larger than the particles shown in Fig. 14(a). Xing et al. [99] used mixed surfactant systems to control the pore structure of mesoporous metal oxides. Glucopyranoside surfactant with a cationic surfactant that readily forms liquid crystalline mesophases was used in their work. A large area of hexagonal phase, as well as cubic, lamellar and solid surfactant phases was found. The ternary phase diagram was used to predict the synthesis of thick mesoporous silica films via a direct liquid crystal templating technique.

Nanocasting may be combined with other micro/nanofabrication techniques, for example, lithography and nanoimprinting to form fine structures [100] or to carry out fine pattern transfer [101]. Sogo et al.[100] proposed using fluorinated polymer and nanocasting lithography to reproduce fine patterns through the procedures shown in Fig. 15(a). The reproduced anti-reflection structures were examined by scanning electron microscope (SEM). A typical SEM image is shown in Fig. 15(b). The fine feature has the size of about 250 nm. Nanocasting lithography (NCL) has the advantage of using conventional contact printing tool. The processing condition is under ambient atmosphere and at room temperature. Thus, it is easy to implement. Hirai et al. [101] used this process to perform fine pattern transfer using various polymers such as polycarbonate (PC), poly methyl methacrylate (PMMA), and poly lactic acid (PLA). The substrates used were quartz PLA and PC plates. High aspect ratio patterns were demonstrated on these substrates as illustrated in Fig. 16.

3. Applications of Nanocasting Materials

Nanocasting materials have found wide applications in chemical engineering and bioengineering fields [102-121]. Catalysis is one of the most important applications [102-108]. For example, mesoporous silica was used as the starting template for deposition of carbon from sucrose and the porous carbon mold was used to nanocast titanium silicalite-1 (TS-1) for catalytic oxidation of aromatic thiophene [102]. Rao et al. [103] nanocast $Cs_xH_{3x}PW_{12}O_{40}$ (CsHPW) salt nanocrystals in silica based nanoporous mold (SBA-15). The CsHPW nanocasts showed enhanced catalytic activity in the isopropanol dehydration

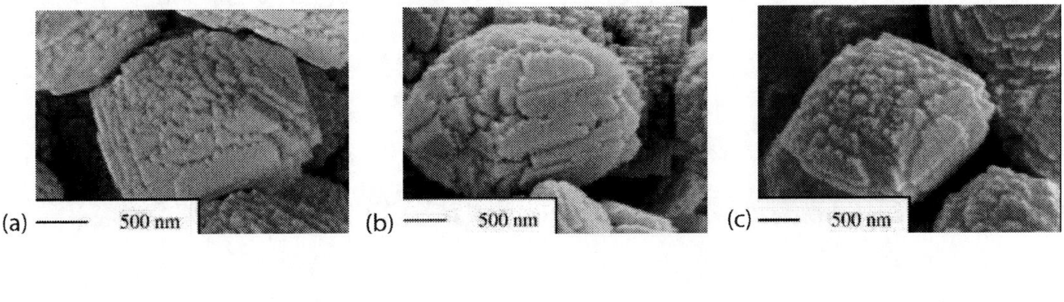

Figure 13. SEM images of: (a) zeolite particle mold, (b) nanocast carbon/zoelite composite materials, (c) carbon replica after zeolite being removed (after Parmentier et al. [94]).

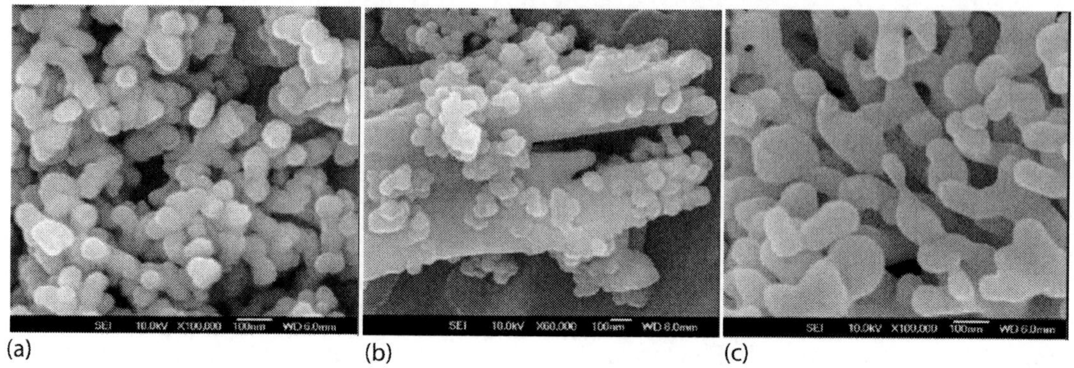

Figure 14. SEM images of: (a) zeolite particle mold, (b) nanocast carbon/zoelite composite materials, (c) carbon replica after zeolite being removed (after Shang et al. [98]).

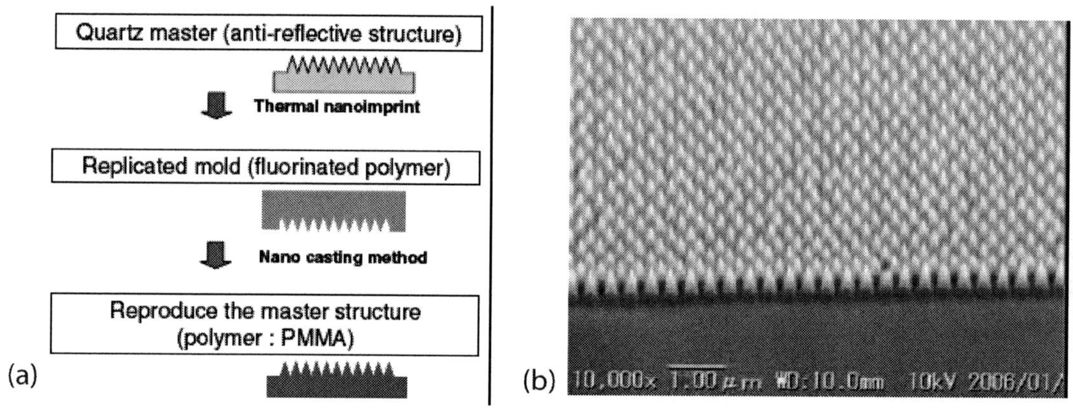

Figure 15. Nanoimprint is combined with nanocasting lithography: (a) procedures for fine pattern reproduction, (b) SEM image of the reproduced anti-reflection structures (after Sogo et al. [100]).

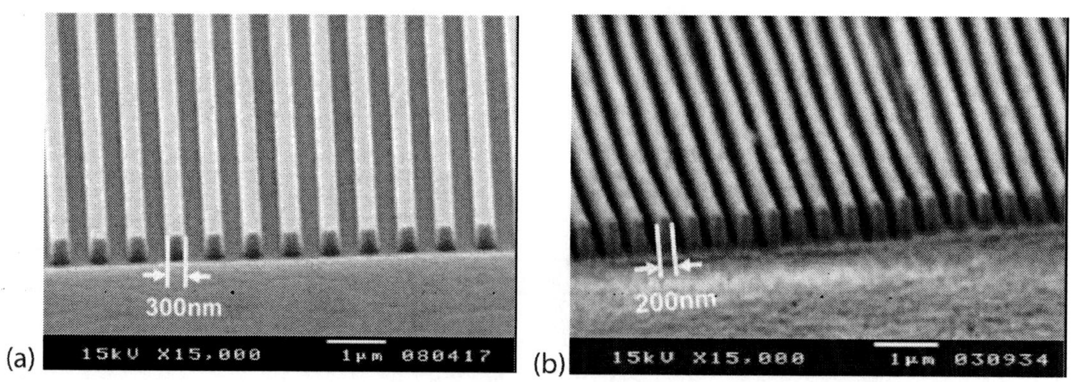

Figure 16. Polymer patterns generated by nanocasting lithography: (a) polycarbonate (PC) on PC plate, (b) poly lactic acid (PLA) on PLA plate (after Hirai et al. [101]).

reactions.

Transition metal ferrites $M_xFe_{3-x}O_4$ (M = Cu, Mn, Co and Ni) were studied as Fenton catalysts for removing organic pollutants from wastewaters through wet oxidation by hydrogen peroxide. Nanocasting was used to synthesize $MnFe_2O_4$ nanoparticles in silica gel porous template [104]. The diameter of the magnetic nanoparticles, is less than 10 nm. The surface area is around 100 m^2/g. It shows a much higher catalytic activity than previously reported heterogeneous catalysts. Nanocasting of ordered mesoporous CeO_2 was performed by Zhang et al. [105]. The ceria was obtained by decomposing $Ce(NO_3)_3$ in the porous carbon mold of CMK-3. The obtained ceria has a 2-D hexagonal ordered structure. The surface area is as high as 167 m^2/g. Its pore size is in the range of 35 nm. The ceria oxides was tested as the support for highly active gold catalyst in CO oxidation and an enhanced catalytic activity was observed. The increased activity is believed due to the nano-sized gold particles and narrow pore size of the ordered mesoporous ceria.

Zhu and Gao [106] made mesoporous spinels of MCo_2O_4 (M = Cu, Mn and Ni) with crystalline walls using concentrated solutions of metal nitrates as precursors and mesoporous silica SBA-15 as the hard template. They found that both $CuCo_2O_4$ and MCo_2O_4 have high catalytic activities and robust stabilities in CO oxidation. Nanocasting was used to improve the order and the density of sites of self-assembled nonionic alkyl glycoside surfactants [107]. Compared with co-precipitated bulk MnCe-oxide catalysts, nanocast oxide-phases of Mn_2O_3 and CeO_2 have reduced and equalized crystal size (at the level of 23 nm) [108]. The full availability of their surface for reacting molecules can be achieved.

Another important application of nanocasting is for fuel cell electrode preparation. Joo et al. [109] prepared ordered mesoporous carbons with controlled particle sizes as catalyst supports for direct methanol fuel cell cathodes. Ordered particles with the sizes of 100, 300, and 700 nm formed the mesoporous silica (OMS) templates. Nanoporous carbon with the uniform mesopores of 45 nm were prepared using the templates. Pt particles with the size of 3 nm were supported by the porous carbon. Ordered and disordered arrays of carbon nanofibers were made by nanocasting for applications such as the highly efficient anode electrode materials for direct methanol fuel cell [110]. Fabrication of highly ordered and disordered arrays of carbon nanofibers (CNFs) may be achieved by replication through nanocasting of SBA-15 silica templates with pretreatment procedures. The structure of the CNFs prepared is special in that they are not strictly tubular, but have filled cores as illustrated in Fig. 17. Fig. 18(a) is the transmission electron microscope (TEM) image of the ordered array of carbon nanofibers. Fig. 18(b) shows the TEM image of the disordered array of carbon nanofibers. The CNFs were found to be excellent catalyst supports. Pt50Ru50 catalyst was supported on them to prepare anode electrode materials. As shown in Fig. 18(c), Pt-Ru nanoparticles are dispersed uniformly on ordered carbon nanofibers after catalyst loading. On disordered carbon nanotubes, the Pt-Ru nanoparticles distributed sporadically, but no significant segregation or clusters of the catalyst were found in Fig. 18(d). The diameter of the PtRu particle measured directly from the TEM micrographs at randomly selected regions was in the range of 2.0-4.5 nm.

As compared with the commercially available E-TEK anode, the nanocast CNF anodes loaded with Pt50-Ru50 showed significant increase in the performance of methanol oxidation in DMFC. This was confirmed by the test results shown in Figs. 19(a) and (b) [110]. Fig. 19(a) presents the polarization curves of three electrodes at 30°C. The three electrodes include the E-TEK anode, the disordered array of carbon nanofiber anode loaded with Pt50-Ru50 (PtRu-DCNFs) and the ordered array of carbon nanofiber anode loaded with Pt50-Ru50 (PtRu-OCNFs). It can be seen that the PtRu-DCNFs anode has the highest power density. PtRu-OCNFs anode ranks the second, and E-TEK anode ranks the third. At higher temperature of 70°C, the same trend was found as shown in Fig. 19(b).

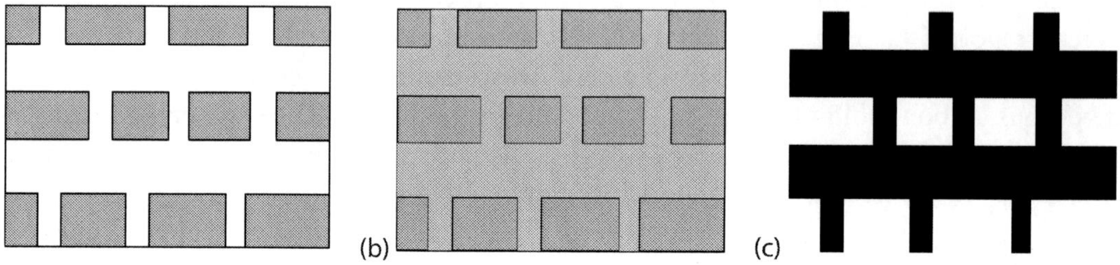

Figure 17. Schematic showing the synthesis procedures for nanocasting carbon nanofibers with cores: (a) starting porous template, (b) impregnating polymer and surfactant in the pore, (c) product of a carbon nanofiber with cores after calcination and carbonization.

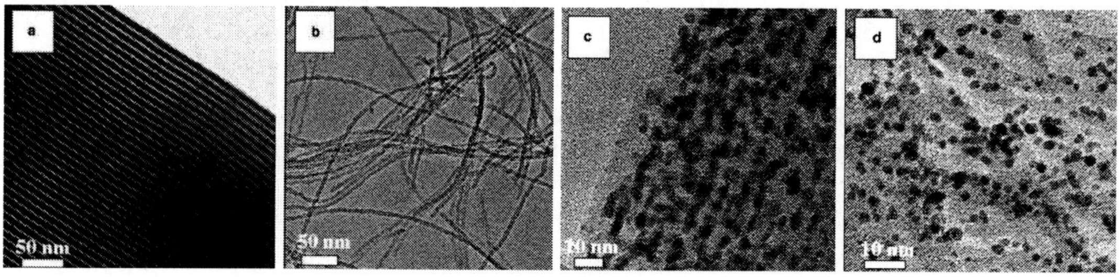

Figure 18. TEM images showing: (a) nanocast ordered carbon nanofibers, (b) nanocast disordered carbon nanofibers, (c) nanocast ordered carbon nanofibers loaded with Pt50-Ru50 nanoparticles, (d) nanocast disordered carbon nanofibers loaded with Pt50-Ru50 nanoparticles (after Chai et al. [110]).

4. Trends in Nanocasting Research

The present research on nanocasting is expected to provide new strategies for preparing novel multifunctional materials with promising applications. Exploring the new applications of nanocast materials are underway. For example, nanocast metal oxide spheres have been considered for macromolecular separation [111]. Highly-ordered nanocast porous thin films were studied for electrochemical analysis and/or electrochemical sensing [112]. Nanocast tin dioxide microspheres are considered for biological signaling because of their phosphopeptide enrichment [113]. New templates or molds have been used for nanocasting, e.g. block co-polymers or multi-block polymers [114] and ionic liquids [115]. The formation mechanisms of mesoporous alumina film on the co-polymer F-127 ($EO_{106}PO_{70}EO_{106}$,

Mav = 12600, BASF) with PPO and PEO blocks were studied by Wan et al. [116]. Fig. 20 is the schematic illustration of the formation mechanisms. First the copolymer was dissolved into EtOH solution, Then $NH_3 \cdot OH$ was added. The inorganic precursors ($AlCl_3 \cdot 6H_2O$ and/or $Al(NO_3)_3 \cdot 9H_2O$) were added into the solution mixture at 70°C. After aging at room temperature for several hours, nanocasting on silicon or glass substrate was performed. The as-prepared samples were calcined at different temperatures (500-900°C) for 4 h in air. It was found that the addition of $NH_3 \cdot OH$ caused the product to have formed as the bulk sediment (Fig. 20, route 1). The explanation for the formation of various structures as shown in Fig. 20 is as follows: First, the organic surfactant template was separated from the condensation of the inorganic species. Then, the volatile solvents evaporated gradually. Finally the micelles assembled into different ordered structures.

Superhydrophobic functionalized $Cu(OH)_2$ nanotube arrays were cast on copper foil by Chen et al. [117]. The wettability of the surface was changed from surperhydrophilicity to superhydrophobicity by chemical modification with 1H,1H,2H,2H–perfluorodecyltriethoxysilane (FAS). They believed that the high roughness of the nanotube arrays along with the generated C-F chains by chemical modification contributed to the enhanced superhydrophobicity. Hard templates, soft templates and reactive templates were used to cast inorganic hollow structures [118]. The formation mechanisms of the hollow structures via different kinds of templates were introduced. The promising and broad applications of hollow materials for controlled drug delivery, catalysis, sensors, lightweight fillers, low-dielectric-constant thin films, photonic crystals, confined-space chemical reactors, biomedical diagnosis and therapy were identified. Calvillo et al. [119] investigated controlling the textural and structural properties of SBA-15 silicas by changing the mass ratio of their precursors. This allows nanocasting different carbon materials with controlled properties. Li et al. [120] also studied texture control of thin films. Liquid crystalline templating mixtures were used in the electrodeposition of PbTe thin films. The hexagonal liquid crystal phase of $C_{16}EO_8$, a non-ionic polyoxyethylene surfactant was used as the template for the formation of the nanostructure. The films have a hexagonal nanostructure and high crystalline rock salt structure. The direct band gap of the films is 0.37 eV which

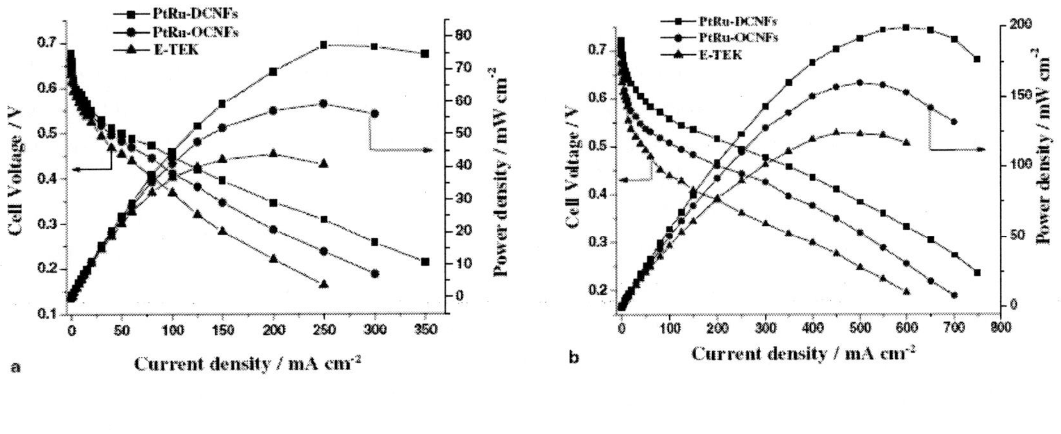

Figure 19. Polarization curves of different fuel cell anodes: (a) at 30°C, (b) at 70°C (after Chai et al. [110]).

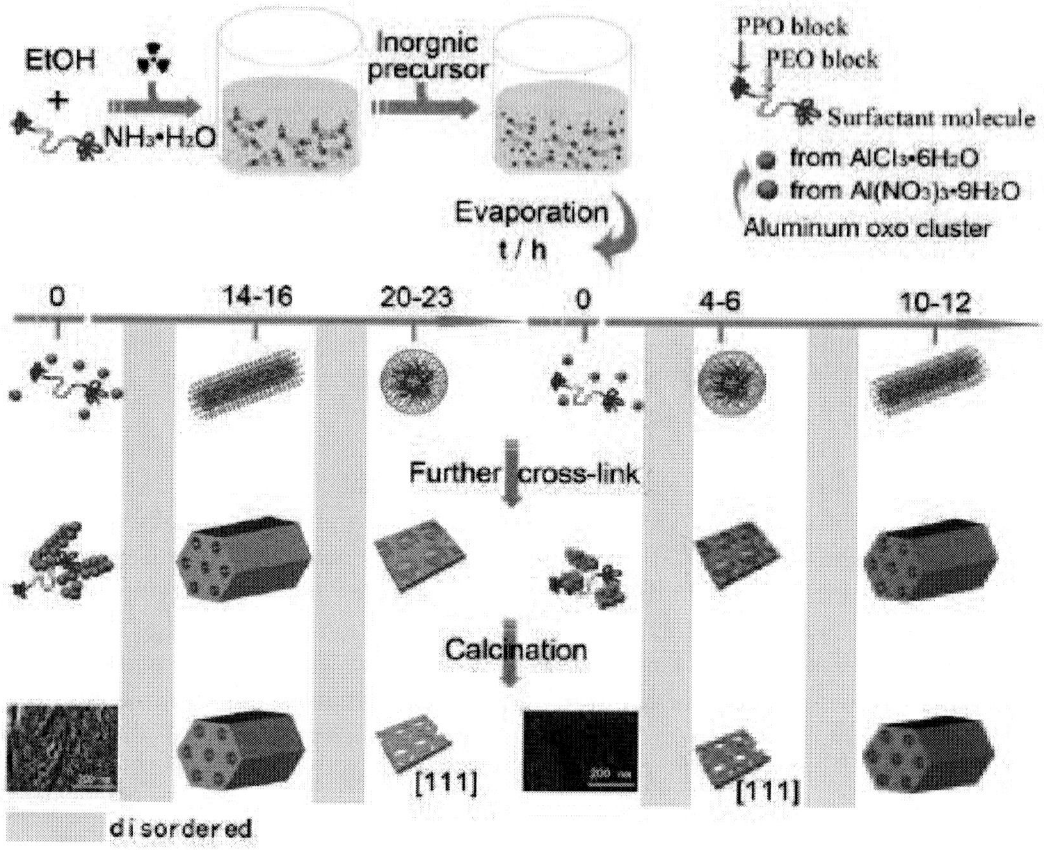

Figure 20. Schematic showing the formation mechanisms of ordered nanocrystalline mesoporous alumina films via nanocasting on the multi-block polymer (after Wan et al. [116]).

is blue shifted in comparison to bulk films, 0.32 eV. The optical measurement results reveal that the nanostructured PbTe films are highly birefringent and of good quality.

Recently, research work on nanocasting biophotoelectrochemical fuel cell has been initiated in the Energy Conversion Research Laboratory at The University of Toledo. The objective of the project is to fabricate a novel biophotofuel cell via nanocasting and develop fundamental understanding of bio- photo- and electrochemical processes related to sustainable energy generation in such a fuel cell. Although photobiological process for hydrogen production has been studied for a long time [121-127], there are many problems remaining to be solved. For example, the efficiency of energy conversion is very low because the performance of the electrodes is not good enough.

Studies have shown that both the quantum efficiency and the hydrogen generation rate of biophotofuel cells can be increased by using new materials such as semiconducting oxides [128-131]. The advantage of using nanomaterials [132, 133] especially nanoparticles has been found because nanoparticles have much higher surface areas than bulk materials. Nevertheless, the agglomeration of particles is a challenging problem to be solved in order to keep the high surface area of the electrode.

We have proposed two ways to potentially solve the agglomeration problem based on the studies of nanoscale thermoelectric energy conversion materials [134, 135]. The first way is to let nanoscale phases grow out-of-the-plane along some preferential directions to form nanoscale fractals or dendrites, which prevents the agglomeration of nanoparticles on substrates. The other method is to use nanoporous substrates or templates because they can confine the growth of nanocrystals either inside the wall of the nanopores or on the surface of the substrates to prevent the agglomeration of nanoparticles. In this work, anodic aluminum oxide nanoporous membrane was used as the nanocasting template or mold to make biophotofuel cell anode with high surface area. Fig. 21(a) shows the morphology of the anodic aluminum oxide (AAO) nanoporous template. Fig. 21(b) shows the nanocast materials on the template.

(a) (b)

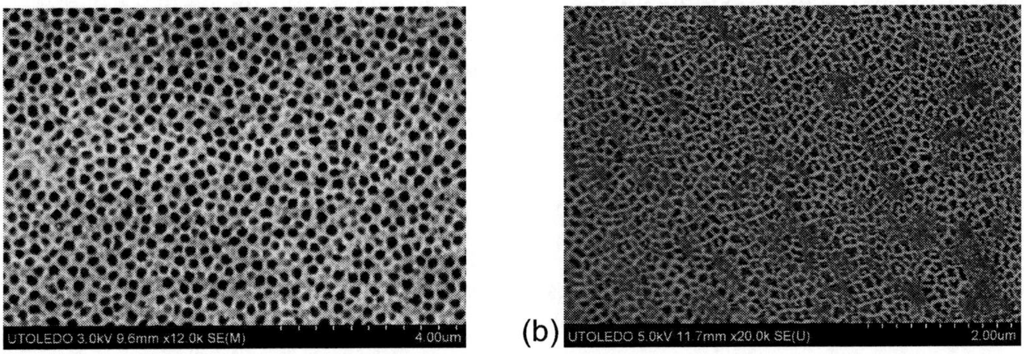

Figure 21. Anode structure: (a) porous template, (b) nanocast materials on the template.

The biophotofuel cell, as schematically shown in Fig. 22, consists of a nanoporous electrode as the anode for biomass decomposition, and a low-hydrogen overpotential metal such as Ni or Pt, as the cathode for hydrogen production. The uniqueness of the fuel cell is its multiple functionalities. It can generate electricity and produce hydrogen from biodegradable fluids under the sun. Meanwhile, noxious gases such as ammonia released from biowastes can be decomposed at the biophotosensitive anode and pure water can be generated for reuse at the cathode.

To prepare such an anode which is able to generate electron under irradiation, and decompose biomass, photosensitive semiconducting substances (e.g. a metal sulphite and a metal oxide) were cast into anodic aluminum oxide templates with regularly aligned nanopores. The generated nanoporous array structures with high surface areas were transferred onto a transparent substrate such as indium tin oxide coated glass to form the anode. The metal thin film was then sputter coated on the top of the electrode in order to establish the electron collecting path.

To prepare a high efficiency cathode, chemical or electrochemical etching was performed to selectively extract metallic elements from alloys to form porous structures. Fig. 23(a) is the morphology of the nickel based alloy cathode in net-like structure. Fig. 23(b) is a higher magnification SEM image of the nickel alloy. It shows the morphology of the dealloyed structure of a single filament of the material as shown in Fig. 23(a). Electrocatalytic property of the porous electrode was characterized. Comparative studies on the electrochemical activity of the nanoporous metallic cathode with the bulk metal cathode were performed.

The work on biophotofuel cell research as described here is preliminary, and continued efforts will be made to refine the nanocasting technology for making the anode. Systematic studies on biophotoelectrochemical behavior of the fuel cell will be carried out. It is expected that through both experimental and theoretical studies, how nanoporous structures with high electroactive surface areas are obtained through controlled electrochemical processing will be unveiled. It will be clarified that at what extent the photoelectron oxidation current density of the nanoporous material increases as compared with that of the bulk electrode with similar chemical compositions. Whether fermentation enhances the photoelectron chemical oxidation of biodegradable fluids at the nanoporous electrode or not will also be revealed. All these fundamental studies will help us to evaluate the performances of the biophotofuel cell in view of electricity generation from biomass under solar rays, hydrogen production, noxious gas decomposition, and clean water regeneration.

Another research work in the Energy Conversion Research Laboratory at The University of Toledo is related to the nanocasting of cobalt sulphide on polymer and ceramic templates. Cobalt sulphide has recently attracted great interest because of its promising applications in catalysis, cathode material of lithium battery, H_2S gas removal, photothermal and photovoltaic energy conversions [136-156]. The cobalt sulphide cast networks prepared in this work consist of nanoparticles with very high surface areas, which are expected to have novel optical and electrical properties.

The nanocasting was performed using a U-shaped reactor as shown in Fig. 24. In Fig. 24, the left-hand side arm of the U-tube was filled with solution *A* (cobalt nitrate solu-

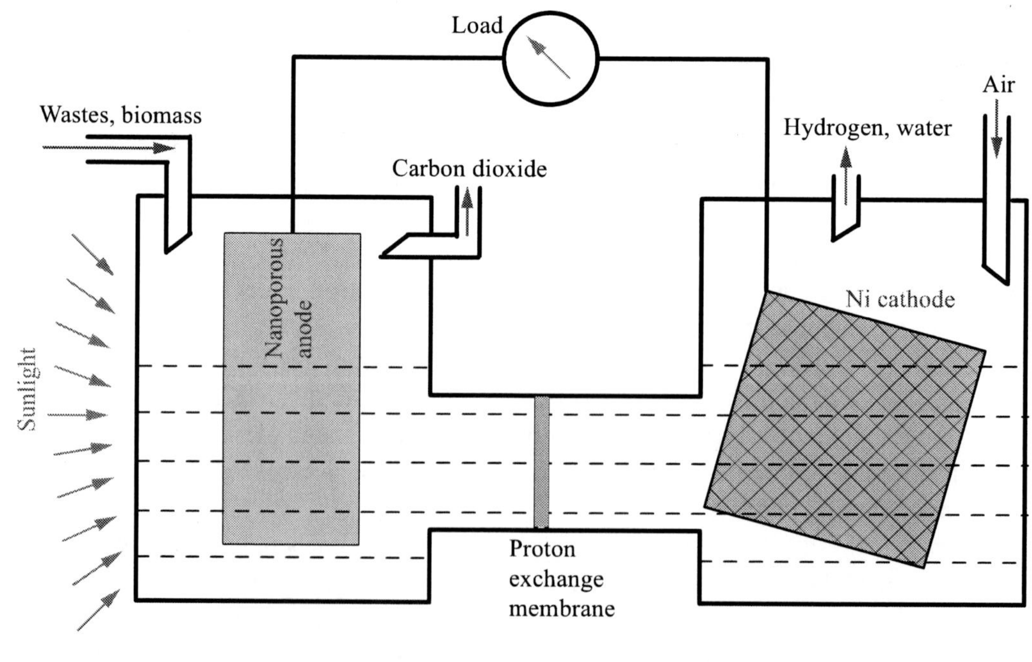

Figure 22. Illustration of the biophotofuel cell.

tion, 0.01 M $Co(NO_3)_2 \cdot 6H_2O$). The right-hand side arm of the tube was filled with solution *B* (sodium sulphide solution, 0.01 M $Na_2S \cdot 9H_2O$). Three different membranes were used in this work. They are regenerated cellulose membrane with the pore size of 40 to 45 μm, nanoporous anodic aluminum oxide membrane with the pore size of 200 nm, and track-etched polycarbonate membrane with the pore size of 800 nm. In order to make sure that the surface of the membrane is the preferred site for casting CoS, a thin layer of gold with the thickness of about 0.4 nm was deposited on these membranes through sputtering. The product from the reaction, CoS nanoparticles, was cast on the surface of the membranes.

Figs. 25(a) and (b) show the SEM images of CoS on the track-etched porous polycarbonate membrane. The morphology of the cast products is featured by microscaled clusters as can be seen in Fig. 25(a). Elongated CoS spheres in different sizes are found in Fig. 25(b). The surface of these spheres is nanoporous and the wall of the nanopores is assembled by many nanoflakes. The average thickness of the nanoflakes is about 25 nm.

In Figs. 26(a) and (b), the SEM images of CoS on the regenerated porous cellulose membrane are presented. The cast CoS products are agglomerates of particles. The size of the particles ranges from 300 nm to 1 μm. These particles consist of nanorods with an average diameter of 30 nm. The aspect ratio of the nanorods is about 5.

The SEM images of CoS on the nanoporous alumina membrane are shown in Figs. 27(a)

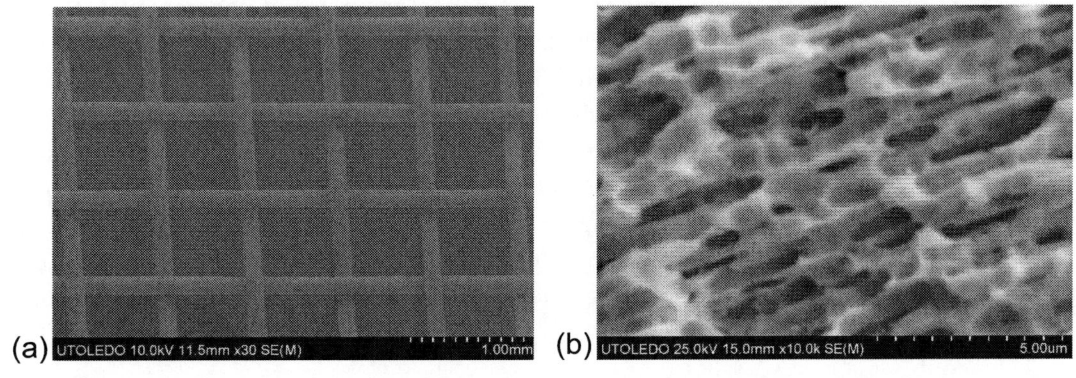

Figure 23. Cathode structure: (a) nickel-based alloy net, (b) porous feature.

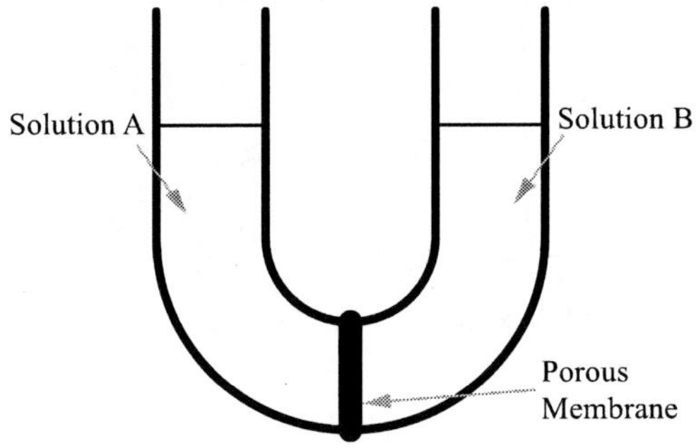

Figure 24. Illustration of the reactor for nanocasting CoS on porous membranes.

and (b). At lower magnification, some clusters of CoS can be seen. CoS ribbons are uniformly distributed on the surface of the aluminum oxide membrane as shown in Fig. 27(a). At higher magnification, the nanoribbons formed the honeycomb structure as shown in Fig. 27(b). Such a feature is the replica of the nanopore within the AAO membrane.

The formation mechanisms of CoS on these porous membranes may be explained by the sketches shown in Figs. 28(a) and (b). Fig. 28(a) shows the formation mechanism of nanocast CoS on porous membrane when S^{2-} diffusion is the dominating mass transfer mechanism. S^{2-} diffused through the pore of the membrane and reacted with Co^{2+} from solution *A*. The heterogeneous nucleation of CoS resulted in the initiation of deposition of the CoS nanostructure at the junction of the pore wall and the surface of the membrane. Nanoribbons or nanofibers are the major products at the initial stage of nanocasting of CoS. With longer deposition times, the nanofibers and ribbons aggregated to form CoS clusters. Similarly, if Co^{2+} migration played a major role in mass transfer, the nanofibers and ribbons initiated on the right-hand side corners of the pore within the membrane as shown in Fig. 28(b).

5. Summary and Conclusions

Nanocasting technology has attracted much attention for its applicability for making a wide range of inorganic, organic and polymeric nanostructures and nanomaterials. The molds consist of porous silica, porous carbon, oxide template, polymer template, organic scaffold, and even plant leaves. In addition, nanocasting process is fast and cost effective, which makes it especially suitable for high rate and large scale production. Nanocasting could be implemented through solgel deposition, electroplating, hydrothermal synthesis, solvent-template-reactant all-in-one synthesis, microwave heating, and emulsions or microemulsions synthesis. Also, nanocasting may be conducted in combination with other micro/nanofabrication technologies such as lithography, soft imprinting etc. Nanocasting has a wide range of applications including catalysis, adsorption, photochemical, photonic nanomaterials and structures.

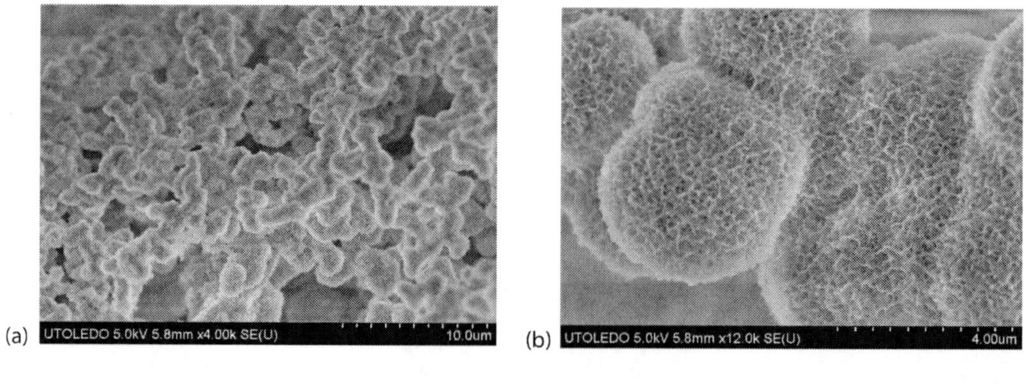

Figure 25. SEM images showing nanocast CoS on track-etched porous polycarbonate membrane: (a) microscaled CoS clusters, (b) nanoporous and nanoflake surface morphological features of CoS crystal networks at a higher magnification.

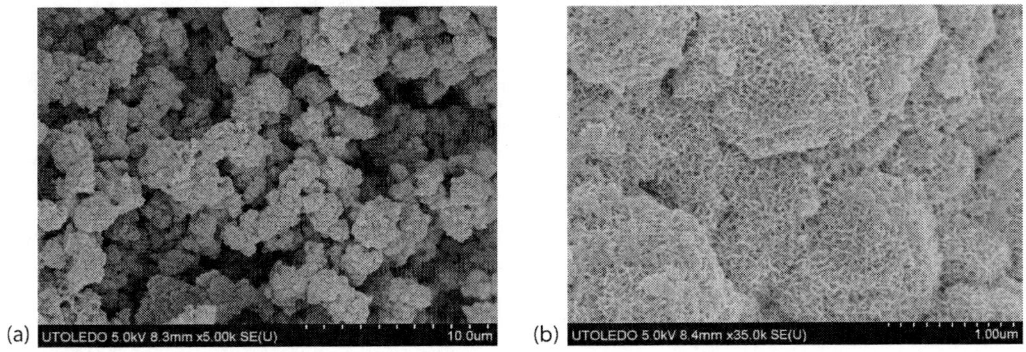

Figure 26. SEM images showing nanocast CoS on regenerated porous cellulose membrane: (a) microscaled CoS clusters, (b) nanoporous and nanorod surface morphological features of CoS crystal networks at a higher magnification.

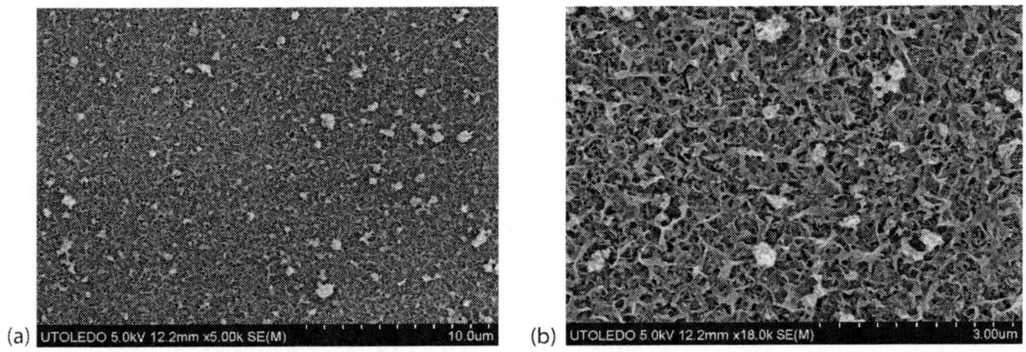

Figure 27. SEM images showing nanocast CoS on porous alumina membrane: (a) at lower magnification, (b) nanoporous and nanoribbon surface morphological features of CoS crystal networks at a higher magnification.

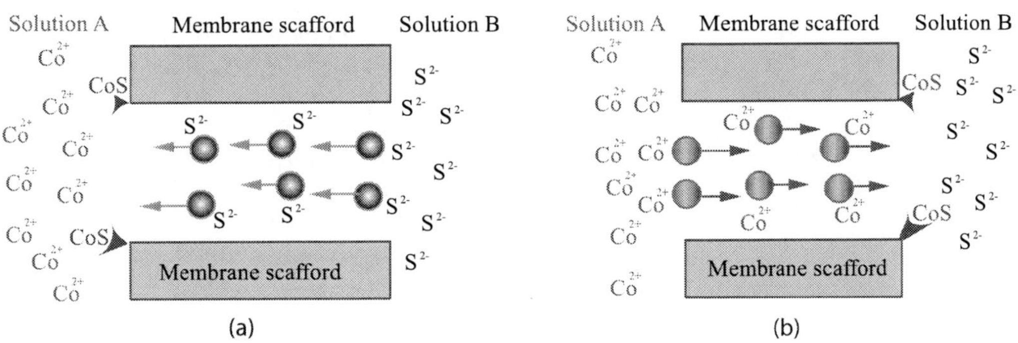

Figure 28. Schematic showing the formation mechanisms of nanocast CoS on porous membrane: (a) S^{2-} transfer dominated mechanism, (b) Co^{2+} transfer dominated mechanism.

Acknowledgments

The research start-up fund from University of Toledo and Summer Faculty Research Fellowships provided by the Office of Research and Development, University of Toledo are gratefully acknowledged. We also acknowledge the permissions from Elsevier B.V. Ltd through Copyright Clearance Center's Rightslink service for using Figures 2–4, 7–16, 18–20.

Contributions from Each Author

YXG planned the work, supervised the research, did nanocasting experiments and wrote the paper. DKI and LY took part in nanocasting of CoS. BJG did biophotofuel cell experiments, researched and organized literatures. JGL did scanning electron microscopy analysis of porous templates and the nanocast CoS material.

References

[1] Shackelford, J. F. *Introduction to Materials Science for Engineers*, 2005, Pearson Prentice Hall, Upper Saddle River, New Jersey, p.423.

[2] Callister, W. D. *Materials Science and Engineering: An Introduction*, 2003, John Wiley & Sons, New York, p.355.

[3] Gan, Y. X.; Overfelt, R. A. *J. Mater. Sci.* 2006, 43, 7537-7544.

[4] Attard, G. S.; Glyde, J. C.; Göltner, C. G. *Nature* 1995, 378, 366-368.

[5] Göltner, C. G.; Antonietti, M. *Adv. Mater.* 1997, 9, 431-436.

[6] Göltner, C.G.; Henke, S.; Weiβenberger, M. C.; Antonietti, M. *Angew. Chem.* 1998, 110, 633-636.

[7] Kramer, E.; Förster, S.; Göltner, C.; Antonietti, M. *Langmuir* 1998, 14, 2027-2031.

[8] Göltner, C.G.; Berton, B.; Krämer, E.; Antonietti, M. *Adv. Mater.* 1999, 11, 395-398.

[9] Han, B. H.; Smarsly, B.; Gruber, C.; Wenz, G. *Microporous Mesoporous Mater.* 2003, 66, 127-132.

[10] Xia, X. H.; Tu, J. P.; Zhang, J.; Xiang, J. Y.; Wang, X. L.; Zhao, X. B. *ACS Appl. Mater. Interfaces* 2010, 2, 186192.

[11] Cruz, H. S.; Spino, J.; Grathwohl, G. *J. European Ceram. Soc.* 2008, 28, 1783-1791.

[12] Chai, F.; Li, D.; Wu, H.; Zhang, C.; Wang, X. *J. Solid State Chem.* 2009, 182, 1661-1665.

[13] Jiang, P.; Bertone, J. F.; Hwang, K. S.; Colvin, V. L. *Chem. Mater.* 1999, 11, 21322140.

[14] Wang, Z.; Li, F.; Ergang, N. S.; Stein, A. *Chem. Mater.* 2006, 18, 5543-5553.

[15] Wang, Z.; Li, F.; Ergang, N. S.; Stein, A. *Carbon* 2008, 46, 1702–1710.

[16] Zhang, F.; Wong, S. S. *Chem. Mater.* 2009, 21, 4541-4554.

[17] Kumar, A.; Whitesides, G. M. *Appl. Phys. Lett.* 1993, 63, 2002- 2004.

[18] Wang, X.; Han, G. R. *Microelectronic Eng.*, 2003, 66, 166–170.

[19] Yu, W. H.; Fei, G. T.; Chen, X. M.; Xue, F. H.; Xu, X. J. *Phys. Letts. A*, 2006, 350, 392–395.

[20] Hu, H.; He, D. *Mater. Letts.*, 2006, 60, 1019–1022.

[21] Kim, K. T.; Cho, S. M. *Mater. Letts.*, 2006, 60, 352–355.

[22] Pellin, M. J.; Stair, P. C.; Xiong, G.; Elam, J. W.; Birell, J.; Curtiss, L.; George, S. M.; Han, C. Y.; Iton, L.; Kung, H.; Kung, M.; Wang, H. H. *Catalysis Letts.*, 2005, 102, 127–130.

[23] Li, N.; Li, X.; Yin, X.; Wang, W.; Qiu, S. *Solid State Comm.*, 2004, 132, 841–844.

[24] Zhao, Y.; Chen, M.; Zhang, Y.; Xu, T.; Liu, W. *Mater. Letts.*, 2005, 59, 40–43.

[25] Hwang, S. K.; Jeong, S. W.; Lee, O. J.; Lee, K. H. *Microelectronic Eng.*, 2005, 77, 2–7.

[26] Wang, X.; Wang, X.; Huang, W.; Sebastian, P. J.; Gamboa, S. *J. Power Sources*, 2005, **140**, 211–215.

[27] Kong, L. B. *Solid State Comm.*, 2005, 133, 527–529.

[28] Chen, P. L.; Chang, J. K.; Pan, F. M.; Kuo, C. T. *Diamond Related Mater.*, 2005, 14, 804–809.

[29] Lee, O. J.; Hwang, S. K.; Jeong, S. H.; Lee, P. S.; Lee, K. H. *Synthetic Metals*, 2005, 148, 263–266.

[30] Sun, X. Y.; Xu, F. Q.; Li, Z. M.; Zhang, W. H. *Mater. Chem. Phys.*, 2005, 90, 69–72.

[31] Yang, Z.; Huang, Y.; Dong, B.; Li, H. L. *J. Solid State Chem.*, 2005, 178, 1157–1164.

[32] Singh, R. A.; Yoon, E. S.; Kim, H. J.; Kim, J.; Jeong, H. E.; Suh, K. Y. *Mater. Sci. Eng. C* 2007, 27, 875-879.

[33] Yuan, Z.; Chen, H.; Zhang, J. *Appl. Surf. Sci.* 2008, 254, 1593-1598.

[34] Lu, A. H.; Schüth, F. *C. R. Chimie* 2005, 8, 609-620.

[35] Puertolas, B.; Solsona, B.; Agouram, S.; Murillo, R.; Mastral, A. M.; Aranda, A.; Taylor, S. H.; Garcia, T. *Applied Catalysis B: Environmental* 2010, 93, 395-405.

[36] Wu, Z.; Meng, Y.; Zhao, D. *Microporous Mesoporous Mater.* 2010, 128, 165-179.

[37] Lepoutre, S.; Smätt, J. H.; Laberty, C.; Amenitsch, H.; Grosso, D.; Lindén, M. *Microporous Mesoporous Mater.* 2009, 123, 185-192.

[38] Escax, V.; Impéror-Clerc, M.; Bazin, D.; Davidson, A. *C. R. Chimie* 2005, 8, 663-677.

[39] Valdés-Solís, T.; Fuertes, A. B. *Mater. Res. Bulltin* 2006, 41, 2187-2197.

[40] Vantomme, A.; Surahy, L.; Su, B. L. *Colloids Surf. A: Physicochem. Eng. Aspects* 2007, 300, 65-69.

[41] Rigby, S. P.; Beanlands, K.; Evbuomwan, I. O.; Watt-Smith, M. J.; Edler, K. J.; Fletcher, R. S. *Chem. Eng. Sci.* 2004, 59, 5113-5120.

[42] Parmentier, J.; Saadhallah, S.; Reda, M.; Gibot, P.; Roux, M.; Vidal, L.; Vix-Guterl, C.; Patarin, J. *J. Phys. Chem. Solids* 2004, 65, 139-146.

[43] Roggenbuck, J.; Schäfer, H.; Tsoncheva, T.; Minchev, C.; Hanss, J.; Tiemann, M. *Microporous Mesoporous Mater.s* 2007, 101, 335-341.

[44] Satishkumar, G.; Titelman, L.; Landau, M.V. *J. Solid State Chem.* 2009, 182, 2822-2828.

[45] Xing, R.; Rankin, S. E. *Microporous Mesoporous Mater.* 2008, 108, 65-76.

[46] Yan, J.; Wang, A.; Kim, D. P. *Microporous Mesoporous Mater.* 2007, 100, 128-133.

[47] Lebeau, B.; Parmentier, J.; Soulard, M.; Fowler, C.; Zana, R.; Vix-Guterl, C.; Patarin, J. *C. R. Chimie* 2005, 8, 597-607.

[48] Wang, Y.; Wang, Y,; Ren, J.; Mi, Y.; Zhang, F.; Li, C.; Liu, X.; Guo, Y.; Guo, Y.; Lu, G. *J. Solid State Chem.* 2010, 183, 277-284.

[49] Sampieri, A.; Pronier, S.; Blanchard, J.; Breysse, M.; Brunet, S.; Fajerwerg, K.; Louis, C.; Pérot, G. *Catalysis Today* 2005, 107-108, 537-544.

[50] Fajula, F.; Galarneau, A.; Di Renzo, F. *Microporous Mesoporous Mater.* 2005, 82, 227-239.

[51] Zhu, J.; Gao, Q.; Chen, Z. *Applied Catalysis B: Environmental* 2008, 81, 236-243.

[52] van der Meer, J.; Bardez, I.; Bart, F.; Albouy, P. A.; Wallez, G.; Davidson, A. *Microporous Mesoporous Mater.* 2009, 118, 183-188.

[53] Lopes, I.; Davidson, A.; Thomas, C. *Catalysis Comm.* 2007, 8, 2105-2109.

[54] Roggenbuck, J.; Waitz, T.; Tiemann, M. *Microporous Mesoporous Mater.* 2008, 113, 575-582.

[55] Kim, Y. S.; Guo, X. F.; Kim, G. J. *Catalysis Today* 2010, 150, 91-99.

[56] Xing, W.; Zhuo, S. P.; Gao, X. *Mater. Lett.* 2009, 63, 1177-1179.

[57] Zhu, S.; Zhou, Z.; Zhang, D.; Wang, H. *Microporous Mesoporous Mater.* 2006, 95, 257-264.

[58] Fuertes, A. B.; Sevilla, M.; Alvarez, S.; Valdes-Solís, T. *Microporous Mesoporous Mater.* 2008, 112, 319-326.

[59] Lu, A. H.; Li, W. C.; Schmidt, W.; Schüth, F. *Microporous Mesoporous Mater.* 2005, 80, 117-128.

[60] Abellán, G.; Carrillo, A. I.; Linares, N.; Serrano, E.; García-Martínez, J. *J. Solid State Chem.* 2009, 182, 2141-2148.

[61] Pevida, C.; Snape, C. E.; Drage, T. C. *Energy Procedia* 2009, **1**, 869-874.

[62] Sun, J.; Ma, D.; Zhang, H.; Bao, X.; Weinberg, G.; Su, D. *Microporous Mesoporous Mater.* 2007, 100, 356-360.

[63] Álvarez, S.; Valdés-Solís, T.; Fuertes, A. B. *Mater. Res. Bulletin* 2008, 43, 1898-1904.

[64] Lu, A. H.; Li, W. C.; Schmidt, W.; Kiefer, W.; Schüth, F. *Carbon* 2004, 42, 2939-2948.

[65] Xiang, L.; Royer, S.; Zhang, H.; Tatibouet, J. M.; Barrault, J.; Valange, S. *J. Hazardous Mater.* 2009, 172, 1175-1184.

[66] Göltner-Spickermann, C. *Current Opinion in Colloid and Interface Sci.* 2002, 7, 173-178.

[67] Yue, W.; Zhou, W. *Progress in Natural Sci.* 2008, 18, 1329-1338.

[68] Wagner, T.; Sauerwald, T.; Kohl, C. D.; Waitz, T.; Weidmann, C.; Tiemann, M. *Thin Solid Films* 2009, 517, 6170-6175.

[69] de Lima, R. K. C.; Batista, M. S.; Wallau, M.; Sanches, E. A.; Mascarenhas, Y. P.; Urquieta-González, E. A. *Appl. Catalysis B: Environmental* 2009, 90, 441-450.

[70] Lu, A. H.; Tuysuz, H.; Schüth, F. *Microporous Mesoporous Mater.* 2008, 111, 117-123.

[71] Lai, X.; Wang, H.; Mao, D.; Yang, N.; Yao, J.; Xing, C.; Wang, D.; Li, X. *Mater. Lett.* 2008, 62, 3868-3871.

[72] Wu, D.; Liang, Y.; Yang, X.; Li, Z.; Zou, C.; Zeng, X.; Lu, G.; Fu, R. *Microporous Mesoporous Mater.* 2008, 116, 91-94.

[73] Pacula, A.; Robert Mokaya, R. *Microporous Mesoporous Mater.* 2007, 106, 147-154.

[74] Li, S.; Liang, Y.; Wu, D.; Fu, R. *Carbon* 2010, 48, 839-843.

[75] Bai, P.; Xing, W.; Wu, P.; Liu, X.; Yan, Z.; Zhao, X. S. *Mater. Lett.* 2007, 61, 4231-4234.

[76] Nakamura, T.; Yamada, Y.; Yano, K. *Microporous Mesoporous Mater.* 2009, 117, 478-485.

[77] Xu, L. Y.; Shi, Z. G.; Feng, Y. Q. *Microporous Mesoporous Mater.* 2007, 98, 303-308.

[78] Pérez-Cabero, M.; Esteve-Turrillas, F. A.; Beltrán, D.; Amorós, P. *Solid State Sci.* 2010, 12, 15–25.

[79] Lu, A. H.; Smătt, J. H.; Backlund, S.; Lindén, M. *Microporous Mesoporous Mater.* 2004, 72, 59-65.

[80] Dibandjo, P.; Chassagneux, F.; Bois, L.; Sigala, C.; Miele, P. *Microporous Mesoporous Mater.* 2006, 92, 286-291.

[81] Liu, Q.; Wang, A.; Xu, J.; Zhang, Y.; Wang, X.; Zhang, T. *Microporous Mesoporous Mater.* 2008, 116, 461-468.

[82] Li, H.; Zhu, S.; Xi, H. Wang, R. *Microporous Mesoporous Mater.* 2006, 89, 196-203.

[83] Ania, C. O.; Pernak, J.; Stefaniak, F.; Raymundo-Piñero, E.; Béguin, F. *Carbon* 2009, 47, 3158-3166.

[84] Yang, Y. X.; Bourgeois, L.; Zhao, C.; Zhao, D.; Chaffee, A.; Webley, P. A. *Microporous Mesoporous Mater.* 2009, 119, 39-46.

[85] Zhou, L.; Li, H.; Yu, C.; Zhou, X.; Tang, J.; Meng, Y.; Xia, Y.; Zhao, D. *Carbon* 2006, 44, 1581-1616.

[86] Jaroniec, M.; Gorka, J.; Choma, J.; Zawislak, A. *Carbon* 2009, 47, 3034-3040.

[87] Zhou, J.; He, J.; Zhang, C.; Wang, T.; Sun, D.; Di, Z.; Wang, D. *Mater. Charact.* 2010, 61, 31-38.

[88] Popp, A.; Engstler, J.; Schneider, J. J. *Carbon* 2009, 47, 3208-3214.

[89] Pérez-Cabero, M.; Puchol, V.; Beltrán, D.; Amorós, P. *Carbon* 2008, 46, 297-304.

[90] Athens, G. L.; Shayib, R. M.; Chmelka, B. F. *Current Opinion in Colloid and Interface Sci.* 2009, 14, 281-292.

[91] Li, H.; Xi, H.; Zhu, S.; Wang, R. *Mater. Lett.* 2006, 60, 943-946.

[92] Kim, J. H.; Fang, B.; Kim, M.; Yu, J. S. *Catalysis Today* 2009, 146, 25-30.

[93] Srinivasu, P.; Vinu, A.; Hishita, S.; Sasaki, T.; Ariga, K.; Mori, T. *Microporous Mesoporous Mater.* 2008, 108, 340-344.

[94] Parmentier, J.; Valtchev, V.; Gaslain, F.; Tosheva, L.; Ducrot-Boisgontier, C.; Möller, J.; Patarin, J.; Vix-Guterl, C. *Carbon* 2009, 47, 1066-1073.

[95] Ducrot-Boisgontier, C.; Parmentier, J.; Patarin, J. *Microporous Mesoporous Mater.* 2009, 126, 101-106.

[96] Fang, Y.; Hua, H.; Chen, G. *Microporous Mesoporous Mater.* 2008, 113, 481-489.

[97] Deshpande, A. S.; Niederberger, M. *Microporous Mesoporous Mater.* 2007, 101, 413-418.

[98] Shang, M.; Wang, W.; Zhou, L.; Sun, S.; Yin, W. *J. Hazardous Mater.* 2009, 172, 338-344.

[99] Xing, R.; Rankin, S. E. *J. Colloid and Interface Sci.* 2007, 316, 930-938.

[100] Sogo, K.; Nakajima, M.; Kawata, H.; Hirai, Y. *Microelectronic Eng.* 2007, 84, 909-911.

[101] Hirai, Y.; Yoshikawa, T.; Morimatsu, M.; Nakajima, M.; Kawata, H. *Microelectronic Eng.* 2005, 78-79, 641-646.

[102] Fang, Y.; Hu, H. *Catalysis Communications* 2007, 8, 817-820.

[103] Madhusudhan Raoa, P.; Goldberg-Oppenheimer, P.; Kababya, S.; Vegab, S.; Landau, M.V. *J. Molecular Catalysis A: Chemical* 2007, 275, 214-227.

[104] Valdés-Solÿs, T.; Valle-Vigón, P.; Álvarez, S.; Marbán, G.; Fuertes, A. B. *Catalysis Comm.* 2007, 8, 2037-2042.

[105] Zhang, H.; Yan, X.; Li, W. *Chin. J. Catal.* 2009, 30, 1085-1090.

[106] Zhu, J.; Gao, Q. *Microporous Mesoporous Mater.* 2009, 124, 144-152.

[107] Rahman, M. S. Rankin, S. E. *J. Colloid Interface Sci.* 2010, 342, 33-42.

[108] Abecassis-Wolfovich, M.; Landau, M. V.; Brenner, A.; Herskowitz, M. *J. Catalysis* 2008, 247, 201-213.

[109] Joo, S. H.; Lee, H. I.; You, D. J.; Kwon, K.; Kim, J. H.; Choi, Y. S.; Kang, M.; Kim, J. M.; Pak, C.; Chang, H.; Seung, D. *Carbon* 2008, 46, 2034-2045.

[110] Chai, G. S.; Yoon, S. B.; Yu, J. S. *Carbon* 2005, 43, 3028-3031.

[111] Smătt, J. H.; Schüwer, N.; Järn, M.; Lindner, W.; Lindén, M. *Microporous Mesoporous Mater.* 2008, 112, 308-318.

[112] Walcarius, A,; Kuhn, A. *Trends in Analytical Chem.* 2008, 27, 593-603.

[113] Leitner, A.; Sturma, M.; Smătt, J. H.; Järn, M.; Linden, M.; Mechtler, K.; Lindner. W. *Analytica Chimica Acta* 2009, 638, 51-57.

[114] Soler-Illia, G. J. A. A.; Crepaldi, E. L.; Grosso, D.; Sanchez, C. *Current Opinion in Colloid Interface Sci.* 2003, 8, 109-126.

[115] Li, Z.; Jia, Z.; Luan, Y.; Mu, T. *Current Opinion in Solid State and Mater. Sci.* 2008, 12, 1-8.

[116] Wan, L.; Fu, H.; Shi, K.; Tian, X. *Microporous Mesoporous Mater.* 2008, 115, 301-307.

[117] Chen, X.; Kong, L.; Dong, D.; Yang, G.; Yu, L.; Chen, J.; Zhang, P. *Appl. Surf. Sci.* 2009, 255, 4015-4019.

[118] Ma, Y.; Qi, L. *J. Colloid Interface Sci.* 2009, 335, 1-10.

[119] Calvillo, L.; Celorrio, V.; Moliner, R.; Cabot, P. L.; Esparbé, I.; Lázaro, M. J. *Microporous Mesoporous Mater.* 2008, 116, 292-298.

[120] Li, X.; Nandhakumar, I. S.; Attard, G. S.; Markham, M. L.; Smith, D. C.; Baumberg, J. J. *Microporous Mesoporous Mater.* 2009, 118, 403-407.

[121] Rao, K. K.; Hall, D. O. *Trends in Biotechnology* 1984, 2, 124–129.

[122] Gassanova, L. G.; Netrusov, A. I.; Teplyakov, V. V.; Modigell, M. *Desalination* 2006, 198, 56–66.

[123] Weaver, P. F.; Lien, S.; Seibert, M. *Solar Energy* 1980, 24, 3–45.

[124] Benemann, J. R.; Miyamoto, K.; Hallenbeck, P.C. *Enzyme Microbial Technol.* 1980, 2, 103–111.

[125] Miura, Y. *Process Biochemistry* 1995, 30, 1–7.

[126] Lambert, G. R.; Smith, G. D. *FEBS Lett.* 1977, 83, 159–162.

[127] Modigell, M.; Holle, N. *Renewable Energy* 1998, 14, 421–426.

[128] Kaneko, M.; Ueno, H.; Saito, R.; Yamaguchi, S.; Fujii, Y.; Nemoto, J. *Appl. Catalysis B: Environmental* 2009, 91, 254–261.

[129] Kaneko, M.; Ueno, H.; Saito, R.; Nemoto, J. *Catal. Lett.* 2009, 131, 184–188.

[130] Kanekoa, M.; Uenoa, H.; Saito, R.; Suzuki, S.; Nemoto, J.; Fujii, Y. *J. Photochem. Photobio. A: Chem.* 2009, 205, 168–172.

[131] Antoniadou, M.; Lianos, P. *J. Photochem. Photobio. A: Chem.* 2009, 204, 69–74.

[132] Antoniadou, M.; Bouras, P.; Strataki, N.; Lianos, P. *Int. J. Hydrogen Energy* 2008, 33, 5045–5051.

[133] Ueno, H.; Nemoto, J.; Ohnuki, K.; Horikawa, M.; Hoshino, M.; Kaneko, M. *J. Appl. Electrochem.* 2009, 39, 1897–1905.

[134] Madhavaram, R.; Sander, J.; Gan, Y. X.; Masiulaniec, C. K. *Mater. Chem. Phys.* 2009, 118, 165–173.

[135] Gan, Y. X.; Sweetman, J.; Lawrence, J. G. *Mater. Lett.* 2010, 64, 449–452.

[136] Bao, S. J.; Li, C. M.; Guo, C. X.; Qiao, Y. *J. Power Sources* 2008, 180, 676-681.

[137] de Bont, P. W.; Vissenberg, M. J.; Hensen, E. J. M.; de Beer, V. H. J.; van Veen, J. A. R.; van Santen, R. A.; van der Kraan, A. M. *Appl. Catalysis A: General* 2002, 236, 205-222.

[138] Wang, J.; Ng, S. H.; Wang, G. X.; Chen, J.; Zhao, L.; Chen, Y.; Liu, H. K. *J. Power Sources* 2006, 159, 287-290.

[139] Kadono, T.; Kubota, T.; Hiromitsu, I.; Okamoto, Y. *Appl. Catalysis A: General* 2006, 312, 125-133.

[140] Zhang, Z.; Ma, J.; Yang, X. *J. Molecular Catalysis A: Chemical* 2003, 195, 189-200.

[141] Morgenstern, K.; Lægsgaard, E.; Besenbacher, F. *Surf. Sci.* 2008, 602, 661670.

[142] Sanders, A. F. H.; de Jong, A. M.; de Beer, V. H. J.; van Veen, J. A. R.; Niemantsverdriet, J. W. *Appl. Surf. Sci.* 1999, 144-145, 380-384.

[143] Sohrabnezhad, S.; Pourahmad, A.; Sadjadi, M. S.; Sadeghi, B. *Physica E* 2008, 40, 684-688.

[144] Trikalitis, P. N.; Kerr, T. A.; Kanatzidis, M. G. *Microporous Mesoporous Mater.* 2006, 88, 187-190.

[145] Choi, J. S.; Maugé, F.; Pichon, C.; Olivier-Fourcade, J.; Jumas, J. C.; Petit-Clair, C.; Uzio, D. *Appl. Catalysis A: General* 2004, 267, 203-216.

[146] Okamoto, Y. *Catalysis Today* 1997, 39, 45–59.

[147] Eze, F. C.; Okeke, C. E. *Mater. Chem. Phys.* 1997, 47, 31–36.

[148] Hu, Q. R.; Wang, S. L.; Zhang, Y.; Tang, W. H. *J. Alloys Compounds* 2010, 491, 707-711.

[149] Liu, X. *Mater. Sci. Eng. B* 2005, 119, 19-24.

[150] Lélias, M. A.; Le Guludec, E.; Mariey, L.; van Gestel, J.; Travert, A.; Oliviero, L.; Maugé, F. *Catalysis Today* 2010, 150, 179-185.

[151] Vissenberg, M. J.; de Bont, P. W.; Gruijters, W.; de Beer, V. H. J.; van der Kraan, A. M.; van Santen, R. A.; van Veen, J. A. R. *J. Catalysis* 2000, 189, 209-220.

[152] Rodríguez-Castellón, E.; Jiménez-López, A.; Eliche-Quesada, D. *Fuel* 2008, 87, 1195-1206.

[153] Qian, X. F.; Zhang, X. M.; Wang, C.; Tang, K. B.; Xie, Y.; Qian, Y. T. *J. Alloys Compounds* 1998, 278, 110-112.

[154] Mijoin, J.; Thévenin, V.; Garcia Aguirre, N.; Yuze, H.; Wang, J.; Li, W. Z.; Pérot, G.; Lemberton, J. L. *Appl. Catalysis A: General* 1999, 180, 95-104.

[155] Toba, M.; Miki, Y.; Kanda, Y.; Matsui, T.; Harada, M.; Yoshimura, Y. *Catalysis Today* 2005, 104, 64-69.

[156] Ledoux, M. J.; Peter, A.; Blekkan, E. A.; Luck, F. *Appl. Catalysis A: General* 1995, 133, 321–333.

INDEX

D

E

F

G

H

I

J

K

L

M

N

O

P

Q

R

S

T

U

V

W

X

Z